Climate-Resilient Crops

Climate-Resilient Crops: Challenges and Opportunities offers a clear and concise guide to the development of crops that can withstand the increasing threats of climate change. With rising global temperatures and more frequent extreme weather events, this book highlights innovative approaches in breeding, biotechnology, and crop management to meet future food demands.

Covering a range of important crops—including cereals, oilseeds, fiber, and vegetables—the book integrates current research on stress tolerance, genetic resource utilization, root systems, and the role of plant metabolites. It also explores cutting-edge tools such as omics and genome editing for crop improvement.

Designed for students, young researchers, and agricultural professionals, this book delivers practical insights into how science can contribute to climate-smart agriculture. By bridging gaps between disciplines and highlighting new breeding strategies, it serves as a valuable resource for those working to build sustainable food systems in an era of climate uncertainty.

Climate-Resilient Crops

Challenges and Opportunities

Edited by

Muhammad Tehseen Azhar, Zulfiqar Ali, and

Muhammad Azam Khan

CRC Press

Taylor & Francis Group

Boca Raton London New York

CRC Press is an imprint of the

Taylor & Francis Group, an **informa** business

Designed cover image: Shutterstock

First edition published 2026
by CRC Press
2385 NW Executive Center Drive, Suite 320, Boca Raton FL 33431

and by CRC Press
4 Park Square, Milton Park, Abingdon, Oxon, OX14 4RN

CRC Press is an imprint of Taylor & Francis Group, LLC

ISBN: 9781032712239 (hbk)
ISBN: 9781032720159 (pbk)
ISBN: 9781032720142 (ebk)

DOI: 10.1201/9781032720142

Typeset in Times
by Newgen Publishing UK

Contents

Foreword

Pakistan is one of the countries in the world where the majority of the population is engaged with agriculture; contributing 24% to the national GDP. Wheat is the main staple food, though rice, corn, sorghum, and some minor cereals are also used as staple foods. Yield and quality are drastically affected by multiple factors, however abiotic stresses including heat, drought, and salinity are of paramount importance among others. Climate change has exacerbated these stresses manyfold in recent years. The use of genetics, genomics, and other modern tools may result in the development of climate-resilient crops to ensure local and global food security. According to reports from NASA and other environmental authorities, the occurrence and duration of heat and salinity stress have increased significantly in the last two decades, and their effects will be more drastic in the near future. Pakistan is one of the top 10 countries affected by climate change. Hence, there is a serious threat to agriculture from climate change in Pakistan. It is worth mentioning that the effects of climate change adversely affect food security. In particular, wheat, rice, and corn are affected by stresses posed by climate change. Floods, droughts, heavy rainfall, heat, and salinity have become natural calamities, encouraging plant researchers to develop climate-resilient crops. One of the ways to do this is through the exploitation of available variations, and second way is to learn new tools of plant breeding, biotechnology, and agronomy. This book, titled *Climate-Resilient Crops: Challenges and Opportunities*, addresses the risks of an ever-changing climate that poses serious threats to food security and how to exploit various resources to eliminate hunger. Climate-resilient crops are the only sustainable way to cope with the challenge of increasing food production for the rapidly growing global population. To address this issue, the latest success stories, achievements, challenges, opportunities, and breeding strategies are discussed in this book. At present, few, if any, books have a combination of up-to-date knowledge of crop breeding and biotechnology, agronomy, and other techniques to mitigate climate change. This information is vital to plant scientists to cope with challenges posed by extreme weather events, such as droughts, floods, heat waves, and cold snaps. This book will add a critical body of knowledge to these areas. In addition, this book can be used as a guide to develop climate-ready field crops, i.e. cereals, fiber crops, and oilseeds. Several strategies can be used to ensure food safety for the growing population and to address the related issues in the provision of quality food. To cope with this challenge, a sound understanding of genetics of the concerned traits in crops, climatic risks, management practices, and mitigation strategies is of primary importance. The editors, Dr Muhammad Tehseen Azhar, Prof. Dr Zulfiqar Ali, and Dr Muhammad Azam Khan have compiled several chapters on various aspects of climate change mitigation in collaboration with national and international authors working on various aspects of the subject. This book explains recent developments with reference to climate change; and their efficient transfer and utilization through modern tools of plant breeding and genetics, biotechnological, and genomics approaches. In addition, the book is designed to facilitate early- to mid-career scientists to assess upcoming risks, designing of strategies for solutions, and the implementation of research activities to boost yields in important crops under the climatic change scenario.

Dr Iqrar Ahmad Khan (S.I., H.I)
Professor Emeritus
Former Vice Chancellor
University of Agriculture, Faisalabad (Pakistan)

About the Editors

Muhammad Tehseen Azhar is working as Associate Professor at the Department of Plant Breeding and Genetics, University of Agriculture, Faisalabad, Pakistan, his primary responsibility is teaching to graduate and post-graduate classes. In addition, his focus is on the screening and development of segregating populations resulting in the development of several lines of cotton with higher yield and improved fiber quality traits. Dr Azhar has specific interest in the development of cotton germplasm with tolerance to biotic and abiotic stresses. These advanced strains of cotton have showed good performance for yield of cotton in national trials with genotypes suitable for cultivation in high-temperature and water stress conditions. Being a Borlaug Alumnus, he has worked with various cotton geneticists, namely, Dr Richard Percy, Dr Lori L. Hinze, and Dr Jame Frelichowsk at Texas A&M University and USDA-ARS, USA. Dr Azhar has completed various research projects funded by the Higher Education Commission, Pakistan, CAS-PARB and the Cooperation Organization Partnership for Science and Technology, China. In addition, Dr Azhar is Endeavour Alumnus and appointed as Adjunct Lecturer at the School of Plant Biology, UWA. Recently he was appointed as Adjunct Associate Professor, School of Agriculture Sciences, Zhengzhou University, Zhengzhou, China. In recognition of his services, Dr Azhar has been appointed as a Chair of Germplasm and Genetic Stock in the International Cotton Genome Initiative (ICGI).

Zulfiqar Ali is a Professor at the Department of Plant Breeding and Genetics, University of Agriculture, Faisalabad, Pakistan. Prof. Ali completed his PhD from the University of Agriculture, Faisalabad, Pakistan, and postdoctoral studies from the Plant Breeding Institute at University of Sydney (2007–2008) and Jiangsu Academy of Agricultural Sciences Nanjing (2009–2011). He has published more than 80 books/chapters/research papers in highly popular journals including *Plant Breeding*, *Nature's Scientific Report*, *Rice*, *Field Crop Research*, and *PLOS One*, and has been serving as the section editor of *Pakistan Journal of Agricultural Sciences*. Prof. Ali has successfully completed various research projects. Currently, he is running six research projects funded by national and international funding agencies including HEC, PARB, ADP, USAID, DFID, ACIAR, etc. He has been granted awards, including the Gold Medal 2017 in Biotechnology from the Pakistan Academy of Sciences. Dr. Ali has also been awarded with the "National Level HEC Best Teacher Award 2021" by the Higher Education Commission (HEC), Pakistan. For the past 21 years, Prof. Ali has been engaged in the fields of plant biotechnology, breeding, genetics, and molecular biology. His research focuses on climate-smart breeding using functional genomics tools.

Muhammad Azam Khan is working as Assistant Professor at the Department of Plant Breeding and Genetics, University of Agriculture, Faisalabad, Pakistan. He is primarily engaged in the teaching and supervision of undergraduate and postgraduate students. He studied genetics and genomics of host plant resistance to fungal diseases in his PhD. He has experience working with genetics of abiotic stress tolerance in wheat and brassica, exploiting the genetic potential of vegetables using wastewater and the selection of bio-energy feedstocks. This hands-on experience explains why he is interested in developing climate-ready crops. More specifically, he is interested in developing brassica types with higher yields while maintaining/improving their quality in the face of a changing climate. Being a graduate from The University of Western Australia, he is ambitious to explore new avenues in research.

Contributors

Furqan Ahmad
Institute of Plant Breeding and Biotechnology, Muhammad Nawaz Sharif University of Agriculture Multan, Multan, Pakistan

Muhammad Qadir Ahmad
Department of Plant Breeding and Genetics, Bahauddin Zakariya University, Multan, Pakistan

Ali Ijaz Ahmed
Department of Plant Breeding and Genetics, University of Agriculture Faisalabad, Pakistan

Hafiz Muhammad Wasif Ali
Department of Plant Breeding Genetics, Bahauddin Zakariya University, Multan, Pakistan

Mubashir Ali
Institute of Plant Breeding and Biotechnology, Muhammad Nawaz Sharif University of Agriculture Multan, Multan, Pakistan

Zulfiqar Ali
Department of Plant Breeding and Genetics, University of Agriculture Faisalabad, Pakistan

Sue J. Armstrong
University of Birmingham, United Kingdom

Sawaira Ashraf
Department of Plant Breeding Genetics, Bahauddin Zakariya University, Multan, Pakistan

Rana Muhammad Atif
Department of Plant Breeding and Genetics, University of Agriculture, Faisalabad, Pakistan

Nadia Ayub
Institute of Plant Breeding and Biotechnology, Muhammad Nawaz Sharif University of Agriculture Multan, Multan, Pakistan

Asad Azeem
Department of Plant Breeding & Genetics, Faculty of Agricultural Sciences & Technology, University of Layyah, Pakistan

Farrukh Azeem
Department of Bioinformatics & Biotechnology, Government College University, Faisalabad, Pakistan

Muhammad Tehseen Azhar
Department of Plant Breeding and Genetics, University of Agriculture Faisalabad, Pakistan; School of Agriculture Sciences, Zhengzhou University, Zhengzhou, P.R. China

Allah Bakhsh
University of Punjab, Lahore, Pakistan

Muhammad Bilal
Department of Environmental Sciences, Government College University, Faisalabad, Pakistan

Ioannis Ganopoulos
Department of Molecular Breeding at Institute of Plant Breeding

and Genetic Resources, ELGO-DIMITRA, Greece

Sadia Hakeem
Cotton Research Institute Multan, Ayub Agriculture Research Institute, Faisalabad, Pakistan

Akhtar Hameed
Institute of Plant Protection, MNS University of Agriculture, Multan, Pakistan

Zeshan Hassan
Department of Plant Breeding & Genetics, Faculty of Agricultural Sciences & Technology, University of Layyah, Pakistan

Sidra Iqbal
Department of Plant Breeding and Genetics, University of Agriculture, Faisalabad, Sub Campus Depalpur Okara, Pakistan

Adil Jamal
Sciences and Research, College of Nursing, Umm Al Qura University, Makkah, Saudi Arabia

Ayesha Javed
Center of Agricultural Biochemistry and Biotechnology, University of Agriculture, Faisalabad, Pakistan

Muhammad Arshad Javed
Department of Plant Breeding and Genetics, University of the Punjab, Pakistan

Muhammad Junaid
Department of Microbiology, Faculty of Pharmacy, Mahidol University, Salaya, Thailand

Apostolos Kalivas
Hellenic Agricultural Organization DIMITRA, Institute of Plant Breeding and Genetics Resources, Greece

Mehwish Kanwal
Pakistan Tobacco Board, Ministry of National Food Security & Research, Government of Pakistan, Pakistan

Çetin Karademir
Department of Field Crops, Siirt University, Turkey

Emine Karademir
Faculty of Agriculture, Siirt University, Turkey

Zulqurnain Khan
Institute of Plant Breeding and Biotechnology, Muhammad Nawaz Sharif University of Agriculture Multan, Multan, Pakistan

Azhar Abbas Khan
Department of Entomology, Faculty of Agricultural Sciences & Technology, University of Layyah, Pakistan

Muhammad Ahsan Khan
Department of Plant Breeding & Genetics, University of Agriculture Faisalabad, Pakistan

Muhammad Azam Khan
Department of Plant Breeding and Genetics, University of Agriculture Faisalabad, Pakistan

Sultan Habibullah Khan
Center of Agricultural Biochemistry and Biotechnology, University of Agriculture, Faisalabad, Pakistan

Rabia Nasir Khan
Department of Bioinformatics and Biotechnology, Government College University, Faisalabad, Pakistan

Qiang Li
Xuzhou Institute of Agricultural Sciences in Jiangsu Xuhuai District/Key Laboratory of Biology and Genetic Breeding of Sweet Potato, Ministry of Agriculture and Rural Affairs/Sweet Potato Research Institute, CAAS, Xuzhou, Jiangsu, China.

Sajid Majeed
Federal Seed Certification and Registration Department, Islamabad, Pakistan

Zahid Manzoor
Key Laboratory of Crop Genetic Improvement, College of Plant Science and Technology, Huazhong Agricultural University, Wuhan, China

Hamna Maqsood
Department of Bioinformatics and Biotechnology, Government College University, Faisalabad, Pakistan

Sammina Mahmood
Department of Botany, Division of Science and Technology, University of Education, Lahore, Pakistan

Mahlka Mukhtar
Institute of Plant Breeding & Biotechnology, MNS-University of Agriculture Multan, Multan, Pakistan

Habibullah Nadeem
Department of Bioinformatics and Biotechnology, Government College University, Faisalabad, Pakistan

Iqrar Ahmad Rana
Center of Agricultural Biochemistry and Biotechnology, University of Agriculture, Faisalabad, Pakistan

Asima Rasheed
Department of Bioinformatics and Biotechnology, Government College University, Faisalabad, Pakistan

Kashif Rashid
Vegetable Research Institute, Ayub Agricultural Research Institute, Faisalabad, Pakistan

Muhammad Abdul Rehman Rashid
Department of Bioinformatics & Biotechnology, Government College University, Faisalabad, Pakistan

Fiaz Rasul
Bio Atlantis Ltd., Clash Industrial Estate, Tralee, Co. Kerry, Ireland

Ijaz Rasul
Department of Bioinformatics and Biotechnology, Government College University, Faisalabad, Pakistan

Husnain Rauf
University of Georgia, Department of Horticulture, Athens, GA, USA

Ahmed Raza
Plant Virology Lab, Crop Diseases Research Institute, National Agricultural Research Centre, Islamabad, Pakistan

Aneeq ur Rehman
Department of Plant Breeding and Genetics, University of Agriculture Faisalabad, Pakistan

Shoaib ur Rehman
Institute of Plant Breeding and Biotechnology, Muhammad Nawaz Sharif University of Agriculture Multan, Multan, Pakistan

Muhammad Abu Bakar Saddique
Institute of Plant Breeding and Biotechnology, Muhammad Nawaz Sharif University of Agriculture Multan, Multan, Pakistan

Fatima Sajal
Institute of Plant Breeding and Biotechnology, Muhammad Nawaz Sharif University of Agriculture Multan, Multan, Pakistan

Attiqa Saleem
Plant Genetic Resources Program, Plant Genetic Resources Institute, National Agriculture Research Center, Islamabad, Pakistan

Muhammad Rizwan Shafiq
Department of Plant Breeding and Genetics, University of Agriculture Faisalabad, Sub Campus Depalpur Okara, Pakistan

Asad Ali Shah
Department of Bioinformatics and Biotechnology, Government College University, Faisalabad, Pakistan

Muhammad Naveed Shahid
Department of Botany, University of Education, Lahore, Pakistan

Tooba Shahid
Department of Bioinformatics and Biotechnology, Government College University, Faisalabad, Pakistan

Zeeshan Shamim
Mirpur University of Science and Technology, Mirpur Azad Kashmir, Pakistan

Zeba Shehnaz
Department of Plant Breeding & Genetics, University of Agriculture Faisalabad, Faisalabad, Pakistan

Muhammad Shehzad
Institute of Crop Science, College of Agriculture and Biotechnology, Zhejiang Provincial Key Laboratory of Crop Germplasm, Zhejiang University, Hangzhou, China

Umbreen Shehzad
Department of Horticulture, Faculty of Agricultural Sciences & Technology, University of Layyah, Pakistan

Muhammad Ali Sher
Institute of Plant Breeding and Biotechnology, Muhammad Nawaz Sharif University of Agriculture Multan, Multan, Pakistan

Muhammad Hussnain Siddique
Department of Bioinformatics and Biotechnology, Government College University, Faisalabad, Pakistan

Daniel K.Y. Tan
Department of Agronomy at the Sydney Institute of Agriculture, The University of Sydney, Australia

Wei Tang
Xuzhou Institute of Agricultural Sciences in Jiangsu Xuhuai District/Key Laboratory of Biology and Genetic Breeding of Sweet Potato, Ministry of Agriculture and Rural Affairs/Sweet Potato Research Institute, CAAS, Xuzhou, Jiangsu, China

Muhammad Tayyab
Center of Agricultural Biochemistry and Biotechnology, University of Agriculture, Faisalabad, Pakistan

Eleni Tsaliki
Hellenic Agricultural Organization DIMITRA, Institute of Plant Breeding and Genetics Resources, Greece

Dandan Wang
Xuzhou Institute of Agricultural Sciences in Jiangsu Xuhuai District/ Key Laboratory of Biology and Genetic Breeding of Sweet Potato, Ministry of Agriculture and Rural Affairs/ Sweet Potato Research Institute, CAAS, Xuzhou, Jiangsu, China

Saba Yaseen
Institute of Plant Breeding and Biotechnology, Muhammad Nawaz Sharif University of Agriculture Multan, Multan, Pakistan

Sadia Zainab
Department of Plant Breeding & Genetics, Faculty of Agricultural Sciences & Technology, University of Layyah, Pakistan

Muhammad Zubair
Department of Bioinformatics and Biotechnology, Government College University, Faisalabad, Pakistan

Section I

Important Cash Crops

1 Breeding Climate-ready Rice

Muhammad Abdul Rehman Rashid, Farrukh Azeem, Tooba Shahid, and Sammina Mahmood

1.1 INTRODUCTION

The term "climate-ready" can refer to the varieties that have been developed or bred to withstand the challenges posed by climate change. These challenges include biotic stresses such as diseases and abiotic stresses such as extreme weather conditions. It may also include various management practices to cope with climatic stresses. The development of climate-ready crops involves traditional breeding techniques such as germplasm introduction, selection, and hybridization, as well as advanced biotechnological methods such as genetic modification and genome editing.

Rice is a primary source of calories and nutrition for a substantial portion of the global population. It provides a significant share of daily energy intake, particularly in countries like China, India, Indonesia, and Bangladesh where it is a staple food (Figure 1.1). For millions of people, particularly smallholder farmers in developing countries, rice farming is a vital source of livelihood. The rice industry – encompassing cultivation, processing, and distribution – makes a considerable contribution to national economies. Deeply ingrained in cultural traditions and culinary practices, rice plays a prominent role in festivals, rituals, and daily meals. Beyond carbohydrates, rice supplies essential nutrients including minerals, vitamins, and fiber. Fortified rice varieties can combat malnutrition by providing additional micronutrients. Its adaptability to diverse environments, from irrigated lowlands to upland regions, makes rice an exceptionally versatile crop. This adaptability makes it a versatile crop that can be cultivated in various climates and conditions. Ongoing research and development in rice breeding and biotechnology have resulted in high-yielding, disease-resistant, and climate-resilient rice varieties. Given its significance, ensuring the sustainability and resilience of rice production systems is crucial for global food security and economic stability. These advancements are vital to maintain rice production in the face of pressing challenges including climate change, pests, and diseases.

Rice plays a crucial role in food security in developing countries. Hence, the investments and economic planning are crucial to alleviate the poverty and promote sustainable food production. In several rice-producing countries, particularly in rural and sub-rural areas, agriculture forms the backbone of the economy, with

DOI: 10.1201/9781032720142-2

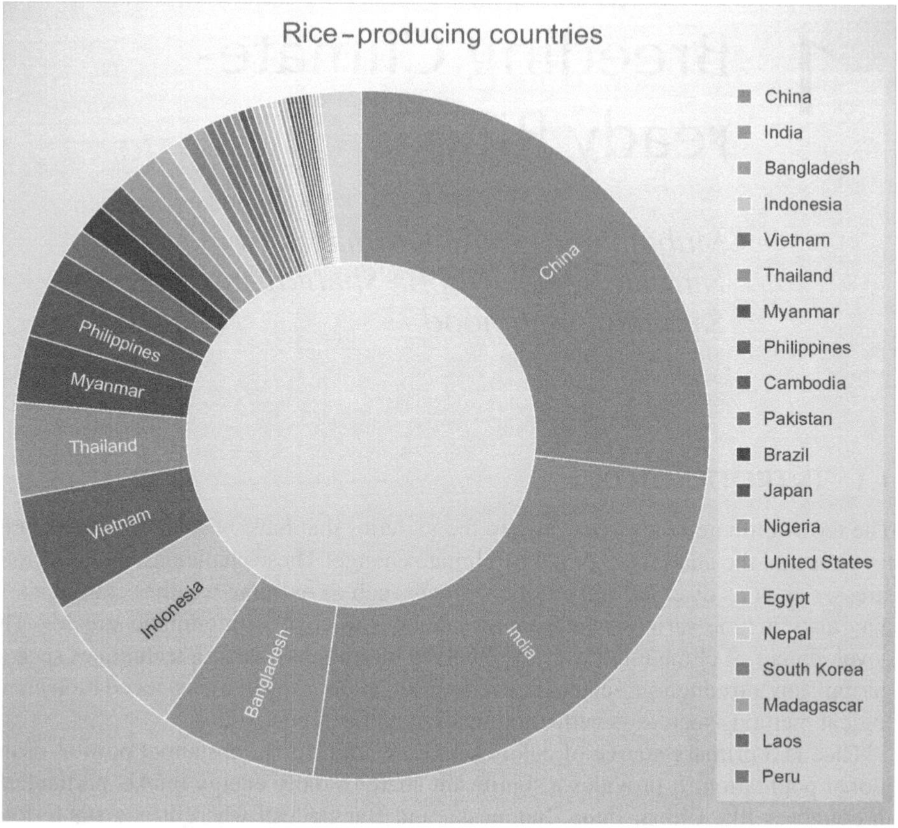

FIGURE 1.1 The top rice-growing areas on the world with the highest rice production in 2023.

high employment rates in the agriculture sector. However, agricultural productivity is heavily dependent on the availability and quality of natural resources, such as irrigation, rainfall, the bio-physical structure of culturable land, and agricultural mechanization. Climatic factors, including rains in monsoon, are fleeting. Thus, a lack of temperature or water control leads to unpredictable agricultural productivity and food security. This uncertainty disproportionately affects rain-dependent communities, perpetuating poverty. To address these adverse climatic challenges, researchers have developed 'climate-ready' rice varieties that can withstand the impact of climate change. The climate-resilient rice varieties are bred to tolerate extreme weather conditions, including drought, flooding, high temperatures, and salinity. They also incorporate management practices to cope with climatic stresses. The development of climate-ready rice employs both traditional breeding methods (introduction, selection, and hybridization) and advanced biotechnological methods (genetic modification and genome editing). Climate-ready rice can be categorized based on the following indicators.

1.1.1 Water Stresses

a. Drought Tolerance: Varieties that thrive with minimal water or survive prolonged drought.
b. Water Logging or Flood Resistance: Varieties that can endure underwater conditions, such as 'scuba rice' developed by the International Rice Research Institute (IRRI).

1.1.2 Temperature Stress

a. Heat Tolerance: Varieties that can maintain high yields under extreme heat.
b. Cold Tolerance: Varieties that germinate and produce grain in low temperatures.

1.1.3 Salinity Stress

Varieties that grow in saline soils, mitigating the effects of rising sea levels and soil salinization.

1.1.4 Biotic Stress

Varieties that are resistant to diseases and pests, which may intensify due to climate change. The ultimate goal of developing climate-ready rice is to ensure food security for populations reliant on rice as a staple food, particularly in regions most vulnerable to climate change.

1.2 CLIMATE-READY RICE: DROUGHT

Rice is crucial for food security in impoverished regions, but its production faces significant challenges in drought-prone lowlands where effective water management is paramount. Rain-fed lowlands, prevalent in countries like India, Thailand, and Bangladesh, are constrained by biophysical factors that limit crop options and productivity. To address these challenges, key management strategies include: (1) introducing rice varieties which are essentially drought tolerant, for instance Sahbhagi Dhan; (2) enhancing water management through irrigation and rainwater harvesting; and (3) and optimizing nutrient use to boost yields. Additional resilience measures include: (a) adjusting the cropping system, such as using shorter-duration varieties and diversified post-rice crops; and (b) implementing advanced establishment techniques like dry direct seeding. Effective soil and field management, including land leveling and subsoil compaction, can mitigate water stress and improve crop performance in vulnerable regions (Figure 1.1, Table 1.1).

Rural economies, where agricultural income supports high employment, improving productivity is vital. Rain-fed lowlands face significant bio-physical constraints, including monsoon patterns that concentrate over 80% of rainfall, limiting crop options and preventing rain-fed crops during dry seasons. Controlling climatic factors and natural resources, particularly water, is critical for enhancing productivity and crop management in drought-prone lowland areas. Fortunately, improved varieties and germplasm can reduce this risk and increase productivity, as demonstrated by recent research (1, 2).

TABLE 1.1
A comprehensive list of areas vulnerable to climate change worldwide, classified by geographical region and the type of vulnerability to climate change*

Continent	Country/region	Risk type	
Asia			
	Maldives	Rising sea level, floods	High
	Bangladesh	Rising sea level, cyclones, drought	High
	Philippines	Typhoons, rising sea level	
	Indonesia	Deforestation, forest fires	
	Mekong River Delta	Rising sea level, flooding	
Europe			
	Mediterranean coastal regions	Rising sea level, heatwaves	Medium
	Alpine regions	Glacier melting, water scarcity	Low medium
	Balkans	Floods, landslides	Low medium
	UK and Ireland	Rising sea level, storm surges	Low medium
	Scandinavian Arctic regions	Permafrost thaw, climate disruption	Low medium
North America			
	Gulf Coast	Rising sea level, hurricanes	Low medium
	California	Wildfires, drought	Low medium
	Arctic regions	Permafrost thaw, climate disruption	Low medium
	Florida	Rising sea level, floods	Low medium
	Great Plains	Drought, tornadoes	Low medium
Latin America			
	Amazon rainforest	Deforestation, drought	Medium
	Andean glaciers	Glacier melting, water scarcity	Medium
	Caribbean islands	Hurricanes, drought, famine, rising sea level	Medium high
	Central American dry corridor	Drought, famine	
	Patagonia	Glacier melting, rising sea level	
Africa			
	Sahel region	Drought, desertification	Medium high
	Horn of Africa	Drought, famine	Medium high
	Nile River Delta	Rising sea level, floods	
	African coastal cities	Rising sea level, coastal erosion	
	Madagascar	Cyclones, deforestation	

TABLE 1.1 (Continued)
A comprehensive list of areas vulnerable to climate change worldwide, classified by geographical region and the type of vulnerability to climate change*

Continent	Country/region	Risk type	
Oceania			
	Great Barrier Reef	Coral bleaching, rising sea level	
	Pacific Islands	Rising sea level, floods	
	Australian bushfire regions	Wildfires, drought	
	New Zealand	Glacier melting, rising sea level	
	Tuvalu	Rising sea level, displacement	High

*Note: Vulnerability and its intensity may affected by economic resilience, infrastructure, and adaptive capacity. Sources: IPCC (Intergovernmental Panel on Climate Change) reports, UNFCCC (United Nations Framework Convention on Climate Change), National Oceanic and Atmospheric Administration (NOAA) National Aeronautics and Space Administration (NASA), World Bank climate data.

1.3 DROUGHT-READY RICE VARIETIES

Several drought-tolerance rice varieties, such as Sahbhagi Dhan, have been successfully released in Southern and Southeastern Asia. These varieties consistently perform well under rain-fed conditions, yielding 0.8–1.0 ton per hectare in drought-prone areas, making it a possible ideal choice for cultivation in drought-prone regions like Bihar, India (3). Notably, Sahbhagi Dhan demonstrated a significant (108%) yield increase. Moreover, during a 10-day drought spell in Eastern Uttar Pradesh (UP) and Bihar, it showed no yield penalty, whereas control varieties suffered yield losses. With just one to two irrigations, it yielded 4.7 tonnes per hectare. Drought severely impacts rain-fed lowland rice by reducing transcription, causing spikelet sterility, and reducing grain filling (4). Water scarcity also decreases soil nutrient uptake, disrupting soil chemistry, and reduces soil nutrient availability (5, 6). Traditional flooding-based management practices often prove ineffective (7). Considering the risk factors, farmers hesitate to invest in quality seed and fertilizers, even with sufficient water availability. Hence, traditional varieties limit potential yields and economic gains throughout the year.

Modern rice varieties, however, can mitigate drought effects in rain-fed lowland rice cultivation by interacting with management interventions and addressing limiting characteristics of rain-fed lowlands.

1.4 CHARACTERIZATION OF DROUGHT-AFFECTED AREAS

The world's largest rain-fed lowlands suitable for rice cultivation are found in Bangladesh, India, and Thailand. Other countries, including China, Myanmar,

Vietnam, Cambodia, Indonesia, Tanzania, Nigeria, Nepal, and the Philippines, have smaller areas of less than 3 million hectares of lowland rice cultivation (7) (Figure 1.1, Table 1.1).

Rain-fed lowland areas can be further categorized into shallow and intermediate rain-fed areas (8, 9). Shallow rain-fed lowlands are characterized by undulating landscapes located outside of the larger flood plains (10, 11). In contrast, intermediate rain-fed areas are situated in deltas, flood plains, and near larger lakes or rivers. Both of these types of sites are prone to drought, and often experience submergence or stagnating flood water (12, 13).

Geography and soil characteristics significantly impact water and nutrient availability to plants (14–16). For example, higher terraces have coarser soil, reducing water retention and fertility. Water movement down slopes further depletes natural resources. Medium terraces often have underground water levels below the rooting zone, causing imbalanced water and nutrient losses. In contrast, lower terraces in valley bottoms experience fewer nutrient losses.

To cope with the climate effects, soils can be classified into four fertility categories: good, poor, very poor, and problematic soils (11, 17, 18). These classifications are based on soil characteristics, weathering effects, and the chemical constraints for rice cultivation (19). The prevalence of 'very poor' soils for rice cultivation is observed in Cambodia, Thailand, the Philippines, Indonesia, and Tanzania. China, Bangladesh, and Myanmar have no major soil constraints, while Nigeria and India have 'poor soils' with weathered characteristics.

Rain-fed rice is primarily grown in humid to sub-humid tropical regions, where uneven rainfall often causes stagnating water and flash floods. Drought is a significant limiting factor for lowland rice due to strong seasonality, sandy soils, heavy tropical rainfall, and undulating topo-morphology. Compared to wheat, lowland rice is more drought-susceptible, with yield losses occurring at 50 kPa soil water tension (20). Soil oxidation causes aluminum toxicity, affecting phosphorus (P) mobilization (21). This negatively impacts rice plants on poor or problematic soils, even during short drought spells.

1.5 MANAGEMENT OF CLIMATIC FACTORS IN RICE-GROWING AREAS

1.5.1 Water Management Options

Irrigation is the most effective way to avoid drought in rain-fed lowland systems, significantly increasing productivity. However, irrigation schemes have limited potential, with a focus on maintenance and rehabilitation (22). Smaller surface reservoirs and tube wells are crucial for rice farming in regions like Bangladesh and India (7). These water sources are used for irrigation during droughts or near the end of the season. Sustainable groundwater consumption is essential for rice cultivation and crop diversification in drought-prone areas (13). Excessive groundwater use can lead to rapid water level drops and salinity issues, with or without heavy metals like arsenic (23).

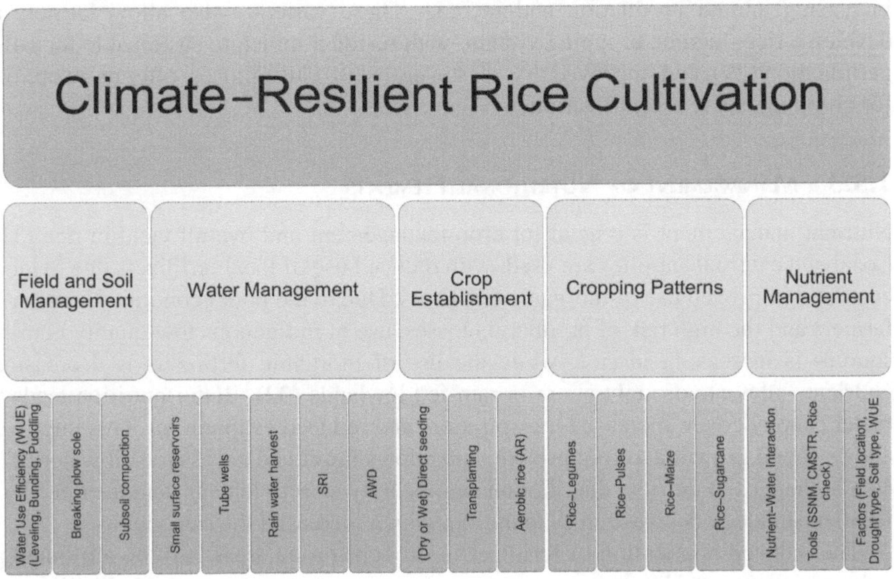

FIGURE 1.2 Various management practices to cope with climate changes for rice cultivation.

Rainwater harvesting is another method to increase water resources, as seen in northeast Thailand, Indonesia, India, and the Philippines (24, 25). These reservoirs serve not only to help with crop-saving irrigation in minor droughts but also generate income through other uses including fish production. Recent water-saving techniques, such as 'System of Rice Intensification' (SRI), 'Alternate Wetting and Drying' (AWD), and controlled drying at various plant growth stages, have been developed to address water scarcity, high costs, and methane emissions from flooded rice fields without any negative effect on rice grain yield (26, 27) (Figure 1.2).

1.5.2 Upgradation of the Cropping System

The adjustment of cropping seasons to optimal rainfall or reducing cropping season length with shorter-duration varieties can help avoid drought. However, farmers' variety choices often do not align with optimal conditions, as drought is just one of various abiotic stresses crops face. Some drought-prone rain-fed lowlands lack sustainable short-duration material, such as traditional photoperiod-sensitive varieties in Cambodia (28). Providing short-duration varieties can reduce drought damage in various rain-fed lowland areas. Shifting from deep-water rice to flood recession or shallow tube well-based irrigated dry season rice has improved the productivity in regions like Bangladesh, Myanmar, Thailand, and Cambodia (7).

Improvements and developments in rain-fed areas are normally influenced by hydrological conditions (7). Crops like sugar cane and maize may be a better option to cultivate in lowlands with poor drainage. Short-duration rice varieties, such as IR-64 drought-1 and Sahbhagi Dhan, offer diversification through post-rice crops, increasing annual production and income. These varieties also allow farmers to develop a rice–legume cropping system, with residual moisture sustainable for pulse germination (29), and improved dry-season crops (30), like pulses, offer more options for cropping, such as lentil varieties (Figure 1.2).

1.5.3 Management of Nutritional Elements

Nutrient management is crucial for crop management and overall yield in rice (31). Local and external supplies are used, with reduced use of local fertilizers due to labor costs and improved use of inorganic fertilizers. Due to the poor economic situation of farmers and the high risk of production losses, use of indigenous low-quality nutrient sources is increasing in rice, while the use of inorganic fertilizers is decreasing. Nutrients alleviate drought stress in rain-fed lowlands (32). Better nutrition leads to better canopy cover, increased transpiration, and reduced evaporation. This improves aboveground growth and root growth, and allows improved access to soil water (27). Minor changes in nutrient distribution significantly affect yield, as demonstrated by a fertilizer-induced decrease in evaporation, which increased the rice yield.

The reduced application of fertilizer in drought-prone areas is often attributed to reduced efficiency. However, water stress doesn't necessarily decrease fertilizer use efficiency, except during extreme drought or flowering. The low yield potential of traditional varieties is further constrained by their relatively low optimal nitrogen requirements compared to irrigated conditions. Nutrient–water interactions could improve crop management in a rice rain-fed system (33). However, the use of fertilizer may increase risk in drought-prone rice environments (34). A larger canopy during mid-season droughts affects total evapotranspiration, but when the soil surface becomes dry, higher transpiration leads to faster water decline and drought stress (7). The application of fertilizer also increases access to soil water sources, to fulfill the higher transpiration requirements (35).

Multiple factors influence fertilizer application in water-limited areas, including field location, drought type, and soil type. For instance, organic fertilizers are more effective on sandy soils, whereas inorganic fertilizer is less effective on fine-textured soil (36). When combined with topographical factors, these variables contribute to widespread adaptation of traditional varieties in rain-fed lowlands. However, the uniform fertilizer application rates across large regions can compromise fertilizer use efficiency in rain-fed lowland rice (Figure 1.2).

To improve nutrient management, site-specific nutrient management (SSNM) recommendations (37) have been proposed (38). While farmers have already adapted traditional practices, recent research-based recommendations and advanced decision-support tools can further enhance management approaches. For example, Haefele and Konbon developed a decision-support tool that leverages farmers' knowledge of their fields to recommend optimal fertilizer types and application rates for rice grown in rain-fed areas of Thailand (17). Tools like 'Crop Manager for Stress Tolerant Rice' in Bihar (CMSTR; http://webapps.irri.org/in/br/cmstr/) and 'Rice Check' developed by IRRI and partners are helpful to provide site- and stress-specific recommendations for flood- and drought-tolerant rice genotypes, calculating fertilizer amounts and offering management advice (Figure 1.2). Notably, CMSTR gained high acceptance among farmers, even with paid services (7).

1.5.4 Crop Establishment

Rice establishment in lowland areas employs three fundamental approaches: (1) dry-direct-seeding (DDS); (2) wet-direct-seeding (WDS); and (3) transplanting (TP). DDS grows seeds on moist or dry soils, while WDS sows seeds on water-saturated soils, and TP replants seedlings. The preferred method depends on factors like water control, labor availability, chemical weed control, and production system diversification. DDS significantly reduces initial water consumption for land preparation and crop establishment (17, 39), enabling early establishment compared to transplanted rice (39). However, DDS may offset water saving in the late crop season due to its lack of puddling. DDS rice roots are deeper, finer, and most extensive, making them more drought tolerant. Early-established and short-duration crops can increase intensification (40) and reduce drought risk in late season.

Historically, direct-seeded systems (DDS or WDS) were common rice establishment methods (7), but this has largely shifted to TP and more intensive cropping patterns in lowland areas. TP dominates in Asia (40) due to higher yields and stability. DDS is preferred in areas with limited labor, mechanization, or hydrological constraints. Research on seed drills for direct-seeded systems is growing in Southern and Southeastern Asia (41) (Figure 1.2). Fully developed drills are available in India, primarily for four-wheel tractors. Weeds are among the significant constraints in DDS, and leveling equipment can improve crop, water, and weed management (42). Further adaptation and feasible business models are needed.

Direct seeding of rice seedlings increases weed pressure, leading to crop losses in drought-prone environments. Herbicides are essential for successful direct-seeded rice cultivation, but limited in rain-fed lowlands. Higher seed rates may suppress the weeds but are not a proper solution (43). Standard management practices are vital for good crop establishment, weed control, and yield improvement (44). Selecting suitable morphological traits for direct seeding in rice is a challenge, as current breeding focuses on transplanting conditions. Suitable traits for a direct seeding approach include high germination rate, vigor, tiller numbers, intermediate height, and resistance to herbicides (45). Conventional varieties, such as those in Thailand, exhibit weed competitiveness under water limitation. Aerobic rice (AR) is a novel approach for water-limited areas (27) (Figure 1.2), but faces challenges including yield potential, grain quality, soilborne diseases, and drought.

1.5.5 Field and Soil Management

Soil and field preparation significantly impact water-use efficiency in rain-fed areas. Effective management practices, such as leveling, bunding, and puddling can improve water use efficiency (WUE). These techniques improve water retention, reduce drought stress, and suppress weeds. Bunds help retain water in fields, while cracks can cause water losses (46). Leveling ensures even water distribution, reducing drought stress, suppressing weeds, and promoting yield gain as observed in Cambodia (42). Puddling reduces percolation losses but requires substantial water, potentially delaying planting and increasing late-season drought risk (39). Advances in tractor-based LASER (light amplification by stimulated emission of radiation) leveling have made it more accessible and affordable in many rice-growing regions.

Various management options can improve the water-holding capacity of water in lowland areas of rice cultivation. These include: (a) breaking the plow sole to allow deeper root growth, tapping into deeper soil water or groundwater (Figure 1.2), however, this increases percolation losses, reducing available water resources; and (b) subsoil compaction to reduce percolation, however its effectiveness depends on the site conditions. More studies are needed to determine suitable large areas with clear characteristics and mostly positive results. Soil amelioration measures are generally beneficial, but their use depends on the cost/value ratio.

REFERENCES

1. Kumar A, Dixit S, Ram T, Yadaw R, Mishra K, Mandal N. Breeding high-yielding drought-tolerant rice: genetic variations and conventional and molecular approaches. *Journal of Experimental Botany*. 2014;65(21):6265–78.
2. Verulkar S , Mandal N, Dwivedi J, Singh B, Sinha P, Mahato R, et al. Breeding resilient and productive genotypes adapted to drought-prone rainfed ecosystem of India. *Field Crops Research*. 2010;117(2–3):197–208.
3. Singh A, Yadav A, Matthew K, Varrieur V, Singh S, editors. Success of IRRI stress tolerant rice varieties in Bihar, India-Experience of IRRAS farmers. Poster Presented at Fourth International Rice Conference; Bangkok, Thailand, 2014.
4. O'Toole J, Moya T. Water deficits and yield in upland rice. *Field Crops Research*. 1981;4:247–59.
5. Haefele SM. *Paddy soils: characteristics, productivity and management*; Elsevier, 2023:447–57.
6. Kato Y, Katsura K. Rice adaptation to aerobic soils: physiological considerations and implications for agronomy. *Plant Production Science*. 2014;17(1):1–12.
7. Haefele SM, Kato Y, Singh S. Climate ready rice: augmenting drought tolerance with best management practices. *Field Crops Research*. 2016;190:60–9.
8. Huke RE. *Rice area by type of culture: South, Southeast, and East Asia. International Rice Research Institution*; 1982.
9. Garrity D, Oldeman L, Morris R, Lenka D. Rainfed lowland rice ecosystems: characterization and distribution. *Progress in Rainfed Lowland Rice IRRI*. 1986:3–23.
10. Limpinuntana V. Physical factors related to agricultural potential and limitations in northeast Thailand. In: *Natural resource management Issues in the Korat Basin of Northeast Thailand: an overview*. IRRI; Metro Manila: IRRI (Limited Proceeding 7, ISBN 971-220-161-9, 2001:3–18.
11. Homma K, Horie T, Shiraiwa T, Supapoj N. Evaluation of transplanting date and nitrogen fertilizer rate adapted by farmers to toposequential variation of environmental resources in a mini-watershed (Nong) in northeast Thailand. *Plant Production Science*. 2007;10(4):488–96.

12. Tsubo M, Basnayake J, Fukai S, Sihathep V, Siyavong P, Chanphengsay M. Toposequential effects on water balance and productivity in rainfed lowland rice ecosystem in Southern Laos. *Field Crops Research.* 2006;97(2–3):209–20.
13. Singh S, Mackill DJ, Ismail AM. Tolerance of longer-term partial stagnant flooding is independent of the SUB1 locus in rice. *Field Crops Research.* 2011;121(3):311–23.
14. Inthavong T, Fukai S, Tsubo M. Spatial variations in water availability, soil fertility and grain yield in rainfed lowland rice: a case study from Savannakhet Province, Lao PDR. *Plant Production Science.* 2011;14(2):184–95.
15. Boling AA, Tuong T, Suganda H, Konboon Y, Harnpichitvitaya D, Bouman B, et al. The effect of toposequence position on soil properties, hydrology, and yield of rainfed lowland rice in Southeast Asia. *Field Crops Research.* 2008;106(1):22–33.
16. Hayashi S, Kamoshita A, Yamagishi J, Kotchasatit A, Jongdee B. Genotypic differences in grain yield of transplanted and direct-seeded rainfed lowland rice (Oryza sativa L.) in northeastern Thailand. *Field Crops Research.* 2007;102(1):9–21.
17. Haefele SM, Bouman B. Drought-prone rainfed lowland rice in Asia: limitations and management options. In: Serraj R., Bennett J., and Hardy B., *Drought frontiers in rice: crop improvement for increased rainfed production.* World Scientific; 2009. p. 211–32.
18. Oberthür T, Kam S. Perception, understanding, and mapping of soil variability in the rainfed lowlands of northeast Thailand. In: Tuong T.P., Kam S.P., Wade L., Pandey S., Bouman B.A.M., and Hardy B., *Characterising and understanding rainfed environments.* IRRI; 2000. p. 75–95.
19. Haefele S, Nelson A, Hijmans R. Soil quality in the major rice environments of the world: characterization and distribution. *Geoderma.* 2014;235(236):250–9.
20. Lafitte R, Blum A, Atlin G. Using secondary traits to help identify drought-tolerant genotypes. In: Fischer K.S., Lafitte R., Fukai S., Atlin G., and Hardy B., *Breeding Rice for Drought-Prone Environments.* IRRI, 2003:37–48.
21. Bell R, Ros C, Seng V, editors. Improving the efficiency and sustainability of fertiliser use in drought-and submergence-prone rainfed lowlands in Cambodia. The Impact of Agricultural Research and Development in South East Asia, Proceedings of an International Conference held; 2001: Cambodian Agricultural Research and Development Institute.
22. Wongprasittiporn C. *The future of large rice-based irrigation system in Southeast Asia.* FAO Regional Office for Asia and the Pacific Bangkok; 2007.
23. Van Geen A, Zheng Y, Cheng Z, He Y, Dhar R, Garnier J, et al. Impact of irrigating rice paddies with groundwater containing arsenic in Bangladesh. *Science of the Total Environment.* 2006;367(2–3):769–77.
24. Kam SP, Hoanh CT, Trébuil G, Hardy B. Natural resource managment issues in the Korat basin of Northeast Thailand: an overview. Metro Manila: IRRI (Limited Proceeding 7, ISBN 971-220-161-9, 2001.
25. Pal A, Bhuiyan S, editors. Rainwater management for drought alleviation: opportunities and options for sustainable growth in agricultural productivity. Fragile lives in fragile ecosystems Proceedings of the International Rice Research Conference: International Rice Research Institute Los Baños; 1995.
26. Yang J, Zhang J. Grain filling of cereals under soil drying. *New Phytologist.* 2006;169(2):223–36.
27. Bouman B, Peng S, Castaneda A, Visperas R. Yield and water use of irrigated tropical aerobic rice systems. *Agricultural Water Management.* 2005;74(2):87–105.

28. Fukai S, Ouk M. Increased productivity of rainfed lowland rice cropping systems of the Mekong region. *Crop and Pasture Science*. 2012;63(10):944–73.
29. Wani SP, Jakkula VS, Singh D. Doubling Farmers' Income: KISAN–MITrA. Proceedings of National Workshop on Doubling Farmers' Income through Scalingup: KISAN–MITrA (Knowledge-based Integrated Sustainable Agriculture Network–Mission India for Transforming Agriculture): ICRISAT; 2017.
30. Singh S, Singh A, Kumar S, Haris A, Sangle U, Bhatt B, et al. Performance of lentil succeeding rice in puddled and un-puddled transplanted conditions in drought-prone rainfed ecosystem. In: Dixit G.P., Singh J., Singh N.P., *Pulses: Challenges & Opportunities under changing climatic scenario*. Indian institute of pulses research, UP, India, 2017:173.
31. Mueller ND, Gerber JS, Johnston M, Ray DK, Ramankutty N, Foley JA. Closing yield gaps through nutrient and water management. *Nature*. 2012;490(7419):254–7.
32. Joint F. Plant mutation reports. International Atomic Energy Agency (IAEA), Vienna Austria, 2006 May;1(1).
33. Lafitte H. Research opportunities to improve nutrient-use efficiency in rice cropping systems. *Field Crops Research*. 1998;56(1–2):223–36.
34. Prasertsak A, Fukai S. Nitrogen availability and water stress interaction on rice growth and yield. *Field Crops Research*. 1997;52(3):249–60.
35. Jin Z, Yue R, Ma Z, Cheng S, Khan MN, Nie L. Effect of water and nitrogen coupling on energy balance and production efficiency in rice production. *Energy*. 2024;288:129739.
36. Haefele S, Naklang K, Harnpichitvitaya D, Jearakongman S, Skulkhu E, Romyen P, et al. Factors affecting rice yield and fertilizer response in rainfed lowlands of northeast Thailand. *Field Crops Research*. 2006;98(1):39–51.
37. Wijnhoud J, Konboon Y, Lefroy RD. Nutrient budgets: sustainability assessment of rainfed lowland rice-based systems in northeast Thailand. *Agriculture, Ecosystems & Environment*. 2003;100(2–3):119–27.
38. Dobermann A, White P, editors. Strategies for nutrient management in irrigated and rainfed lowland rice systems. Resource Management in Rice Systems: nutrients. Papers presented at the International Workshop on Natural Resource Management in Rice Systems: Technology Adaption for Efficient Nutrient Use, Bogor, Indonesia, 2–5 December 1996; 1999: Springer.
39. Rathore A, Mazid A, Pane H, Romyen P, Haefele S, Johnson D, editors. Challenges and opportunities of direct seeding in rice-based rainfed lowlands of Asia. Limited Proceedings; 2009.
40. Pandey S, Velasco L. Economics of direct seeding in Asia: patterns of adoption and research priorities. In: Pandey S., Mortimer M., Wade L., Tuong T.P., Lopez K., and Hardy B., *Direct Seeding: Research Strategies and Opportunities*. IRRI, 2002:3–14.
41. Esdaile R, Bell R, Hossain I, Haque E, Sinath P, Vial L. *Development of conservation farming implements for two-wheel tractors (power tillers) in Cambodia, Lao PDR and Bangladesh*. Australian Centre for International Agricultural Research (ACIAR): 2009.
42. Rickman J, Pyseth M, Bunna S, Sinath P. *Direct seeding of rice in Cambodia*. 2001.
43. Fukai S, Ouk M. Increased productivity of rainfed lowland rice cropping systems of the Mekong region. *Crop Pasture Science* 63: 944–973. 2012.
44. Chauhan BS. *Weed management in direct-seeded rice systems. International Rice Research Institute*; 2012.

45. Mahajan G, Chauhan BS. The role of cultivars in managing weeds in dry-seeded rice production systems. *Crop Protection*. 2013;49:52–7.
46. Tuong T, Wopereis M, Marquez J, Kropff M. Mechanisms and control of percolation losses in irrigated puddled rice fields. *Soil Science Society of America Journal*. 1994;58(6):1794–803.

2 High-temperature Tolerance in Cotton

Sajid Majeed, Aneeq ur Rehman, Iqrar Ahmad Rana, Çetin Karademir, Emine Karademir, Daniel K.Y. Tan, and Muhammad Tehseen Azhar

2.1 BACKGROUND

Cotton crop has multiple uses and is mainly utilized as a source of natural fibers in the textile industry. It is also used as a vegetable oil, feed, and fuel. Cotton is widely cultivated in sub-tropical and tropical parts of the world. In 2023–24, cotton was cultivated on an area of 31.1 million hectares, with the production of 112.9 million bales (one bale = 480 lb.). India was the largest cultivating country with the devotion of 5.14 million hectares for its production, while China ranked first with the production of 27.3 million cotton bales. Other top cotton-producing countries include India, Brazil, the USA, Pakistan, Australia, and Turkey (USDA, 2024). Cotton cultivation poses various challenges including biotic and abiotic stresses, and harvesting and post-harvest management. Heat stress due to global climate change and warming is the result of excessive carbon emissions and is one the biggest challenges to its productivity, particularly in developing countries which have limited resources to tackle the climate change issues. Increased CO_2 can enhance cotton growth under optimal conditions, but it does not mitigate the negative effects of rising temperature, reducing benefits in future climatic conditions (Broughton et al., 2017). Heat stress is caused by a rise in the mean temperature during cultivation of a crop for a sufficient period of time to cause irreversible damage to its growth and development. The range of optimal temperatures varies for each crop. Temperate crops are more vulnerable to high temperature, but tropical and sub-tropical species also have a negative response toward heat stress (Hall, 2001). However, genetic diversity also exists within the crop in its natural gene pool. Plant breeders take this as an opportunity, and are always exploring and exploiting natural variations to develop stress-resistant cultivars without compromising yield and quality. As a result, per hectare production of cotton has been increased since the last century despite climate change and other challenges. However, there is a dire need for continuous and rapid development of advanced genotypes for which novel genomic and biotechnological tools may also be utilized in plant breeding. Therefore, this chapter aims to provide the basics for understanding of the effects of heat stress on cotton productivity, its mechanisms, and mitigating strategies from the breeding point of view for researchers.

DOI: 10.1201/9781032720142-3

2.2 EFFECT OF HEAT STRESS ON COTTON

High-temperature stress can cause irreversible damage to cotton crops at almost every stage of their growth. The reproductive or flowering stage is considered to be most vulnerable to heat stress. High temperatures at both day and night times result in low productivity and yield (Khan et al., 2020). The following are some of the effects of heat stress on the cotton plant and its productivity.

2.2.1 Effect on Morphology

Heat stress may affect cotton from germination to fiber maturity. The optimal temperature for proper germination of cotton seed ranges between 28–30 °C. An increase or decrease from the optimal range can cause physiological disturbance in the seed. This results in either abnormal seedlings or the inability of seeds to germinate due to disturbed activity of its enzymes (Krzyzanowski and Delouche, 2011). Optimal environmental conditions, particularly temperature, are a prerequisite for normal seedling growth and development. The tap root system of cotton depends upon the soil temperature and conditions. The length of cotton root decreases significantly when the temperature exceeds the optimum. Temperatures more than 45 °C cause irreversible damage to its root system. Overall growth and weight of the seedlings also highly depend upon temperature, while no seedling emergence above 50 °C has been reported (Raphael et al., 2017). Lateral growth stages are also affected by heat stress, particularly high temperature at the time of bud formation which results in a lower number of lateral buds and fruit-bearing nodes. Retention of fruiting bodies declines significantly when the temperature exceeds the optimum, causing a decline in the number of flowers and boll formation. Hence, the ultimate yield of seed cotton decreases (Tariq et al., 2017). Not only the boll numbers, but the size and weight of the bolls also decrease significantly during heat stress. Overall, cotton yield depends on various factors, among which boll size and number are the most important.

2.2.2 Effect on Fiber Traits

Besides yield, the quality of the cotton fibers is also an important component from an industrial point of view. Cotton fibers require an optimal period of time and environmental conditions for their proper growth, development, and maturity. Partially developed or immature fibers are categorized as low-quality raw material in the textile industry and have a lower market value. Among various fiber quality traits, three important ones from a plant breeding point of view are fiber length, strength, and fineness. Further, these quality traits are highly influenced by the environment due to being polygenic in nature (Ali et al., 2008). Researchers have found that maintaining a canopy temperature of 25 °C to 28 °C from flowering to maturity optimizes lint yield and fiber quality in cotton (Conaty et al., 2015). Heat stress led to early maturity of the plant that results in incomplete development of the fibers within the boll. A shorter period of heat stress is less damaging to fiber quality, however, an increase in the duration of high temperature for more than one week can cause physiological damage to the cellulose synthesis process and a reduction in the quality of cotton fibers. Cotton fibers developed under heat stress are coarse in nature due to insufficient time for their maturity. Similarly, smaller-sized bolls developed under high-temperature stress conditions possess shorter fiber length with decreased strength (Bo et al., 2017).

2.2.3 Effect on Reproductive Parts

The reproductive or flowering stage and parts are considered to be the most sensitive to heat stress in cotton. The ultimate outcome of this crop is based on the quantity and quality of fibers which come from the bolls (fruits) formed from the flowers after successful fertilization. A high number of bolls set under optimal conditions is positively correlated with more seed cotton yield per plant (Majeed et al., 2024). Cotton pollen becomes non-viable under heat stress due to a change in the rate of enzymatic activities and denaturation of vital enzymes, particularly those involved in redox reactions such as cellular respiration and photosynthesis. Heat exposure at 40 °C during the tetrad stage reduced pollen size and boll production by 65% (Masoomi-Aladizgeh et al., 2020). Enzymes belonging to the dehydrogenase group are directly affected by a change in temperature. It causes a decrease in the number of available viable and normal pollen grains for the germination process. The ovaries fertilized from non-viable or abnormal pollen either result in poor boll formation or abnormal transition to subsequent developmental stages. Such fruiting bodies usually shed before reaching maturity (Najeeb et al., 2017; Abro et al., 2023). An optimal length and size of pollen tube is essential to ensure the successful transfer of pollen grains from stigma to ovary to ensure successful fertilization. Not only does the length of the pollen tube decrease but some viable pollen fails to form pollen tubes under heat stress (Burke et al., 2004). A negative effect of high-temperature stress on the length of filament, receptivity of stigma, and shape/size of the ovary has been also reported (Snider and Oosterhuis, 2012). All these factors decrease the amount of seed setting, boll formation, and boll retention on the cotton plant, which ultimately results in a poor yield as well as reduced fiber quality.

2.2.4 Effect on Molecular Homeostasis

An optimal concentration of specific micro- and macromolecules is essential for normal activities of cells for their proper development, functioning, and multiplication. These requirements of molecules change with the growth and development stages at a cellular level and are highly influenced by the external environment and stresses (Suzuki et al., 2012). Such disturbances at a cellular level change the permeability of the membrane, leading to leakage of electrolytes from the cell. They also disturb the photosynthetic efficiency by damaging chlorophyll content (Cottee et al., 2014). A change in the electrolyte and water concentrations also changes the rate of conductance of stomata. Higher conductance usually results in rapid loss of moisture and increased concentration of cellular $CO_{2.}$ This causes a rapid loss of cellular moisture content due to the increased rate of transpiration. The rates of energy production and consumption of the cell also change in such circumstances, which creates metabolic imbalances by disturbing redox reactions. Changes in the concentration of molecules cause a rapid increase in the reactive oxygen species (ROS) in the cell. The production of ROS is part of the plant's defense mechanism in response to stresses to balance the concentration of molecules. However, a significant increase in the cellular concentration of ROS causes oxidative stress in plants. Oxidative stress oxidizes proteins and cause lipid peroxidation in the cell. Both lipids and proteins are main components of cellular organelles involved in various activities for normal functioning of plant cells (Majeed et al., 2019).

2.3 MITIGATION STRATEGIES FOR HEAT TOLERANCE IN COTTON

2.3.1 Conventional Breeding

Breeding novel germplasm and developing stress-resilient cultivars is considered to be the most suitable among various strategies including a change in the cropping period, identification of new areas for cotton production, and providing continuous irrigation during the heat stress period. Heat stress tolerance in cotton is a complex feature which is controlled by a combination of various morphological and physiological attributes. Hence, it is necessary to have a suitable gene pool with significant diversity among traits for initiating a breeding program for this purpose. Germplasm storage and characterization is a continuous activity which is being carried out in many countries at a national or regional level. Furthermore, breeders also have their own genetic resources which they evaluate from time to time against certain conditions to achieve their breeding objectives. Significant diversity has been reported in cotton germplasm for heat tolerance by various researchers (Majeed et al., 2019; Jaconis et al., 2021). Cotton belongs to the genus *Gossypium*, which comprises more than 50 species. Most of the germplasm screening has been carried out for its four cultivable species only, i.e., *G. hirsutum*, *G. barbadense*, *G. arboreum*, and *G. herbaceum*, among which most of the literature is related to one species (*G. hirsutum*) only. Wild relatives of crops have evolved naturally and possess significant genetic diversity to be utilized in breeding programs. Some wild Australian cotton species were more heat resistant than cultivated cotton, with different degrees of resistance to high daytime temperatures (Dubey et al., 2024). Wild species of cotton could potentially be used to transfer heat-tolerant genes to commercial cultivars of cotton, but it is difficult to utilize their genetic potential due to their cross incompatibilities with domesticated species. However, several techniques have been developed to utilize them in cross combination, e.g. through embryo rescue, in vitro pollination methods, and bridge cross-breeding methods (Majeed et al., 2021; Saleem et al., 2021). The use of physical and chemical mutagens to produce novel genetic combinations and traits has been widely explored in cotton. A promising variety (NIAB-78), developed through mutation breeding, was released in Pakistan covering more than half of the total cotton cultivating area for years. The advancement of technologies has enabled scientists to induce the mutation at a specific site in the genome rather than randomly mutating and screening for desirable plants/traits. The use of techniques such as CRISPR/Cas9 has been increased in the last few years.

2.3.2 Molecular Basis of Heat Tolerance in Cotton

Cotton crops respond to high-temperature stress through a variety of mechanisms, including physiological, biochemical, and molecular processes. These are some of the main mechanisms that have been found in studies. Cotton cultivars that can withstand high temperatures usually have more stable cell membranes. This stability reduces cell damage by maintaining cellular integrity and function at high temperature. The lipid makeup of cell membranes has a major impact on their fluidity, flexibility, and stability. Fatty acid profile adjustments can improve membrane resilience. In order to maintain the appropriate membrane fluidity at various temperature, plants modify the 'unsaturation degree' (the quantity of double bonds) in the fatty acid (FA) tails inside their cell membranes. Membranes can become excessively fluid at higher temperature and excessively hard at low temperature. Due to their double bonds, unsaturated fatty acids cause 'kinks' in the membrane, which keep it flexible and avoid tight packing. Fatty acid desaturase (FAD) mutant plants exhibit this adaptation because their increased unsaturation levels make them particularly sensitive to temperature variations (He and Ding, 2020). This adaptation makes membrane structures more robust and flexible, which helps the cells to properly function when they are under stress. Plants that experience heat stress may form lipid rafts; these are specific areas of the membrane that are enriched in particular sterols and lipids. These rafts can improve the function and stability of membranes and are essential for signaling pathways that encourage heat tolerance (Cacas et al., 2016). The interaction between the cell membrane and the cell wall also affects the membrane's stability. Increases in the amount of lignin and pectin in the cell wall can provide the membrane with more support and boost its resistance to heat damage (Yun et al., 2013). A stronger structure that supports the membrane under heat stress may result from these modifications.

Heat shock proteins (HSPs) are essential for protecting cells from heat stress. These are divided into several groups based on their molecular weight, such as HSP60, HSP70, HSP90, and HSP100 (Khan and Shahwar, 2020). According to Usman et al. (2014), HSPs in cotton mostly act as molecular chaperones and are responsible for protein folding, refolding denatured proteins, transport, accumulation, and degradation. For example, Hsp90 plays a role in the translocation of signaling proteins such as kinases and actin. It also plays a role in the hormonal abscisic acid signaling pathway that controls stomatal closure and saves water in hot weather. Improved heat tolerance in cotton has been associated with increased HSP expression. This increased expression of certain HSPs in cotton plants has been linked to increased yield and growth. Higher levels of HSP expression are frequently seen in heat-tolerant cotton cultivars, which helps them perform better overall under heat stress (Ijaz et al., 2024). The control of gene expression in response to heat stress is one of its functions. They can improve the plant's capacity to withstand high temperatures by influencing the expression of other stress-responsive genes. The gene that encodes the production of the HSP101 protein was detected in *Arabidopsis thaliana* and transformed into the cotton plant. This transgenic cotton plant, therefore, increases the rate of pollen germination and the length of the pollen tube under high-temperature stress (Nadeem et al., 2018). More reactive oxygen species (ROS) can be produced at high temperature, which can harm cellular constituents. In order to prevent oxidative damage to plants, HSPs help detoxify ROS by controlling the activity of antioxidant enzymes (Driedonks et al., 2015). Some other antioxidants may also play a role in scavenging ROS and reducing oxidative stress, including ascorbate peroxidase, catalase, and superoxide dismutase (Mishra et al., 2023). These mechanisms collectively contribute to the ability of cotton plants to withstand high temperature, ensuring their growth and productivity in challenging environmental conditions.

REFERENCES

Abro, A.A., M. Anwar, M.U. Javwad, M. Zhang, F. Liu, R. Jiménez-Ballesta, E.A. Salama and M.A. Ahmed. 2023. Morphological and physio-biochemical responses under heat stress in cotton: Overview. *Biotechnology Reports.* 40:e00813.

Ali, M.A., M.T. Seyal, S.I. Awan, S.N. Shahid Niaz, S.A. Shiraz Ali and A.A. Amjad Abbas. 2008. Hybrid authentication in upland cotton through RAPD analysis. *Australian Journal of Crop Science* 2(3):141–149.

Bo, X., Z.-G. Zhou, L.-T. Guo, W.-Z. Xu, W.-Q. Zhao, B.-L. Chen, Y.-L. Meng and Y.-H. Wang. 2017. Susceptible time window and endurable duration of cotton fiber development to high temperature stress. *Journal of Integrative Agriculture.* 16:1936–1945.

Broughton, K.J., R.A. Smith, R.A. Duursma, D.K.Y. Tan, P. Payton, M.P. Bange and D.T. Tissue. 2017. Warming alters the positive impact of elevated CO_2 concentration on cotton growth and physiology during soil water deficit. *Functional Plant Biology.* 44:267–278.

Burke, J.J., J. Velten and M.J. Oliver. 2004. In vitro analysis of cotton pollen germination. *Agronomy Journal.* 96:359–368.

Cacas, J.-L., C. Buré, K. Grosjean, P. Gerbeau-Pissot, J. Lherminier, Y. Rombouts, E. Maes, C. Bossard, J. Gronnier and F. Furt. 2016. Revisiting plant plasma membrane lipids in tobacco: A focus on sphingolipids. *Plant Physiology.* 170:367–384.

Conaty, W.C., J.R. Mahan, J.E. Neilsen, D.K.Y. Tan, S.J. Yeates and B.G. Sutton. 2015. The relationship between cotton canopy temperature and yield, fibre quality and water-use efficiency. *Field Crops Research.* 183:329–341.

Cottee, N.S., I.W. Wilson, D.K.Y. Tan and M.P. Bange. 2014. Understanding the molecular events underpinning cultivar differences in the physiological performance and heat tolerance of cotton (*Gossypium hirsutum*). *Functional Plant Biology.* 41:56–67.

Driedonks, N., J. Xu, J.L. Peters, S. Park and I. Rieu. 2015. Multi-level interactions between heat shock factors, heat shock proteins, and the redox system regulate acclimation to heat. *Frontiers in Plant Science.* 6:999.

Dubey, G., A.L. Phillips, D.J. Kemp and B.J. Atwell. 2024. Physiological and structural traits contribute to thermotolerance in wild Australian cotton species. *Annals of Botany. mcae* 098.

Hall, D.A.E. 2001. *Heat Stress and Its Impact. Botany and Plant Sciences.* Department University of California, Riverside.

He, M. and N.-Z. Ding. 2020. Plant unsaturated fatty acids: Multiple roles in stress response. *Frontiers in Plant Science.* 11:562785.

Ijaz, A., Z. Anwar, A. Ali, A. Ditta, M.Y. Shani, S. Haidar, B. Wang, L. Fang, S.M.-U.-D. Khan and M.K.R. Khan. 2024. Unraveling the genetic and molecular basis of heat stress in cotton. *Frontiers in Genetics.* 15:1–15. doi: 10.3389/fgene.2024.1296622.

Jaconis, S.Y., A.J. Thompson, S.L. Smith, C. Trimarchi, N.S. Cottee, M.P. Bange and W.C. Conaty. 2021. A standardised approach for determining heat tolerance in cotton using triphenyl tetrazolium chloride. *Scientific Reports.* 11:5419.

Khan, A.H., L. Min, Y. Ma, Y. Wu, Y. Ding, Y. Li, S. Xie, A. Ullah, M. Shaban and H. Manghwar. 2020. High day and night temperatures distinctively disrupt fatty acid and jasmonic acid metabolism, inducing male sterility in cotton. *Journal of Experimental Botany*. 71:6128–6141.

Khan, Z. and D. Shahwar. 2020. Role of heat shock proteins (HSPs) and heat stress tolerance in crop plants. In: Rajib Roychowdhury, Shuvasish Choudhury, Mirza Hasanuzzaman, Sangeeta Srivastava, *Sustainable Agriculture in the Era of Climate Change*. Springer Nature, 211–234.

Krzyzanowski, F.C. and J.C. Delouche. 2011. Germination of cotton seed in relation to temperature. *Revista Brasileira de Sementes*. 33:543–548.

Majeed, S., M.T. Chaudhary, A.M. Hulse-Kemp and M.T. Azhar 2021. *Introduction: Crop Wild Relatives in Plant Breeding. Wild Germplasm for Genetic Improvement in Crop Plants*. Elsevier.

Majeed, S., M.T. Chaudhary, M.S. Mubarik, I.A. Rana, M. Shaban, D.K. Tan, Y. Jia, X. Du, L. Hinze and M.T. Azhar. 2024. Genetics of biochemical attributes regulating morpho-physiology of upland cotton under high temperature conditions. *Journal of Cotton Research*. 7:3.

Majeed, S., T.A. Malik, I.A. Rana and M.T. Azhar. 2019. Antioxidant and physiological responses of upland cotton accessions grown under high-temperature regimes. *Iranian Journal of Science and Technology, Transactions A: Science*. 43:2759–2768.

Masoomi-Aladizgeh, F., U. Najeeb, S. Hamzelou, D. Pascovici, A. Amirkhani, D.K. Y. Tan, M. Mirzaei, P.A. Haynes and B.J. Atwell. 2020. Pollen development in cotton (*Gossypium hirsutum*) is highly sensitive to heat exposure during the tetrad stage. *Plant, Cell & Environment*. 44:2150–2166.

Mishra, N., C. Jiang, L. Chen, A. Paul, A. Chatterjee and G. Shen. 2023. Achieving abiotic stress tolerance in plants through antioxidative defense mechanisms. *Frontiers in Plant Science*. 14:1110622.

Nadeem, M., J. Li, M. Wang, L. Shah, S. Lu, X. Wang and C. Ma. 2018. Unraveling field crops sensitivity to heat stress: Mechanisms, approaches, and future prospects. *Agronomy*. 8:128.

Najeeb, U., M. Sarwar, B.J. Atwell, M.P. Bange and D.K. Tan. 2017. Endogenous ethylene concentration is not a major determinant of fruit abscission in heat-stressed cotton (*Gossypium hirsutum* L.). *Frontiers in Plant Science*. 8:1615.

Raphael, J.P., B. Gazola, J.G. Nunes, G.C. Macedo and C.A. Rosolem. 2017. Cotton germination and emergence under high diurnal temperatures. *Crop Science*. 57:2761–2769.

Saleem, H., M.T. Chaudhary, A. Shakeel, S.H. Wani, X. Du and M.T. Azhar. 2021. *Wild Cotton Genepool: An Unopened Treasure. Wild Germplasm for Genetic Improvement in Crop Plants*. Elsevier.

Snider, J.L. and D.M. Oosterhuis. 2012. *Heat Stress and Pollen-Pistil Interactions. Flowering and Fruiting in Cotton*. The Cotton Foundation, 59–78.

Suzuki, N., S. Koussevitzky, R. Mittler and G. Miller. 2012. ROS and redox signalling in the response of plants to abiotic stress. *Plant, Cell & Environment*. 35:259–270.

Tariq, M., A. Yasmeen, S. Ahmad, N. Hussain, M.N. Afzal and M. Hasanuzzaman. 2017. Shedding of fruiting structures in cotton: factors, compensation and prevention. *Tropical and Subtropical Agroecosystems*. 20:251–262.

USDA. 2024. *Cotton: World Markets and Trade December 2024*. United States Department of Agriculture; Foreign Agricultural Service, Government of United States.

Usman, M.G., M. Rafii, M. Ismail, M. Malek, M.A. Latif and Y. Oladosu. 2014. Heat shock proteins: Functions and response against heat stress in plants. *International Journal of Science and Technology Research.* 3:204–218.

Yun, Z., H. Gao, P. Liu, S. Liu, T. Luo, S. Jin, Q. Xu, J. Xu, Y. Cheng and X. Deng. 2013. Comparative proteomic and metabolomic profiling of citrus fruit with enhancement of disease resistance by postharvest heat treatment. *BMC Plant Biology.* 13:1–16.

3 Hemp as An Environmentally Friendly Crop

E. Tsaliki, A. Kalivas, and I. Ganopoulos

3.1 INTRODUCTION

Hemp cultivation, known from ancient times for a variety of products, is currently undergoing a revival, and through the holistic approach of the crop value chain, many new environmentally friendly applications have been revealed. Cannabis is positioned as one of the most significant crops in the bioeconomy since the entire value chain of hemp, from cultivation to processing along with usage, recycling, reuse, and waste disposal, meets the goals of the circular economy action plan and addresses the issues of combating climate change.

3.2 HISTORY AND TAXONOMY OF CANNABIS

The exact origin of hemp (*Cannabis sativa* L.) is an issue of conflict; however, there much archeological evidence since the Neolithic period has been described, and both historians and archeologists agree that hemp was among the first non-food crops to be cultivated (1). The hemp plant originated from temperate central Asia and migrated around the world, because of its adaptability and importance to people who were continuously on the move. The Greeks and Romans used hemp for sailing ropes and fabrics, while Scythians were the first to use hemp as a drug. Hemp was likely used also in Persia and Arabia because the word 'hashish' is Arabic in origin and comes from 'hashish al kief,' which translates to 'dried herb of pleasure.' European explorers pushed hemp production throughout America, Australia, and western Africa in the 1500s and 1600s in order to meet the high demand for rope and sails in the shipping sector. During the 19th and 20th centuries, the decreased demand, competition from alternative fiber sources, and the fact that hemp cultivation is prohibited in many nations, led to a decrease in hemp production (2).

In the 1970s, botanists debated the classification of cannabis species – whether all varieties belong to one species *C. sativa* L. or should be divided into more distinct species. Currently, according to most scientific conventions (3), the plant is classified under the *Cannabis* genus as one species, *C. sativa* L., with three subspecies (*C. sativa* L., C. *indica* Lam., and *C. ruderalis* Jan) and many varieties (1, 4–6). Various chemical variations of cannabis exhibit morphological as well as chemical differences. The proposed morphological differences of cannabis include that *sativa* is tall with

DOI: 10.1201/9781032720142-4

narrow leaves, suitable as a hemp type; *indica* is short with wide leaves, mainly used as a drug type; and *ruderalis* is the shortest with the sparsest leaves, considered as a wild-type. These are of limited use for botanical classification because most cannabis strains encountered today are hybrids between *sativa* and *indica* that cannot be easily distinguished from their morphological characteristics (7).

The relative proportions of the cannabinoids tetrahydrocannabinol (THC) and cannabidiol (CBD) in hemp buds have been used for distinguishing three predominant chemotypes since the plant's life cycle has no impact on the ratio of CBD to THC (8). 'Drug type' was defined as the first one (chemotype I) having a low CBD/THC content ratio because the predominant cannabinoid is THC. The second intermediate type (chemotype II) has both THC and CBD, the two main cannabinoids, in a content ratio close to unity (typically ranging from 0.5 to 3.0), but CBD is slightly more prevalent. The third one, also-called 'fiber' or 'non-drug' type (chemotype III), has primarily CBD, and a THC content lower than 0.3%, and therefore has a high CBD/THC ratio (9). Other researchers describe five chemotypes. The first two chemotypes are considered as drug types with relatively high THC. The other three chemotypes have potential for both textile and pharmaceutical purposes, where THC is undesirable (10). In the late 1990s, hemp was rediscovered as a sustainable source of raw materials. Currently, there are thousands of hemp products available on the global market, while novel uses for hemp are constantly arising. Cannabis was mostly used for three commodities: bast fiber (for cordage and textiles), seed (for food and seed oil), and flowering tops (for medicinal and psychoactive drugs) (11).

Most states in the United States of America permit the cultivation of hemp as a crop, and according to data gathered by the USDA in 2021, the whole area planted with hemp was 54,152 acres, the overall harvested area was 33,480 acres, while the value of hemp production was USD 824 million (12). Growers of hemp in Europe are entitled to area-based direct payments, but they must also satisfy the standard requirements for eligibility for direct payments. Specifically, they must use only certified seed of hemp varieties listed in the EU Common Catalogue of Varieties of Agricultural Plant Species, and the hemp variety they are cultivating must have a THC content of less than 0.3% (13). However, the national limits may be different, such as Switzerland where the limit is 1.0% (14). With 70% of the EU's total production, France is the top producer, followed by the Netherlands (10%) and Austria (4%) (13). Other countries with active hemp grower and/or consumer markets are Australia, where the limit of THC is 1.0% (15), New Zealand, India, Japan, Korea, Turkey, Egypt, Chile, and Thailand (16). China is regarded as a global leader in the manufacturing of consumer textiles and hemp, but regarding China's hemp production and cultivation, official statistics are lacking (17). Hemp plants are diploid (2N D 2X D 20) and typically dioecious (having male and female flowers on separate plants). Male (staminate) plants in dioecious varieties are taller and thinner, have fewer leaves enclosing the flowers, and fade away quickly after pollen is shed. The shorter female (pistillate) plants persist till seed maturity and produce a greater number of leaves at the terminal inflorescences. X and Y chromosomes determine a cannabis plant's sex, and while monoecious plants are uncommon in nature (6), they have been developed by breeders and the genetic control of monoecy has proved that monoecious plants have XX chromosomes (2). In the modern era, monoecious varieties have been chosen to reduce the agronomic challenges related to the sexual vegetative dimorphism that occurs in dioecious varieties (18).

Although X/Y sexual inheritance, wind pollination, and dioecious sexuality are all rather uncommon in plants, cannabis includes all three of these features (6), and it has yet to benefit from the advanced level of breeding applied to other commercial crops. Cannabis plants are cross-fertile, making manual pollination of female plants an easy aspect of conventional breeding. In the natural world, pollen causes fertile female plants to stop flowering and set seeds. For cannabinoid production, female inflorescences are kept entirely isolated from pollen. This allows the plants to grow abnormally large flowers and produce more active ingredients. However, even the windborne pollen from a single male plant can destroy the crop, so it is essential for drug production to leave out all male plants. There are various ways to achieve exclusion, such as: (a) pulling out male plants prior to anthesis, or flower maturity; (b) using hormonally feminized seeds to produce mostly female plants; and (c) using vegetative propagation, or cloning, to produce uniform stands of female plants (4).

3.3 RECENT GENETICS AND GENOMICS ADVANCES IN HEMP (*CANNABIS SATIVA* L.)

Recent progress in genome-editing tools, particularly for non-transgenic genome-edited crops, offers significant opportunities in hemp cultivation. These advancements can address regulatory challenges faced by genetically modified crops and improve traits like cannabinoid production, CO_2 to biomass conversion, and soil health enhancement. High-quality cannabis genome sequences from various strains have been identified, aiding in the production of valuable bioactive molecules (19). Advances in *Agrobacterium*-mediated transformation and CRISPR/Cas9 technology have enabled the successful generation of gene-edited hemp plants. This includes manipulating genes related to cannabinoid biosynthesis, highlighting a significant step in the functional validation of gene roles in hemp (20). A review of the advances and challenges in metabolic engineering strategies for industrial hemp highlights the potential for increasing the production of desirable phytochemicals. This includes the use of CRISPR/Cas systems and transgenic methods for improving hemp's pharmaceutical and industrial applications (21). A genome-wide association study (GWAS) has revealed the genetic architecture underlying fiber quality in hemp. This research identified quantitative trait loci (QTLs) and candidate genes related to cellulose, hemicellulose, and lignin biosynthesis, crucial for developing high-quality hemp fiber cultivars (22). Mapping of QTLs associated with various agronomic and biochemical traits in hemp has provided insights into the genetic control of these traits. This research has identified key loci and candidate genes, significantly modernizing the understanding of complex traits in hemp (23).

GWAS studies have elucidated the genetic mechanisms of flowering time and sex determination in hemp. Identifying genes related to light perception, transcription factors regulating flowering, and phytohormone balance offers valuable tools for hemp breeding (22). Furthermore, research on PCR primers for early identification of male hemp plants showed that specific primer sets could be effective tools for determining plant sex at the seedling stage, which is crucial for hemp cultivators focusing on CBD production (24). A study using whole-genome resequencing of various hemp accessions provided insights into its domestication history. The research identified candidate genes associated with traits differentiating hemp and drug cultivars and shed light on the evolution of hemp and its cannabinoids (25). Multi-omics studies have contributed significantly to understanding the molecular mechanisms in hemp. These comprehensive approaches help to identify correlations between biological processes and metabolic pathways, crucial for elucidating the relationships between cannabis sub-species (26).

In conclusion, the advancements in genetic and genomic research in hemp are paving the way for improved strains with desired traits, enhancing both agricultural and industrial applications of this versatile crop.

3.4 AGRONOMY OF HEMP

Hemp crops are high yielding and do not need additional fertilizers and almost no pesticides. These characteristics in most fields, allow their implementation in intensive agricultural systems and possibly as a long-term strategy for conservation farming systems that are particularly appreciative of environmental and climate policy objectives (18). As an annual crop, hemp can reach heights of one to five meters. It is a herbaceous, wind-pollinated crop with both dioecious and monoecious varieties that have been developed through breeding and selection (4). Optimal growth occurs in well-drained, medium-textured soils (pH 6.0–7.0) that maintain high moisture levels and experience average temperatures of 16–27°C. Soil preparation before sowing is similar to that for spring crops, and the time required from emerging to harvesting the plant is 3–4 months. The final yield is determined by harvest timing, nitrogen content, and seeding density (27). Compared to other crops, hemp requires less water because of its well-developed root system, which prevents soil erosion (28).

Soil preparation for hemp sowing should be smooth and not compacted, as with other spring crops. The planting season of industrial hemp in the northern hemisphere is from the second half of April (29) to mid-May (30). These planting dates enable the period of vegetative growth to align with the optimal growing temperatures and the longest days, which is necessary to maximize stem growth and delay flowering of inflorescences (31). Earlier or later plantings in the northern hemisphere reduce crop growth and yield, mainly in response to short day lengths, low temperatures, and reduced solar radiation (29). The primary determinants of hemp quality and yield in the field are not only the plant density, the input of fertilizers, and the irrigation schedule, but mainly the interaction of genotype, environment, and agricultural management (9). Temperate climates are ideal for the growth of cannabis plants, and the planting should be done in a well-prepared fertile seedbed that is not compacted. The amount of light that hemp crops receive correlates directly to their biomass output when the plants are healthy and not subject to any restricting conditions, such as water or nutrient deficiencies (32). Early flowering has to be avoided because the plant minimizes its vegetative growth once the blossom begins, and consequently the final production output is affected (30).

The plants have a branched taproot system that typically grows between 30 and 60 cm below the surface, but in loose soils, it can reach a depth of 2.5 meters (33). The hemp stems are usually angular, branching with a woody interior and, occasionally, the internodes are hollow. The interior material is known as hurd, while the outer part is commonly referred to as 'bark' or bast. The height of the stems can differ, but during the vegetative phase, a phyllotaxy with two synchronous helices is obtained. At the beginning of flowering this arrangement becomes asynchronous, producing an alternating leaf pattern. The number of leaflets varies, because of the genetic origin of each cultivar, but is always odd and increases progressively on the green, palmate leaves. The hemp plant remains green during its development, although during maturing, the lower leaves on the stem shed, especially where the sowing density is high. The leaves feature small secretory epidermal glands called resinous trichomes, sporadically located on their upper (adaxial) and lower (abaxial) surfaces. Various types of glandular, or non-glandular epidermal trichomes may develop (34).

Hemp requires a photoperiod of 12–14 hours daylight during the vegetation phase, since is a short-day plant sensitive to photoperiods. A more extended photoperiod increases plant height, delays flowering, and prolongs the vegetative stage, which is preferable to hemp genotypes grown for fiber and biomass production. A longer time of darkness, however, results in early blooming and lower yield (35).

A high amount of soil water is necessary for hemp crops during the first phase of root establishment. After that point, if the root system is well developed, cannabis may be able to tolerate relatively drier situations (36). Nevertheless, the hemp genotype, agroclimatic area, soil properties, meteorological factors, and evapotranspiration significantly affect the quantity of water needed for hemp production. In the semi-arid Mediterranean environment of southern Italy, it was found that 250 mm of water was optimal for monoecious early fiber genotypes, while dioecious late genotypes needed at least 450 mm (30). Additionally, as reported by Di Bari et al. (37), hemp needs more water (435 mm) than sunflower and soybean, but less than sorghum. However, these investigations were carried out in temperate northern latitudes, where evapotranspiration is lower than in southern latitudes. In the Mediterranean climate of northern Greece, six monoecious hemp varieties were tested in a 3-year field experiment and the results showed that the choice of genotype affected hemp production. Therefore, it would be helpful for farmers to grow hemp varieties that are most suitable to their specific environment (38).

Fiber-producing hemp is tall and densely planted, and the harvesting demands differ from those of the shorter, bushier seed crops that are planted less densely (39). The harvesting process is affected not only by the height of the plants and their extremely strong and highly lignified fiber, but also by the handling of enormous amounts of fresh biomass (40). When hemp plants are cultivated for two separate uses, harvesting becomes more complicated since the leftover stalks are taken for fiber production after the seed heads are harvested, and efficient seed and biomass separation is ideal only with modified cereal combine harvesters (41). Harvesting machines that could at the same time harvest and sort straw and inflorescences have been developed recently.

Fluorescents can be picked at the blooming stage to extract cannabinoids, or at seed maturity for grain threshing, depending on the final product (40).

Cannabis plants can be grown in three environments: indoors, outdoors, and greenhouses. The grower's choice of approach is heavily influenced by the way the plant will be used in the end. Outdoors, which is the most common method, is mainly applied for fiber and seed production. For cannabinoid-oriented production, indoor and greenhouse methods are preferable since they allow better control of the growing conditions. Compared to indoor and greenhouse production, the energy required to produce 1 gram of dry flowers is 18 times lower in outdoor cultivation, but outdoor cultivation could be the viable option only in climates like that of southern Europe. Furthermore, compared to outdoor cultivation, indoor cultivation generates around 25 times more carbon dioxide per gram. Concerning the production of extracts from flowering tops, it has been documented that indoor cultivation could result in more grams per square meter, but outdoor crops could provide more grams per kWh of energy (42). Cross-pollination between varieties may be a major problem in outdoor hemp fields where wind and/or insects could promote cross-pollination. Apart from pollination, many stresses, including wind, hail, lack of moisture, pests, and others can raise the total quantity of THC within crops, and therefore producers could face the risk of exceeding the current laws of each country, which might lead to fines and crop damage. These problems are eliminated by indoor cultivation, which offers full control over each phase of the cultivation process and provides the desired quality and quantity of flowering tops. This is the reason why indoor cultivation is exclusively used for medical cannabis production (42).

3.5 SUSTAINABILITY OF HEMP

Being one of the fastest-growing plants, hemp helps not only to protect the environment but also aligns with the objectives of the European Green Deal (EGD), which include decreasing greenhouse gas emissions, developing a circular economy, handling efficiently natural resources, and moving from fossil fuels to renewable energy sources (43).

The European Industrial Hemp Association (44) in 2023 published a comprehensive study realized by the Nova-Institute, titled 'Unlocking Carbon Transfer: Wood and Hemp for Sustainable Construction,' which claims that while hemp takes only 5 months to grow, it can sequester 9–15 tonnes of CO_2 per hectare, a quantity comparable to what a young forest can accomplish. Furthermore, hemp crops support biodiversity, absorb large quantities of carbon dioxide, slow down climate change, and do not require a lot of pesticides or herbicides, maintaining environmental sustainability (45). The crop may be zero-carbon or carbon-negative depending on the cultivation and processing methods used and, as reported by Bouloc et al. (46), 1 ha of a hemp-cultivated field stored 3.06 tons of carbon.

Using life cycle analysis to compare hemp to other potential energy crops, Hanegraaf et al. (47) found that hemp was one of the best crops taking into consideration energy balance, greenhouse gas emissions, resource uses, biodiversity impacts, and economic returns. Using life cycle assessment (LCA) and carbon footprint (CF) methodologies in the Mediterranean region, Campiglia et al. (48) have determined the environmental sustainability of various hemp varieties. From the LCA and CF results, it was found that the most environmentally sustainable variety was Ferimon, while Epsilon 68, Futura 75, and Santhica 27 varieties gave the highest impact. The LCA results for the other agricultural practices demonstrated that while minor CF resulted from using less N fertilizer, the negative effects increased with decreasing plant densities and N fertilizer. The genotypes E68/F75/S27, grown with 50 kg ha^{-1} of N fertilizer and 40 plants m^{-2}, had the largest impact (18.720 kg CO_2 eq), while Ferimon grown with 50 kg ha^{-1} of N fertilizer and 120 plants m^{-2} had the lowest (0.161 kilogram CO_2 eq).

Eco-friendliness has received little attention in the past because monocrops have dominated the agriculture sector and little data are available comparing the ecological benefits of different crops. A report titled 'Evaluating the Ecological Friendliness and Biodiversity of the Hemp Crop' evaluated the ecological friendliness of *C. sativa* L. as a fiber and oilseed crop, comparing hemp with 20 monocrops using crop characteristics that affected biodiversity. According to the author's assessment, alfalfa is the most beneficial to biodiversity, followed by wood trees and hemp. Several significant global crops, such as potatoes, cereals, cotton, sugar cane, sunflowers, rapeseed, soybeans, and tobacco, performed poorly in terms of these environmental parameters (49).

Hemp demonstrates several agronomic functions and can be a great option for environmentally friendly actions. The processing of hemp generates zero waste because all parts of the plant can be further utilized. Hemp crops also help regenerate soils and can significantly increase the benefits of land reclamation and rotations. The extensive root system of industrial hemp, according to Adesina et al. (36), optimizes soil porosity and friability while regenerating soil structure, while numerous studies have proved that hemp could reduce the amount of harmful fungi and parasitic nematodes in the soil (29, 50).

The ability of hemp plants to remove heavy metals from the soil – phytoremediation capacity – can also be significantly efficient for ecosystem recovery and land reclamation. As the cannabis root system grows, it can eventually eliminate heavy metals and other contaminants from deeper soil layers (51). Furthermore, because *C. sativa* L. plants store carbon in their stems and roots through photosynthesis, they can absorb significant amounts of carbon dioxide and help to lessen the consequences of climate change (36, 52).

Weed problems in hemp cultivation are usually limited, however this depends on agronomic practices as well as environmental factors. According to Jankauskienė et al. (53), hemp fields had a greater amount of weeds in a given year with high air temperatures in the late spring. However, since hemp may overcome most weed species in favorable growing conditions – like early planting and adequate moisture – fewer herbicides typically are needed in hemp cropping systems. After emerging from the ground, hemp often covers the area rapidly and outcompetes most weeds, however, at wider spacing, hemp frequently fails to develop canopies dense enough to inhibit the growth of weeds (54).

Hemp is consumed by many insect species. Three distinct insect classes were found in a preliminary study conducted at the Institute of Plant Breeding and

Genomic Resources (IPGRB) in Thermi, Greece. The most common class was Hemiptera, which primarily consisted of herbivores. Among them, many stink bug species (Pentatomidae) were identified, including *Halyomorpha halys*, *Nezara viridula* (L.), and *Dolycoris baccarum* (L.); additionally, some predator species were noted, including *Zelus renardii Kolenati* (Hemiptera: Reduviidae), *Nabis* sp. *Latreille* (Hemiptera: Nabidae), and *Rhynocoris iracundus* (Poda) (Hemiptera: Reduviidae) (55). Feeding brown marmorated stink bugs (BMSBs), *Halyomorpha halys*, in industrial hemp fields of Greece, indicated that BMSB adults may significantly decrease the protein and oil content of hemp seeds (56).

3.6 ENERGY PRODUCTION BY HEMP

Cannabis has also been classified as an energy crop thanks to its high biomass and potential to generate solid fuels like briquettes or pellets for heat production, as well as biodiesel, biogas, and bioethanol (57). Hemp is a promising crop for producing energy and biofuels because of its high content of cellulose as well as its productivity (58). Hemp can produce two types of fuels: hemp biodiesel which is made from the oil of pressed hemp seed and hemp ethanol/methanol mainly made from the fermented stalk. Hemp gas yields, with an energy output of about 296 GJ/ha, were equivalent to those of other annual crops, such as corn and sugar beets (*Beta vulgaris* L.) (59). According to Rice (60) and Zegada-Lizarazu and Monti (61), hemp may be effective in (energy) cropping system rotations at both northern and southern latitudes in terms of disrupting pest cycles and enhancing soil quality. Ahmad et al. (62) reported that conversion rates from hemp oil to fatty acid methyl esters are usually 90% or higher, and in some cases, the product yield has exceeded 97%. However, until large-scale manufacturing generates economies of scale that lower their costs, the current value of these oils for human consumption would probably limit their utilization in industrial bioenergy systems.

Industrial hemp is valuable as a raw material for social sustainability because of its ability to create local and regional supply networks. These regional supply networks are supported by two important factors. First, transporting harvested hemp stalks is expensive due to the low density of stalk bundles and, secondly, local soil and climate factors determine the most suitable hemp varieties in any particular area. Due to these two factors, it is more profitable to grow, process, and produce goods derived from hemp locally, near its source fields (63).

3.7 HEMP AS A PHYTOREMEDIATOR

Hemp is suitable for phytoremediation because it is recognized as a hyperaccumulator – a plant that may accumulate metals or metalloids in its tissues in hundreds or thousands of times greater quantities compared to other plants (64). Degraded land can be converted to productive agricultural land using hemp crops, especially in regions where heavy metal contamination is a problem. According to Mahmood et al. (65), hemp can aid in the stabilization of contaminated environments, is resistant to pollutants, and can also accumulate metals. Furthermore, hemp offers the potential to support a circular economy model, in contrast to the majority of plants used in phytoremediation, since its yield and dead biomass byproducts can be used to create a variety of alternatives for household or industrial products (66). Hemp has an extensive root system, which may penetrate deep soil substrate layers and thereby absorb heavy metals, which is one of its qualities that contribute to its phytoremediation ability. In particular, industrial hemp has a root system that allows it to reach quite deep, which makes it possible to extract metals from it. Due to its increasing durability in high concentrations of heavy metals, and its notable tissue yield potential, hemp permits the accumulation and dilution of heavy metals (66). According to Placido and Lee (67), hemp can be used to clean up areas affected by metals, pesticides, solvents, explosives, crude oil, polyaromatic hydrocarbons, and toxins, while hemp hurd and fiber quality are not affected (68).

3.8 USES OF HEMP FIBER

Hemp's basic and best-known use since ancient times has been in textiles. The fibers of the plant are strong enough to allow its use in clothing fabrics, ropes, shoes, sailing fabrics, etc. Given their low density as well as their specific mechanical properties, hemp fibers can be spun into threads, ropes, woven fabrics, canvas, and other materials. A novel and potential use for hemp fibers is the production of insulating products and fiber-reinforced composites (43). A hemp stem or stalk is made up of two sublayers: the inner layer is made from hurd or shive fiber, which is a less valuable woody (xylem core body) component, and the outer layer is formed from more valuable fiber bundles known as bast fiber (69).

According to Jankaauskiene et al. (53), the outer bast of hemp is composed of 60–70% cellulose, 15–20% hemicellulose, 2–4% lignin, 2–4% pectin, and 1–2% fat and wax. Three different types of fibers make up the outer bast: primary bast fibers, short bast fibers, and intermediate-length secondary bast fibers. Hemp's woody core has a chemical profile that is close to that of hardwood, with around 40% cellulose, 20% hemicelluloses, and 20% lignin (11).

Core fibers are extracted from bark mechanically with a decorticator or by dry or wet retting, as well as by combining the two methods. Hemp fibers that have been dried and baled can then undergo additional mechanical separation (cottonizing, shredding into smaller lengths, and spinning into yarn, for example) after being separated (43). There are four steps involved in the fiber extraction process: retting, breaking, scutching, and hackling. Field retting lasting 4–6 weeks and water retting lasting 1–2 weeks are the most widely used retting techniques. During retting, pectins are broken down, and fungi and naturally occurring microorganisms bond the fiber together. Then, fluted rollers are used to break down the stems. The stems are split by a beater during the scutching process and the fiber is separated and straightened by hackling and combing.

Hemp fiber has a diameter of 16–50 microns and a length of 1–5 meters. Textiles made of hemp have certain unique characteristics that make them ideal for a variety of specialized uses. Hemp textiles have inherent antibacterial qualities, resist mold and bacteria, shade well, and are good heat conductors, while also blocking ultraviolet light (70). In another study, hemp hurd powder exhibited antibacterial effects against *Pseudomonas aeruginosa* and *Escherichia coli* (71). The antibacterial properties of hurd, which contain more lignin than bast fibers, may be attributed to lignin-related substances such as phenolic compounds, alkaloids, and cannabinoids. The market share of textiles, fabrics, and apparel derived from hemp fiber has expanded worldwide due to the growing demand for environmentally conscious and eco-friendly natural products. Because of their potential to reduce the spread of infections in medical and healthcare settings, antibacterial fabrics have drawn a lot of attention recently. Hemp has antibacterial properties that combat a variety of harmful microorganisms (72) mainly because of active antibacterial components such as alkaloids, flavones, and saponins (70). Furthermore, hemp possesses anti-mildew qualities, and hemp fabric is hypoallergenic (73). Socks made from hemp fiber are popular because they are naturally antifungal and antibacterial and they keep the feet odor-free (74).

The advantages of hemp fibers over synthetic fibers which reinforce composites are significant, and their high specific strength, low density, low production cost, bio-renewable nature, and environmentally friendly behavior make them suitable for a wide range of applications (27). Biodegradable mulch, horticultural planting materials, pressed fiber goods, paper and pulp products, building and construction materials, insulating materials, hurd-produced animal bedding, plastic biocomposites, and compressed cellulose plastics are among the other products made from hemp fiber (63). Furthermore, hemp straw was recommended as bedding for horses and cows mostly because of its hydrophilic qualities, which allow it to provide a stable, dust-free surface that does not slide or slip (74).

In China, hemp has been utilized as a raw material for papermaking since 105 AD. The US Constitution and Declaration of Independence, along with the first copy of the Bible have been linked to the wonderful history of hemp paper (75), but over time, the technology vanished as a result of the development of wood-based paper, cutting-edge technology, and consumer demand (76).

3.9 GRAIN OR SEED PRODUCTS

Hemp seed has historically been important in the human diet (77). According to Jeliazkov et al. (78), it contains 20–30% edible oil, 20–30% protein, 20–25% fiber, 20–30% carbohydrates, and several additional nutrients and vitamins for human consumption. Among the goods made from hemp seed are: whole and dehulled cannabis seeds, hemp seed oil, hemp seed flour, hemp seed cake (a result of mechanical oil pressing), hemp seed meal, hemp hulls, and hemp protein isolates and concentrates (41).

Hempseed consists of almost 30% oil, of which more than 80% is polyunsaturated fatty acids. Two important acids that are particularly abundant in hempseed oil are linoleic acid (18:2, omega-6) and alpha-linolenic acid (18:3, omega-3) in a ratio of about 3:1, which are thought to be ideal for a healthy human diet (79), and not usual compared to other common vegetable oils. Because hemp seed has the ideal ratio of omega-3 to omega-6, it can also be beneficial for fish and bird feed (80).

Hempseed oil is often extracted using solvents or cold pressing and the by-product that results is called hempseed meal (HSM). HSM, depending on the type of cannabis variety used, may contain 30–40% of protein in the dry matter (DM) of flour and can serve as a nutritious feed additive, due to its high content of protein and energy (81). Cold-pressed seed oil has more healthy ingredients like natural antioxidants which defend against aging-related illnesses such as cancer, heart disease, and other problems (82), while gamma-linolenic acid, another ingredient in hemp seed oil, has the potential to prevent osteoporosis (83).

Seven industrial hemp varieties were grown in Greece for 3 years and the seeds were studied for antioxidant activity, phytochemical profile, and nutritional value. According to this study, the properties of hemp seeds are influenced by growth year and genotype, but hemp seeds may prove to be a promising food crop, owing to their high nutritional content and their potential as antioxidants (84).

Hemp seeds, depending on the variety, are ovoid or spherical in shape and 3–5 mm in length. The pericarp, which surrounds the seeds, effectively acts as a protective shell of the two cotyledons (embryonic leaves), which are rich in proteins and oil but have less albumen, compared to other oil crops. Following the removal of the bract, the seeds are crushed under intense pressure in order to produce oil and husk, which is additionally utilized as a fertilizer and cow feed (11, 85).

3.10 METABOLITES

In addition to the above products, hemp inflorescence produces a resinous blend from over 100 known cannabinoids and aromatic compounds, mostly terpenes (86). While each cannabinoid has different biochemical properties, they can only be produced and accumulated in glandular trichomes found on capitate stalks. THC, the main psychoactive ingredient in cannabis, is produced by decarboxylation of its acid form and is approved for medical use in numerous countries (77). The review of Caffarel et al. (87) gives a summary of the current knowledge regarding cannabis potential as a breast cancer treatment. The primary byproduct of THC breakdown is cannabinol, or CBN, which is produced when the plant is exposed to light or stored for a long time, but only exists in trace amounts during the formation of hemp buds (88).

Cannabidiol (CBD) is another cannabinoid that can be found in significant concentrations in hemp. While CBD is not psychoactive, several scientific studies have demonstrated its potential as a cure for a wide range of illnesses and symptoms. Among these are anxiety disorders (including post-traumatic stress disorder), diabetes, cancer, obesity, dystonia, epilepsy, neurodermatitis, and Alzheimer's disease. Because of its antibacterial qualities, CBD can be used to reduce inflammation and prevent infection mainly because it acts well against streptococci, staphylococci, and even clinically significant MRSA (methicillin-resistant *Staphylococcus aureus*). At lower dosages, CBD has physiological effects that preserve health along with anti-inflammatory, neuroprotective, and antioxidant properties stronger than tocopherol (vitamin E) or ascorbate (vitamin C). CBD can be used as a cosmetic element to reduce sebum and sebocytes. Studies have demonstrated the efficacy of CBD in curing problems like psychosis and epilepsy, as well as its greater tolerability and fewer adverse effects when compared to traditional treatments for these disorders (89).

Hemp also contains flavonoids, spiroindans, dihydrostilbenes, phenanthrenes, and dihydrophenanthrenes. The distinct flavor of cannabis comes from the volatile terpenes found in essential oils; monoterpenes account for 47.9–92.1% of total terpenes, while sesquiterpenes make up 5.2–48.6%. This particular flavor helps training of dogs to identify illegal marihuana. Although the physiological function of resin and volatile oils is less understood, it is likely that they serve as animal defense mechanisms or to attract insects. Furthermore, at high temperatures, the ether oils reduce the transpiration of plants (90).

3.11 HEMPCRETE

Hempcrete is a composite that is made from a mixture of hemp hurds crushed to a length of 5–40 mm, with lime used as a binder along with water (91), and has gained a lot of attention lately as a natural resource for construction materials (27). Hempcrete is a suitable non-load-bearing material for walls, doors, and roof insulation because of its lightweight, excellent moisture buffering, thermal insulation, and acoustic properties (92). Hempcrete provides long endurance, is thermally stable, makes homes fire-resistant, and can replace concrete effectively in a wide range of buildings and insulation (93).

Many of the currently existing environmental problems could be resolved if hemp was used in the construction process. Hempcrete's ability to absorb more carbon from the air during the growth of hemp plants than it releases during manufacturing is its main benefit when used in buildings. Furthermore, it keeps absorbing carbon after being used in buildings, storing more throughout the building's life than was released during construction (75).

3.12 HEMP AS FEED

The scientific community has raised numerous concerns and doubts about the safety of hemp application as animal feed, not only to the health of animals but also for the safety of humans who consume the produced animal product. Industrial hemp can be used as animal feed in three forms, as seeds, as residue after extraction of the oil, and as hay either from the whole plant or from its upper part. Despite research findings indicating the presence of anti-dietary elements including tannins, phytic acid, and trypsin inhibitors, all three types show a significant nutritional value that makes them a very good feed for both ruminants and monogastric animals. Specifically, for dairy cows, the amount of cannabis considered safe, based on the final content of cannabinoids in the milk, is 0.5–1.5 kg DM of hay per cow per day, having in mind always the daily milk production of the cow and the content of hemp in cannabinoids (94). Kalaitzidis et al. (95) showed that a Holstein dairy cow diet enriched with 3.5% hemp seed cake (HSC), at a daily weight of 1 kilogram per cow, did not affect the amount or composition of milk produced but influenced the fatty acid profile of produced milk. Although research for the use of cannabis in animal production is still at the early stages, its safety has been proven in experimental studies, where it appears that the amounts that can cause a small number of side effects (toxicity) are much higher than the permitted levels that are usually used in animal rations (75, 96).

3.13 NEW HEMP PRODUCTS

Cannabis essential oils have been noted to be a potentially environmentally friendly substitute for pesticides, despite the high degree of heterogeneity found in studies. This is due to the plant's fascinating mechanisms of action, eco-friendly cultivation methods, and the recent regulations that have allowed and supported large-scale hemp cultivation (97). Amaducci et al. (18) stated the potential use of hemp extracts as bio-pesticides in medicinal applications against insects, nematodes, mesophilic fungi, and possibly weeds, poses a new challenge for hemp cultivation. Hesami et al. (98) proposed cannabis as a new ornamental plant because their study indicates that cannabis has high ornamental potential both as a potted and landscaping plant, with high feasibility for sustainable exploitation.

Finally, new developments in technology have increased the applications of hemp fiber and hurds in the creation of plastics, oil-absorbent materials, carbon nanosheets, 3D-printed filaments, and fiber-reinforced polymer composites (11).

3.14 INHIBITORY FACTORS OF HEMP CULTIVATION

Despite the above, industrial hemp cultivation is sometimes accompanied by allergies and breathing difficulties. According to several studies, hemp pollen can cause allergic skin reactions, sometimes accompanied by symptoms similar to asthma (99). Hemp decortication and organic dust have also been linked to allergy symptoms (43). Processing hemp is quite labor-intensive, and it is practiced exclusively in East Europe and Asia because of the lower wages in those countries (100, 101). Hazardous working conditions, noise pollution, and poor posture can further increase the risk of physical harm and musculoskeletal illnesses ,and only with sufficient training and knowledge can these risks can be mitigated.

REFERENCES

1. McPartland JM, Hegman W. Cannabis utilization and diffusion patterns in prehistoric Europe: a critical analysis of archaeological evidence. *Vegetation History and Archaeobotony*. 2018; 27: 627–634. https://doi.org/10.1007/s00334-017-0646-7.
2. Fike J. Industrial hemp: renewed opportunities for an ancient crop. *Critical Reviews in Plant Sciences*. 2016; 35(5–6): 406–424. https://doi.org/10.1080/07352689.2016.1257842.
3. ISTA (International Seed Testing Association). *ISTA list of stabilised plant names*, 6th ed. Published by The International Seed Testing Association (ISTA) Zurichstr. 50, CH-8303 Bassserdorf, Switzerland, 2014. Available from: https://redebiotecnologia.files.wordpress.com/2018/09/ista-lista-de-nomes-estabilizados-2013.pdf [Accessed: 16 January 2024].

4. Rahn B, Pearson JB, Trigiano RD, Gray DJ. The derivation of modern cannabis varieties. *Critical Reviews in Plant Sciences*. 2017; 35(5-6): 328–348. https://doi.org/10.1080/07352689.2016.1273626.
5. Sawler J, Stout JM, Gardner KM, Hudson D, Vidmar J, Butler L et al. The genetic structure of Marijuana and Hemp. *PLoS One*. 2015 Aug 26; 10(8): e0133292. https://doi.org/10.1371/journal.pone.0133292.
6. Clarke RC, Merlin MD. Cannabis domestication, breeding history, present-day genetic diversity, and future prospects. *Critical Reviews in Plant Sciences*. 2017; 35(5–6): 293–327. https://doi.org/10.1080/07352689.2016.1267498.
7. Vergara D, Baker H, Clancy K, Keepers KJ, Mendieta P, Pauli CS, et al. Genetic and genomic tools for *Cannabis sativa*. *Critical Reviews in Plant Sciences*. 2017; 35(5–6): 364–377. https://doi.org/10.1080/07352689.2016.1267496.
8. Staginnus C, Zörntlein S, de Meijer E. A PCR marker linked to a THCA synthase polymorphism is a 426 reliable tool to discriminate potentially THC-rich plants of *Cannabis sativa* L. *Journal of Forensic Science*. 2014; 59: 919–926. https://doi.org/10.1111/1556-4029.12448.
9. Sikora V, Berenji J, Latković D. Influence of agroclimatic conditions on content of main cannabinoids in industrial hemp (*Cannabis sativa* L.). *Genetika-Belgrade*. 2011; 43(3): 449–456. https://doi.org/10.2298/GENSR1103449S.
10. Jin D, Jin S, Chen J. Cannabis classification systems and growth trends of the North American medical Cannabis industry. Paper published on Conference Journal "The Second Annual Academic Conference of the Specialty Committee of TCM Pharmacognosy of the World Federation of Chinese Medicine Societies". 2015; 23–27. Available from: www.researchgate.net/publication/321596927_Cannabis_Classification_Systems_and_Growth_Trends_of_the_North_American_Medical_Cannabis_Industry [Accessed: 16 January 2024].
11. Schluttenhofer C, Yuan L. Challenges towards revitalizing Hemp: a multifaceted crop. *Trends in Plant Science*. 2017; 22(11): 917–929. https://doi.org/10.1016/j.tplants.2017.08.004.
12. USDA. State, Territory, and tribal hemp program contact information. 2023. Available from: www.ams.usda.gov/rules-regulations/hemp/information-producers/grower-contact-list [Accessed: 11 January 2024].
13. EU (European Union) Hemp data. Agriculture and rural development. Available from: https://agriculture.ec.europa.eu/farming/crop-productions-and-plant-based-products/hemp_en [Accessed: 11 January 2024].
14. Willi, C. Cannabis law and legislation in Switzerland: CMS expert guides, Cannabis law and legislation in Switzerland | CMS Expert Guides. 2023. Available from: https://cms.law/en/int/expert-guides/cms-expert-guide-to-a-legal-roadmap-to-cannabis/switzerland [Accessed: 11 January 2024].
15. Carruthers M. Review on the regulation of industrial hemp-based food products and cosmetic products in Australia. *HWL Ebsworth Lawyers*. 2023. Available from: https://hwlebsworth.com.au/review-on-the-regulation-of-industrial-hemp-based-food-products-and-cosmetic-products-in-australia/ [Accessed: 11 January 2024].
16. Johnson R. *Hemp as an agricultural commodity*. Washington, DC, USA: Congressional Research Service; 2013. p. 1–29. Available from https://sgp.fas.org/crs/misc/RL32725.pdf [Accessed: 16 January 2024].
17. USDA. *China: China's hemp and hemp products import policies*. USDA Foreign Agricultural Service; CH2021-0113, 2023. Available from: https://fas.usda.gov/data/china-chinas-hemp-and-hemp-products-import-policies [Accessed: 11 January 2024].

18. Amaducci S, Scordia D, Liu FH, Zhang Q, Guo H, Testa G, et al. Key cultivation techniques for hemp in Europe and China. *Industrial Crops and Products*. 2014; 68: 2–16. https://doi.org/10.1016/j.indcrop.2014.06.041.
19. Shiels D, Prestwich BD, Koo O, Kanchiswamy CN, O'Halloran R, Badmi R. Hemp genome editing–challenges and opportunities. *Frontiers in Genome Editing*. 2022; 4: 823486.
20. Zhang X, Xu G, Cheng C, Lei L, Sun J, Xu Y, Deng C, et al. Establishment of an Agrobacterium-mediated genetic transformation and CRISPR/Cas9-mediated targeted mutagenesis in Hemp (*Cannabis Sativa* L.). *Plant Biotechnology Journal*. 2021; 19(10): 1979–1987. https://doi.org/10.1111/pbi.13611.
21. Deguchi M, Kane S, Potlakayala S, George H, Proano R, Sheri V et al. Metabolic engineering strategies of industrial hemp (*Cannabis sativa* L.): a brief review of the advances and challenges. *Frontiers in Plant Science*. 2020; 11: 580621. https://doi.org/10.3389/fpls.2020.580621.
22. Petit J, Salentijn E, Paulo M, Denneboom C, Trindade L. Genetic architecture of flowering time and sex determination in hemp (*Cannabis sativa* L.): a genome-wide association study. *Frontiers in Plant Science*. 2020; 11: 569958. https://doi.org/10.3389/fpls.2020.569958.
23. Woods P, Campbell B, Nicodemus T, Cahoon E, Mullen J, McKay J. Quantitative trait loci controlling agronomic and biochemical traits in *Cannabis sativa*. *Genetics*. 2019; 219(2). https://doi.org/10.1093/genetics/iyab099.
24. Nalesnik A, Erickson L. Analysis of published PCR primers to determine sex of industrial hemp seedlings. *The FASEB Journal*, 2020; 34. https://doi.org/10.1096/fasebj.2020.34.s1.08695.
25. Ren G, Zhang X, Li Y, Ridout K, Serrano-Serrano M, Yang Y, et al. Large-scale whole-genome resequencing unravels the domestication history of *Cannabis sativa*. *Science Advances*. 2021; 7. https://doi.org/10.1126/sciadv.abg2286.
26. Sirangelo TM, Ludlow RA, Spadafora ND. Multi-omics approaches to study molecular mechanisms in *Cannabis sativa*. *Plants*. 2022; 11(16): 2182. https://doi.org/10.3390/plants11162182.
27. Ahmed ATMF, Islam MZ, Mahmud MS, Sarker ME, Islam MR. Hemp as a potential raw material toward a sustainable world: a review. *Heliyon*. 2022 Jan 13; 8(1): e08753. https://doi.org/10.1016/j.heliyon.2022.e08753.
28. Andre CM, Hausman JF, Guerriero G. *Cannabis sativa*: the plant of the thousand and one molecules. *Frontiers in Plant Science*. 2016 Feb 4; 7: 19. https://doi.org/10.3389/fpls.2016.00019.
29. Van der Werf HMG, van Geel WCA, van Gils LJC, Haverkort AJ. Nitrogen fertilization and row width affect self-thinning and productivity of fibre hemp (*Cannabis sativa* L.). *Field Crops Research* 1995; 42(1): 27–37. https://doi.org/10.1016/0378-4290(95)00017-k.
30. Cosentino SL, Riggi E, Testa G, Scordia D, Copani V. Evaluation of European developed fibre hemp genotypes (*Cannabis sativa* L.) in semi-arid Mediterranean environment. *Industrial Crops and Products*. 2013; 50: 312–324. https://doi.org/10.1016/j.indcrop.2013.07.059.
31. Hall J, Bhattarai SP, Midmore DJ. The effects of different sowing times on maturity rates, biomass, and plant growth of industrial fiber hemp. *Journal of Natural Fibers*. 2013; 10(1): 40–50. https://doi.org/10.1080/15440478.2012.756639
32. Garcia Tejero IF, Duran Zuazo VH, Perez-Alvarez R, Hernandez A, Casano S, Moron M et al. Impact of plant density and irrigation on yield of hemp (*Cannabis sativa* L.) in a mediterranean semi-arid environment. *JAST*. 2014; 16(4): 887–895. http://jast.modares.ac.ir/article-23-9987-en.html.

33. Farag S, Kayser O. The cannabis plant: botanical aspects. *Handbook of Cannabis and Related Pathologies*. 2017; 3–12. https://doi.org/10.1016/b978-0-12-800756-3.00.
34. Sunoj Valiaparambil SJ, Dong X, Trostle C, Pham H, Joshi MV, Jessup RW et al. Hemp agronomy: current advances, questions, challenges, and opportunities. *Agronomy*. 2023; 13: 475. https://doi.org/10.3390/agronomy13020475.
35. Salentijin EMJ, Petit J, Trindade LM. The complex interactions between flowering behavior and fiber quality in hemp. *Frontiers in Plant Science*. 2019 May 16; 10: 614. https://doi.org/10.3389/fpls.2019.00614.
36. Adesina I, Bhowmik A, Sharma H, Shahbazi A. A review on the current state of knowledge of growing conditions, agronomic soil health practices and utilities of hemp in the United States. *Agriculture*. 2020; 10: 129. https://doi.org/10.3390/agriculture10040129.
37. Di Bari V, Campi, P, Colucci R, Mastrorilli M. Potential productivity of fibre hemp in southern Europe. *Euphytica*. 2004; 140: 25–32. https://doi.org/10.1007/s10681-004-4751-1.
38. Tsaliki E, Kalivas A, Jankauskiene Z, Irakli M, Cook C, Grigoriadis I, Panoras I, et al. Fibre and seed productivity of industrial hemp (*Cannabis sativa* L.) varieties under mediterranean conditions. *Agronomy*. 2021; 11: 171. https://doi.org/10.3390/agronomy11010171.
39. Huang S, Li H, Xu, J., Zhou H , Seeram NP, Ma H, Gu Q. Chemical constituents of industrial hemp roots and their anti-inflammatory activities. *Journal of Cannabis Research*. 2023; 5: 1. https://doi.org/10.1186/s42238-022-00168-3.
40. Baraniecki P, Latterini F, Stefanoni W, Frankowski J, Wielgusz K, Pari L. Assessment of the working performance of an innovative prototype to harvest hemp seed in two different conditions of terrain slope. *Agronomy*. 2022; 12: 185. https://doi.org/10.3390/agronomy12010185.
41. Burton RA, Andres M, Cole M, Cowley JM, Augustin MA. Industrial hemp seed: from the field to value-added food ingredients. *Journal of Cannabis Research*. 2022; 4: 45. https://doi.org/10.1186/s42238-022-00156-7.
42. United Nations (UN) Conference on Trade and Development. *Commodities at a glance: special issue on industrial hemp*. UN iLibrary; 2022. Available from: https://doi.org/10.18356/9789210019958 [Accessed: 16 January 2024].
43. Zimniewska M. Hemp fibre properties and processing target textile: review. *Materials*. 2022; 15: 1901. https://doi.org/10.3390/ma15051901.
44. EIHA (European Industrial Hemp Association). *Unlocking carbon transfer: wood and hemp for sustainable construction*; EIHA, 2023. Available from: https://eiha.org/wp-content/uploads/2023/07/Press-release-Carbon-Storage-in-Hemp-and-Wood-raw-materials-for-Construction-Materials.pdf [Accessed: 12 January 2024].
45. Dhondt F, Muthu SS. The environmental and social impacts of Hemp. In *Hemp and sustainability*. Singapore: Springer; 2021. p. 15–35. http://dx.doi.org/10.1007/978-981-16-3334-8_2.
46. Bouloc P, Allegret S, Arnaud L. *Hemp: industrial production and uses*. Wallingford, UK: CABI; 2013. ISBN: 978-1-84593-792–8.
47. Hanegraaf MC, Biewinga EE, Bijl Gvan der. Assessing the ecological and economic sustainability of energy crops. *Biomass Bioenergy*. 1998; 15: 345–355. https://doi.org/10.1016/S0961-9534(98)00042-7.
48. Campiglia E, Gobbi L, Marucci A, Rapa M, Ruggieri R, Vinci G. Hemp seed production: environmental impacts of *Cannabis sativa* L. agronomic practices by Life Cycle Assessment (LCA) and carbon footprint methodologies. *Sustainability*. 2020; 12(16): 6570. https://doi.org/10.3390/su12166570.

49. Montford S, Small E. A comparison of the biodiversity friendliness of crops with special reference to hemp (*Cannabis sativa* L.). *Journal of International Hemp Association*. 1999; 6: 53–63. Available from: www.druglibrary.net/olsen/HEMP/IHA/jiha6206.html [Accessed: 12 January 2024].
50. Kok CJ, Coenen GCM, De Heij A. The effect of fibre hemp (*Cannabis sativa* L.) on selected soil-borne pathogens. *Journal of the International Hemp Association*. 1994; 1(1): 6–9. Available from: www.druglibrary.net/olsen/HEMP/IHA/iha01103.html [Accessed: 16 January 2024].
51. Wu Y, Trejo HX, Chen G, Li S. Phytoremediation of contaminants of emerging concern from soil with industrial hemp (*Cannabis sativa* L.): a review. *Environment, Development and Sustainability*. 2021; 23(10): 14405–14435. https://doi.org/10.1007/s10668-021-01289-0.
52. Liu M, Thygesen A, Summerscales J, Meyer AS. Targeted pre-treatment of hemp bast fibres for optimal performance in biocomposite materials: a review. *Industrial Crops and Products*. 2017; 108: 660–683. https://doi.org/10.1016/j.indcrop.2017.07.027.
53. Jankauskienė Z, Butkutė B, Gruzdevienė E, Cesevičienė J, Fernando AL. Chemical composition and physical properties of dew- and water-retted hemp fibers. *Industrial Crops and Products*. 2015; 75: 206–211. https://doi.org/10.1016/j.indcrop.2015.06.044.
54. Shikanai A, Gage KL. Allelopathic Potential of hemp: implications for integrated weed management. *Frontiers in Agronomy*. 2022; 4: 832471. https://doi.org/10.3389/fagro.2022.832471.
55. Andreadis SS, Kalivas A, Tsaliki E. First report of entomofauna in industrial cannabis cultivation in northern Greece. Proceeding of 18th Panhellenic Entomological Congress, Komotini-Greece; 2019 Oct 15–18. p. 87.
56. Andreadis SS, Moysiadis T, Gogolashvili NE, Koutsogeorgiou EI, Fifis GT, Tsaliki E. Quality degradation of industrial hemp due to infestation by *Halyomorpha halys* (Hemiptera: Pentatomidae). *International Journal of Pest Management*. 2022; *70*(4), 1141–1148. https://doi.org/10.1080/09670874.2022.2089769.
57. Kraszkiewicz A, Kachel M, Parafiniuk S, Zając G, Niedziółka I, Sprawka M. Assessment of the possibility of using hemp biomass (*Cannabis Sativa* L.) for energy purposes: a case study. *Applied Science*. 2019; 9: 4437. https://doi.org/10.3390/app9204437.
58. Salentijn EMJ, Zhang Q, Amaducci S, Yang M, Trindade LM. New developments in fiber hemp (*Cannabis sativa* L.) breeding. *Industrial Crops and Products*. 2015; 68: 32–41. https://doi.org/10.1016/j.indcrop.2014.08.011.
59. Prade T, Svensson SE, Andersson A, Mattsson JE. Biomass and energy yield of industrial hemp grown for biogas and solid fuel. *Biomass Bioenergy*. 2011; 35: 3040–3049. http://dx.doi.org/10.1016/j.biombioe.2011.04.006.
60. Rice B. Hemp as a feedstock for biomass-to-energy conversion. *Journal of Industrial Hemp*. 2008; 13: 145–156. http://dx.doi.org/10.1080/15377880802391274.
61. Zegada-Lizarazu W, Monti A. Energy crops in rotation. A review. *Biomass and Bioenergy* 2011; 35(1): 12–25. https://doi.org/10.1016/j.biombioe.2010.08.001.
62. Ahmad M, Ullah K, Khan MA, Zafar M, Tariq M, Ali S, et al. Physicochemical analysis of hemp oil biodiesel: a promising non edible new source for bioenergy. *Energy Sources, Part A: Recovery, Utilization, and Environmental Effects*. 2011; 33(14): 1365–1374. https://doi.org/10.1080/15567036.2010.499420.
63. Kaur G, Kander R. The sustainability of industrial hemp: a literature review of its economic, environmental, and Social Sustainability, 2023 MDPI. 15, 6457. Available from: https://doi.org/10.3390/ su15086457 [Accessed: 28 December 2023].

64. Linger P, Müssig J, Fischer H, Kobert J. Industrial hemp (*Cannabis sativa* L.) growing on heavy metal contaminated soil: fibre quality and phytoremediation potential. *Industrial Crops and Products*. 2001; 16(1): 33–42. https://doi.org/10.1016/S0926-6690(02)00005-5.
65. Mahmood A, Rashid S, Malik RN. Determination of toxic heavy metals in indigenous medicinal plants used in Rawalpindi and Islamabad cities, Pakistan. *Journal of Ethnopharmacology*. 2013; 148(1): 158–164.
66. Golia EE, Bethanis J, Ntinopoulos N, Kaffe GG, Komnou AA, Vasilou C. Investigating the potential of heavy metal accumulation from hemp. The use of industrial hemp (*Cannabis Sativa* L.) for phytoremediation of heavily and moderated polluted soils. *Sustainable Chemistry and Pharmacy*. 2023; 31: 100961. www.mdpi.com/2223-7747/11/5/595.
67. Placido DF, Lee CC. Potential of industrial hemp for phytoremediation of heavy metals. *Plants*. 2022; 11(5): 595. https://doi.org/10.3390/plants11050595.
68. Xu Y, Deng G, Guo H, Yang M, Yang Q. Accumulation and sub cellular distribution of lead (Pb) in industrial hemp grown in Pb contaminated soil. *Industrial Crops and Products*. 2021; 161: 113220.
69. Lee CH, Khalina A, Lee SH, Liu M. A comprehensive review on bast fibre retting process for optimal performance in fibre-reinforced polymer composites. *Advances in Materials Science and Engineering*. 2020; Article ID 6074063: 27. https://doi.org/10.1155/2020/6074063.
70. Hao XM, Yang Y, An LX, Wang JM, Han L. Study on antibacterial mechanism of hemp fibre. *Advanced Materials Research*. 2014; 887–888: 610–613. https://doi.org/10.4028/www.scientific.net/amr.887-888.610.
71. Khan BA, Warner P, Wang H. Antibacterial properties of hemp and other natural fibre plants: a review. *BioResources*. 2014; 9(2): 3642–3659. https://doi.org/10.15376/biores.9.2.3642-3659.
72. Appendino G, Gibbons S, Giana A, Pagani A, Grassi G, Stavri M, et al. Antibacterial cannabinoids from *Cannabis sativa*: a structure activity study. *Journal of Natural Products*. 2008; 71: 1427–1430. https://doi.org/10.1021/np8002673.
73. Duan Y, Yao J, Zhang Y. Development on the multi-component blending yarn of hemp/anti bacteria ne rayon/micro-porous polyester ber. *Advanced Materials. Research*. 2011; 317–319: 2013–2017. http://doi.org/10.4028/www.scientific.net/AMR.317-319.2013.
74. Crini G, Lichtfouse E, Chanet G, Crini NM. Applications of hemp in textiles, paper industry, insulation and building materials, horticulture, animal nutrition, food and beverages, nutraceuticals, cosmetics and hygiene, medicine, agrochemistry, energy production and environment: a review. *Environmental Chemistry Letters*. 2020; 18(1): 1451–1476. http://doi.org/10.4028/10.1007/s10311-020-01029-2ff.ffhal-02926335.
75. Yano H, Fu W. Hemp: a sustainable plant with high industrial value in food processing. *Foods*. 2023; 12: 651. https://doi.org/10.3390/foods12030651.
76. Malachowska E, Przybysz P, Dubowik M, Kucner M, Przybysz K. Comparison of papermaking potential of wood and hemp cellulose pulps. *Annals of Warsaw University of Life Sciences - SGGW, Forestry and Wood Technology*. 2019; 91. https://api.semanticscholar.org/CorpusID:136128302.
77. Farinon B, Molinari R, Costantini L, Merendino N. The seed of industrial hemp (*Cannabis sativa* L.): nutritional quality and potential functionality for human health and nutrition. *Nutrients*. 2020; 12: 1935. https://doi.org/10.3390/nu12071935.
78. Jeliazkov VD, Noller JS, Angima S, Rondon SI, Roseberg RJ, Summers S, Jones GB, Sikora V. *What is industrial hemp?* Corvallis, OR, USA: Oregon State University

Extension Service; 2019. Available from https://catalog.extension.oregonstate.edu/sites/catalog/files/project/pdf/em9240.pdf [accessed: 8 January 2024].

79. Leizer C, Ribnicky D, Poulev A, Dushenkov S, Raskin I. The composition of hemp seed oil and its potential as an important source of nutrition. *Journal of Nutraceuticals, Functional & Medical Foods*. 2000; 2(4): 35–53. https://doi.org/10.1300/J133v02n04_04.
80. Carus M, Sarmento L. The European Hemp Industry: cultivation, processing and applications for fibres, shives, seeds and flowers. *European Industrial Hemp Association*. 2016; 5: 1–9. Available from: www.researchgate.net/publication/281366876_The_european_hemp_industry_Cultivation_processing_and_applications_for_fibres_shives_and_seeds [Accessed: 11 January 2024].
81. Russo R, Reggiani R. Evaluation of protein concentration, amino acid profile and antinutritional compounds in hempseed meal from dioecious and monoecious varieties. SCIRP; , 2015. Available from: www.scirp.org/journal/paperinformation?paperid=52936 [Accessed: 12 January 2024].
82. Yu LL, Zhou KK, Parry J. Antioxidant properties of cold-pressed black caraway, carrot, cranberry, and hemp seed oils. *Food Chemistry*. 2005; 91(4): 723–729. https://doi.org/10.1016/j.foodchem.2004.06.044.
83. Borhade S. Chemical composition and characterization of hemp (*Cannabis sativa*) seed oil and essential fatty acids by HPLC method. *Archives of Applied Science Research*. 2013; 5(1): 5–8. http://scholarsresearchlibrary.com/archive.html.
84. Irakli M, Tsaliki E, Kalivas A, Kleisiaris F, Sarrou E, Cook CM. Effect of genotype and growing year on the nutritional, phytochemical, and antioxidant properties of industrial hemp (*Cannabis sativa* L.) seeds. *Antioxidants*. 2019; 8: 491. https://doi.org/10.3390/antiox8100491.
85. Citti B, Pacchetti MA, Vandelli, et al. Analysis of cannabinoids in commercial hemp seed oil and decarboxylation kinetics studies of cannabidiolic acid (CBDA). *Journal of Pharmaceutical and Biomedical Analysis*. 2018; 149: 532–540. http://doi.org/10.1016/j.jpba.2017.11.044.
86. Clark S. *Seeding Rates and Harvest Dates on Productivity of 'Tropic Sun' Sunn Hemp in South-Central New York*. 2016.
87. Caffarel MM, Andradas C, Pérez-Gómez E, Guzmán M, Sánchez C. Cannabinoids: a new hope for breast cancer therapy? *Cancer Treatment Reviews*. 2012; 38(7): 911–918. https://doi.org/10.1016/j.ctrv.2012.06.005.
88. Russo EB, Marcu J. Cannabis pharmacology: the usual suspects and a few promising leads. *Advances in Pharmacology*. 2017; 67–134. https://doi.org/10.1016/bs.apha.2017.03.004.
89. EIHA (European Industrial Hemp Association). Position paper on: Reasonable regulation of cannabidiol (CBD) in food, cosmetics, as herbal natural medicine and as medicinal product Hürth (Germany), October 2016, update January 2017 and update October 2018; 2018. [accessed 24 January 2024].
90. Johnson MS, Wallace JG. Genomic and chemical diversity of commercially available high-CBD industrial hemp accessions. *Frontiers in Genetics*. 2021; 12: 682475. https://doi.org/10.3389/fgene.2021.682475.
91. Tronet P, Lecompte T, Picandet V, Baley C. Study of lime hemp concrete (LHC) – Mix design, casting process and mechanical behaviour, *Cement and Concrete Composites*. 2016; 67: 60–72, ISSN 0958-9465. https://doi.org/10.1016/j.cemconcomp.2015.12.004.

92. Delhomme F, Hajimohammadi A, Almeida A, Jiang C, Moreau D, Gan Y, et al. Physical properties of Australian hurd used as aggregate for hemp concrete. *Materials Today Communications*. 2020; 24: 100986. https://doi.org/10.1016/j.mtcomm.2020.100986.
93. Arrigoni A, Pelosato R, Melia P, Ruggieri G, Sabbadini S, Dotelli G, et al. Life cycle assessment of natural building materials: the role of carbonation, mixture components and transport in the environmental impacts of hempcrete blocks. *Journal of Cleaner Production*. 2017; 149: 1051–1061. http://dx.doi.org/10.1016/j.jclepro.2017.02.161.
94. Klir Ž, Novoselec J, Antunović Z. An overview on the use of hemp (*Cannabis sativa* L.) in animal nutrition. *Poljoprivreda*. 2019; 25(2): 52–61. https://doi.org/10.18047/poljo.25.2.8.
95. Kalaitsidis K, Parissi Z, Theodoridis A, Tsaliki E, Vasilopoulou K, Dokou S, et al. Evaluation of hemp cake (*Cannabis sativa*) and other hemp by-products of Greek origin and efficacy in dairy cow nutrition. *Archiva Zootechnica*. 2023; 26(2): 148–169. https://doi.org/10.2478/azibna-2023-0020.
96. Fallahi S, Bobak Ł, Opaliński S. Hemp in animal diets-Cannabidiol. *Animals (Basel)*. 2022 Sep 22; 12(19): 2541. https://doi.org/10.3390/ani12192541.
97. Ona G, Balant M, Bouso JC, Gras A, Valles J, Vitales D, Garnatje T. The use of *Cannabis sativa* L. for pest control: from the ethnobotanical knowledge to a systematic review of experimental studies. *Cannabis and Cannabinoid Research*, 2022; 7(4): 365–387. https://doi.org/10.1089/can.2021.0095.
98. Hesami M, Pepe M, Baiton A, Salami SA, Jones AMP. New insight into ornamental applications of Cannabis: perspectives and challenges. *Plants*. 2022; 11: 2383. https://doi.org/10.3390/plants11182383.
99. Williams C, Thompstone J, Wilkinson M. Work-related contact urticaria to *Cannabis sativa*. *Contact Dermatitis*. 2008 Jan; 58(1): 62–63. PMID: 18154568. https://doi.org/10.1111/j.1600-0536.2007.01169.x.
100. Turunen L, Van der Werf H. The production chain of hempand flax textile yarnand its environmental impacts. *Journal of Industrial Hemp*. 2007; 12(2): 43–66. https://doi.org/10.1300/J237v12n02_04.
101. Van der Werf HMG. The effect of plant density on light interception in hemp (*Cannabis sativa* L.). *Journal of the International Hemp Association*. 1997; 4(1): 8–13. Available from: www.internationalhempassociation.org/jiha/jiha4106.html [Accessed: 11 January 2024].

4 Breeding Climate-resilient Oilseed Crops

Sidra Iqbal, Ahmed Raza, Muhammad Junaid, Mehwish Kanwal, and Muhammad Rizwan Shafiq

4.1 INTRODUCTION

4.1.1 Importance of Oilseed Crops

Oilseed crops stand out as a major source of vegetable oil, offering a substantial role in the global agricultural trade. The demand for refined edible oil products and industrial oil sources has indispensably grown in the current scenario of increasing population and declining fossil fuels [1]. These crops are not only cost-effective but also are able to thrive in diverse agro-climatic landscapes. Among other noteworthy oleiferous crops, such as coconut, palm oil, sesame, safflower, and olive, significant attention and research efforts are dedicated to several major oilseed crops, including soybeans, sunflower, canola, cottonseed, and peanuts.

Globally, efforts are underway to modify oilseeds for enhanced nutritional value, improved oil quality, and favorable composition. Emerging oilseeds, such as Camelina and Pennycress, are gaining prominence due to their adaptability to marginal lands and potential for biofuel production. The defatted meal derived from oilseeds serves as an excellent source of protein in animal nutrition, as it is not only highly valuable as a food source but also contains essential amino acids that positively contribute to human health.

Furthermore, the oil contains polyunsaturated fatty acids, which offer both health benefits for consumers and diverse industrial applications [2]. Flax (linseed) and castor bean are notable examples of industrial oilseed crops, with their seed oils finding applications in various industrial sectors including biodiesel, fertilizer, fibers, and cosmetics [3]. Canola oil, a specific variant of rapeseed oil, is widely embraced for culinary purposes, with its lower saturated fat content (~70 g kg^{-1}). It is categorized into two types: Conventional, featuring an average of almost 600 g kg^{-1} oleic and 110 g kg^{-1} linolenic acids; and high oleic, which contains about 700 g kg^{-1} oleic and 30 g kg^{-1} linolenic acids. Recent research has developed sunflower varieties with oleic acid exceeding 750 g kg^{-1}. The high oleic acid varieties are gaining popularity for their

DOI: 10.1201/9781032720142-5

enhanced oxidative stability, particularly in the context of end-use applications, as an alternative to trans-esterification methods [4].

4.1.2 Impact of Climate Change on Agriculture

The connection between the agricultural sector and climate change is globally intertwined, gaining heightened significance amidst the escalating disparities in population and world food production. The increased seasonal temperature might shorten the duration of many crops, resulting in reduced final yields and outbreaks of pests and diseases, ultimately impacting the overall food security of a nation [5]. Climate change scenarios embrace increased temperatures, alterations in rainfall patterns, and elevated concentrations of atmospheric CO_2. These factors can have profound effects on crop yield, quality, and quantity, as well as growth rates, photosynthesis, and transpiration rates, and are anticipated to diminish agricultural production, particularly in regions with low incomes or in the developing world, with a notable focus on Africa and parts of Asia [6]. Changes in moisture availability, influenced by variations in water use and agricultural inputs, are key considerations. Environmental repercussions, such as alterations in the intensity and frequency of soil drainage, compromise agricultural productivity in terms of nitrogen leaching, soil erosion, and a reduction in crop diversity.

Climate change is primarily attributed to a rise in greenhouse gas (GHG) concentrations in the atmosphere, resulting in an intensification of the greenhouse effect. The impact of the greenhouse effect on agriculture unfolds in three key factors. Firstly, elevated atmospheric CO_2 concentrations can interfere with metabolism and impact the growth rate of both crop plants and weeds. Secondly, climate changes induced by CO_2 variations may bring about shifts in temperature, precipitation, and sunshine levels, exerting an influence on plant productivity. Lastly, the rise in sea levels poses a threat to farmland, leading to inundation and an increase in the salinity of groundwater in coastal areas. These dynamics pose a serious threat to food production, thereby jeopardizing food security. Addressing these challenges necessitates the implementation of special agricultural measures to mitigate the impacts [7].

Conversely, agriculture holds significant potential for both mitigating and adapting to climate change. Mitigation encompasses interventions to reduce emission sources or enhance GHG sinks, while adaptation involves adjustments in natural or human systems in response to prevailing climatic changes, aiming to minimize harm or capitalize on beneficial opportunities. Practices such as organic agriculture and sustainable systems play a crucial role in mitigating agricultural GHG emissions. These approaches involve energy conservation, reduced reliance on carbon-based inputs, less use of synthetic fertilizers, and other strategies designed to lessen GHG emissions. Additionally, they contribute to carbon sequestration in the soil. While engaging in mitigation and adaptation activities related to climate change through agricultural practices, several challenges or barriers may arise, including financial constraints and issues related to policy making and implementation. In essence, agricultural activities have the dual role of being both a source of GHGs and a sink, particularly through the storage of carbon in soil organic matter and biomass [8].

4.1.3 Objectives of Breeding Climate-resilient Oilseed Crops

A key goal in breeding climate-resilient oilseed crops is to bolster stress tolerance. Anticipated climate change is poised to bring about intensified environmental stresses including drought, heat, and salinity, which can substantially impact crop growth and yield. The development of oilseed crops with enhanced resilience to these stresses becomes instrumental in ensuring a more consistent and sustainable food supply. Achieving this objective involves the identification and incorporation of genes conferring stress tolerance into breeding programs, coupled with the application of advanced molecular breeding techniques to expedite the breeding process. In essence, fortifying stress tolerance stands as a pivotal objective in the endeavor to breed oilseed crops that can withstand and thrive in the face of changing climatic conditions [9]. This strategy is not merely a reaction to climate change; it is also a crucial step to safeguard future agricultural productivity.

Enhancing yield and quality is another crucial objective in the breeding of climate-resilient oilseed crops. Resilient crops have the capacity to maintain or even boost yields and quality amid diverse climatic and environmental changes. This goal involves the development of crop varieties that possess traits such as improved photosynthetic efficiency, efficient nutrient uptake, and enhanced resistance to diseases. The improvement of yield and quality not only contributes to a stable and dependable food supply in the context of changing environmental conditions, but also offers benefits to farmers by increasing profitability and curbing food wastes [10]. Establishing resistance to pests and diseases emerges as a pivotal goal in the breeding of climate-resilient oilseed crops. Anticipated changes in climate are expected to alter the dynamics of pest and disease pressures, posing challenges in maintaining crop health and yield. The development of oilseed crops with enhanced resistance to pests and diseases serves to diminish the reliance on harmful pesticides and other chemical inputs. This shift promotes more sustainable and environmentally friendly farming practices, contributing to a stable and dependable food supply. Moreover, the cultivation of oilseed crops with resistance to pests and diseases aids in reducing crop losses and enhancing profitability for farmers [11].

4.2 CLIMATE CHANGE AND ITS IMPACT ON OILSEED CROPS

Escalating temperatures present a significant global concern, with various models predicting a rise in carbon dioxide (CO_2) level, which contributes to increased ambient temperatures and global warming. This phenomenon poses severe consequences for agricultural production systems worldwide. The Intergovernmental Panel on Climate Change (IPCC) has predicted an annual temperature rise of 1.5°C from the year 2030 to 2052 [12]. Heat stress induced by elevated temperatures is defined as a prolonged period with air temperatures surpassing threshold levels, capable of causing injuries or irreversible damage to crop plants. Heat stress accelerates crop growth and hinders the proper completion of growth stages, ultimately leading to immature development because of disrupted carbon assimilation. Thus, heat stress looms as a substantial and imminent threat to productive crop cultivation globally. For instance, a slight increase of only 2°C can lead to accelerated flowering and reduced seed filling in rapeseed, ultimately leading to a 12–25% decline in yield. Similarly, sunflower production may experience a 20% yield reduction, with a 10% drop in oil content under combined drought and heat stress in semi-arid regions [13].

The repercussions of global climate change, including spatial, temporal, and regional patterns, raise significant concerns in agricultural production. This urgency is heightened by the fact that the geographical distribution of plant species is heavily contingent on their adaptation to various temperature zones [14]. Altered precipitation patterns wield a substantial impact on the yields of oilseed crops, with drought being a particularly formidable factor causing significant reductions. In numerous countries, climate change-induced temperature extremes significantly impede wheat production, resulting in yield reductions of up to 6%. Sensitivity to drought is evident in the upper Ganges basin, where crop yield is intricately dependent on the total rainfall received during the relatively brief growing season. The decline in yield attributable to drought is associated with diminished photosynthesis rates and disruptions in assimilate transport. These factors collectively contribute to decreased plant growth and yield [15].

Shifts in precipitation patterns can also induce changes in nutrient availability and soil quality, influencing both the productivity and quality of oilseed crops. Altered seasonality of precipitation may result in excessive water during off-seasons and insufficient water during crucial crop growth phases. This scenario can lead to soil degradation, nutrient leaching and, ultimately, a decline in crop yields. The combined impacts of soil warming and modified precipitation on the productivity and quality of oil crops remain insufficiently studied [16]. Nevertheless, existing research indicates that increased mean precipitation can enhance soil water content and multi-functionality [17]. On the other hand, elevated rainfall and increased humidity create conditions conducive to the prevalence of plant diseases that thrive in moist environments [18]. For instance, higher temperatures coupled with high humidity are correlated with heightened disease severity caused by the potato blight pathogen, *Phytophthora infestans*. This heightened risk of disease poses a significant threat, leading to substantial losses in both crop yield and quality. Furthermore, the evolving weather patterns and rising temperatures could contribute to alterations in the distribution of pests and diseases, exacerbating stress on essential pollinators [19].

Even a minimal increase in environmental temperature can lead to elevated incidence of phytopathogens in oilseed crops [20]. This increased population and pressure of pests demands extensive applications of pesticides, as earlier observed in the United State of America, where increasing temperature of southern regions is directly proportional to insect populations and vice versa in higher latitude regions. The chemical control of Lepidopteran insects in sweet corn is much higher, i.e. 15 to 32, 4 to 8, and 0 to 5 times in Florida, Delaware, and New York, respectively [21]. This diverse application of pesticides poses a negative impact on human health and biodiversity as well [22]. The rise in temperature can also retard the efficiency of pesticides against targeted pests after a rise in post-exposure temperature. The toxicity of pyrethroids (lambda-cyhalothrin and bifenthrin) to *Ostrinia nubilalis* is such an example [23].

Changes in precipitation, either excessive or suppressive, can have prominent impacts on crop–pest interaction, especially on many phytopathogens, whose incidence is significantly influenced by rainfall and humidity [21]. During floods in Iowa in 1993, a prolonged spell of humidity fostered the spread of pathogens like *Puccinia* and *Alternaria* species, leading to plant diseases like rust and leaf blight, respectively [24]. The invasion of desert locust is also encouraged by changes in rainfall ratio of that specific region. In 2004, abnormal rainfall patterns disrupted the ecological balance in northern and western Africa, leading to an increase in vegetation and soil moisture, which was an ideal situation for desert locust breeding and swarm formation. This situation resulted in locust infestation in almost 10 counties including Mauritania, Mali, and Niger [25]. Extreme weather conditions can affect crop–pest interactions by accelerating the life cycle of pests, thereby adversely affecting crop protection strategies. For example, in May 1993, hot and dry weather in Slovakia reduced the populations of the egg parasitoid *Trichogramma evanescens*. Moreover, there was no incidence of lucrative parasitism of European corn borer eggs that year. Prolonged droughts have increased the incidence of spider mites (*Tetranychus urticae*) in several crops globally, including cotton, maize, fruits, and vegetables, as they thrive in hot dry conditions [26, 27].

Furthermore, climate change is threatening the world's natural plant biodiversity food supply [18, 28]. It is anticipated that any potential yield gains in the coming five decades will be due to climate change-mediated altered pressure of disease-causing agents (pathogens) [29]. Climate change may ease plant infection in several ways, i.e. by changing host–pathogen evolution, altering host–pathogen interactions, and facilitating the exposure of newly evolved strains of pathogens that have the ability to break down host-plant resistance [18]. Insects are also major problems in *Brassica* crops, contributing to major yield losses and aesthetic damage. The insects invading oilseed crops belong to the orders of Hymenoptera, Homoptera, Diptera, Lepidoptera, and Coleoptera [30, 31]. Unfortunately, no natural resistance has been found and the only possible way to protect the plants is through direct pesticide application or practicing an integrated pest management program which has resulted in the emergence of new resistant insect pest species [32] and narrative approaches are required to control these insect pests [33].

Diseases of oilseed crops are a significant concern for farmers and can lead to low productivity and economic losses. Seed-borne diseases caused by pathogens present externally or internally on the seed pose a potential threat to the cultivation of annual edible oilseeds such as sesame, peanut, soybean, sunflower, rapeseed-mustard, and safflower [34]. Pests and diseases, including viruses, also contribute to reduced oilseed crop productivity. Integrated pest management (IPM) is suggested as a solution, with various components and recommendations for successful implementation in oilseed crops [35]. In Australia, viral diseases pose a threat to cereal and oilseed crops, with several viruses causing significant yield and quality losses. The need for increased resources and research to address these threats is emphasized in a review

[36]. Disease resistance in oilseed crops is an important consideration in the context of climate change [37, 38]. Studies have revealed that climate change can influence the severity and spread of diseases in oilseed crops such as oilseed rape [39]. It has been predicted that climate change will lead to an increase in the severity and spread of phoma stem canker epidemics in oilseed rape in the UK [40]. Additionally, the major gene for resistance to *Leptosphaeria maculans*, a pathogen causing phoma stem canker, has been found to be temperature sensitive [41]. Furthermore, disease control measures such as fungicide use and cultivar resistance can contribute to reducing greenhouse gas emissions associated with oilseed crop production [39]. Oilseed crops can exhibit successful disease resistance, contributing to their climate resilience. For example, winter oilseed rape cultivars have been found to show disease resistance against pathogens such as *Plasmodiophora brassicae* and *L. maculans*, resulting in a decrease in GHG emissions of 32 and 24 kg CO_2 t^{-1}, respectively [42]. These findings highlight the potential of disease control measures in oilseed crops to mitigate climate change impacts and contribute to sustainable agriculture.

4.3 BREEDING STRATEGIES FOR CLIMATE RESILIENCE

Plant breeding programs for crop improvement mainly rely on the availability of significant genetic variations, which can be exploited by the breeders through effective selection to attain the maximum genetic yield potential of a crop. The phenotypic selection of plants from a population covers both heritable and non-heritable components [43]. The advancement in crop production technology, utile farm machinery, development of effective pesticides and fertilizers, knowledge of crop's molecular and genetic bases has resulted in increased yield over time. However, this also led to adverse environmental and climatic phenomena along with challenges for future food security, such as greenhouse gas effects, uncertain weather conditions, changed growing season dynamics, and the emergence of extreme biotic and abiotic stresses [42, 44]. Unpredictable environments along with global warming have resulted in altered objectives in current breeding programs and compelled breeders to discover novel breeding strategies to overcome environmental adversities.

4.3.1 TRADITIONAL BREEDING APPROACHES

Primarily, oilseed breeding is focused on developing oilseed crops adaptable to new areas, with improved meal and oil quality and quantity using selection and hybridization approaches. The presence of genetic variations in a population enables the breeder to improve the selected traits or develop new varieties via traditional or conventional breeding approaches without disturbing the natural inheritance laws. Plants, on the basis of their superior performances for specific traits, are selected and incorporated into the breeding program to develop new cultivars using hybridization [45]. Conventional breeding presents numerous challenges to gain substantial improvements as it mostly depends on the phenotypic variations in a crop resulting from environmental factors, hence trait improvements become laborious and time consuming. Pure line selection, mass selection, recurrent selection, hybridization, and backcross breeding are some of the common traditional breeding methods used for the improvement of desirable traits in various oilseed crops [43, 46, 47]. The backcross breeding method was successfully applied to introduce climate-resilient genes from somatic hybrids of *Sinapis alba* into *Brassica juncea* [48]. Interspecific hybrids (F_1) and backcross progenies were developed using *Ogura* cytoplasmic male sterile broccoli lines and rapeseed [49]. The backcross breeding strategy helped to obtain resistance against *Sclerotinia* stem rot in *B. juncea* and *S. alba* [48, 50]. Pure line selection was implied in sesame, *Sesamum indicum*, to improve the seed yield and nutritional quality [51]. Similarly, soybean cultivars were improved for various agronomic and chemical parameters using a pure line selection technique [52]. High-throughput (HT) precision phenotyping was applied in different crops including canola, sunflower, castor bean, cotton, soybean, jatropha, and maize to identify improved crop plants for agronomic traits along with higher seed yield, oil mass, and oil content [53]. Recurrent selection was applied in soybeans to improve selected traits and develop improved cultivars [54]. Conventional breeding has helped humankind to introduce new crop varieties with improved seed yield with higher oil and fatty acid compositions and its significance is undeniable, yet the drastic changes in the global environmental conditions and increased population around the globe have necessitated the need to develop new breeding strategies that are more robust in achieving the desirable outcomes.

4.3.2 Marker-assisted Selection

Selection of superior plants obtained from sexual mating is a classical plant breeding approach where selection is based on the whole plant level, i.e. phenotypic appearance, resulting from both genotype and environment and their interactions. This plant appearance-based selection has contributed significantly to crop improvement but has limited efficiencies when plant traits are mostly derived by the environment [55, 56]. Moreover, the selection of more complex traits controlled by multiple alleles makes phenotypic selection more difficult and unreliable. Molecular and genetic markers based on indirect selection in such situations present more accurate and effective breeding approach. Through molecular markers, allele/genes controlling specific traits can be identified and linked, and this linked marker locus can be used as a substitute for the targeted trait, allowing significant breeding efficacy [55].

Marker-assisted selection (MAS) utilizes DNA markers that enable the breeder to specify the genetic regions that control the expression of targeted traits [57]. DNA markers comprise of short DNA segments that are used to detect variations in genes/alleles within the same gene pool [58, 59]. Various DNA markers are easily available, efficient, and effective for use in different breeding programs. Moreover, DNA markers are not influenced by inheritance or environmental factors, thus the original phenotypic expression of plants is maintained [60].

MAS is quite effective in selecting different agronomic and chemical characteristics in both early and late developmental stages of a plant, not only phenotypically but also at genotypic levels [61]. MAS enables breeders to select less heritable traits that are rather expensive and tedious to improve in the breeding population, but the MAS approach can be utilized more effectively when the selected quantitative traits have a significant amount of variation within the breeding population [45]. Among different available DNA markers, simple sequence repeat (SSR) markers are mostly utilized in breeding programs as SSRs are easier to use and cheaper. Other examples of DNA markers widely used for MAS are amplified fragment length polymorphism (AFLP), restriction fragment length polymorphism (RFLPs), inter simple sequence repeats (ISSR), random amplified polymorphic DNA (RAPD), and single nucleotide polymorphism (SNP) [62].

Different MAS and genomics approaches for seed quality, oil yield, and disease resistance in *Brassica napus* L. have been comprehensively reviewed [63]. Similarly, for all the major *Brassica* crop species, use of molecular markers including SNP markers has been reviewed in detail [64, 65]. MAS was applied to develop *B. napus* L. lines with higher oleic and lower linolenic acid contents [66]. In *Brassica* vegetables, use of marker-assisted breeding for resistance to clubroot and *Fusarium* wilt disease has been reported [67]. Safflower (*Carthamus tinctorious* L.) lines with higher oleic acid content were developed [68]. In various groundnut species, use of RFLPs, SSRs, ISSR, and RAPD has been reviewed in detail for improved yield and oil content [69]. The application of molecular markers for the improvement of various economically important traits, identification of markers linked to monogenic characters, and markers for disease and pest resistance in sunflower has been thoroughly reviewed [70]. Disease-resistant sunflower, especially for downy mildew, was selected through MAS and SSR markers [71–74]. Kompetitive allele-specific PCR (KASP) markers to restore fertility and genetic purity test have been successfully applied in sunflower [75].

4.3.3 Modern Biotechnology and Genetic Engineering

Climate change has affected the production and quality of almost all major crops, including oilseed crops. The oil content, fatty acid composition, seed yield, and other agronomic parameters are facing global climate change consequences, and conventional breeding is not fully able to address these challenges [76]. Hence, modern biotechnological techniques, genetic engineering-based approaches, and molecular breeding strategies can be a useful tool for breeders. By integrating these technologies and approaches, researchers can develop oilseed crops that are better equipped to combat the impacts of climate change and ensure global food security [77]. Modern technology is crucial for developing climate-resilient oilseed crops. Advanced production technologies are needed to boost oilseed production under changing climate conditions [38]. Genetic engineering enables breeders to directly transfer the beneficial genes and/or manipulate the existing genes to develop and improve desirable traits and phenotypes. Additionally, novel alleles can be introduced in the existing germplasm through genetic engineering to obtain improved crop plants [78].

Genomics-assisted breeding (GAB) is a promising technology for developing climate-resilient oilseed crops. GAB involves using genomic analysis to identify genes or quantitative trait loci that confer stress tolerance in crop species. This approach has been successful in improving the capacity of crop plants to adapt to various environmental stresses, including pests, diseases, heat, cold, salt, floods, submersion, and drought [42]. Molecular breeding techniques, combined with advancements in phenotyping and high-throughput sequencing technologies, have transformed traditional breeding into GAB, enabling more efficient and targeted crops.

Additionally, breeding for phenotypic plasticity and resilience to multiple abiotic stresses, such as temperature and precipitation changes, can help oilseed crops adapt to the challenges posed by climate change [79]. Modern biotechnology, including genetic improvement and omics approaches, can enhance crop productivity, increase yields, and ensure food security [42]. Understanding the phenotypic plasticity of oilseed crops and breeding for climate resilience are important strategies. Phenotypic plasticity is the ability of a plant to alter its phenotype under varying environmental conditions, enabling it to thrive under diverse and fluctuating conditions. Genome-wide association studies and high-throughput plant phenotyping can aid in the discovery of new traits and improve phenotypic prediction. For instance, breeding rapeseed varieties with wide flowering windows might mitigate the drastic effects of drought during critical growth stages. Emphasis on traits such as plasticity in soybean root architecture, leaf morphology in sunflower, stomatal conductance in mustard, pod setting and filling in peanuts under erratic rainfall patterns, and drought-induced leaf rolling in sunflower have been well presented in a review [80]. Phytoremediation, a novel and environment-friendly technology, can be used to treat oilseed crops contaminated by heavy metals [81]. Perennial cropping systems, along with adaptive management strategies and digital technologies, offer potential for adapting agriculture to a changing climate and supporting multiple environmental benefits. Successful development and deployment of climate-resilient oilseed crops depend on grower adoption, scalability, and modifications to markets and supply chains.

Genomic selection (GS) is a more targeted and less time-consuming approach for grain and yield improvement. GS uses genome-wide markers to determine the breeding value of a specific trait by predicting the quantitative gene loci effects [82]. While MAS has been revolutionary for improving traits under influence of major genes, genes with multiple quantitative trait loci (QTL) and small-effect genes can be mapped effectively through GS. In contrast to phenotyping, GS is a more effective and economical approach to improve quantitative parameters including yield in a breeding population under biotic or abiotic stresses [83, 84]. The role of GS and phenomics in soybean, groundnut, and *Brassica* species breeding along with other crops until 2020 are discussed comprehensively in reviews [85, 86]. In Chinese semi-winter rapeseed (*B. napus* L.) inbred lines, GS was successfully applied to obtain improved quantitative traits [87]. Similarly, GS was applied in rapeseed for improving the yield, plant growth, and other agronomic parameters [76]. GS has played an effective role in improving the agronomic seed yield, oil content, and climate resilience in crops like soybean, sunflower, safflower, oil palm (*Elaeis guineensis* Jacq.), and groundnut [60, 88–92).

4.4 CHALLENGES AND LIMITATIONS

Conventional breeding approaches have been applied to improve oilseed production and yields. However, they have certain limitations due to laborious and imprecise results. On the other hand, crops developed using transgenic approaches or genetically modified (GM) crops could not get public acceptance due to their negative perception and extreme regulatory concerns. Although transgenic crops have been proved to have enormous potential to fulfill the needs of humanity in terms of yield, higher oil, and fatty acid content, their acceptance in society remains debatable [93].

4.4.1 Regulatory and Ethical Concerns

There are several concerns related to the effects of transgenic crops on human health, environment, biodiversity, ecosystem, and traditional farming. There is a negative perception about GM crops and transgenic plants that their cultivation may lead to a rise of more extreme biotic and abiotic stresses and affect beneficial plants and insects. Pests and pathogens may develop resistance to genetically manipulated crops over time and have adverse effects on the ecosystem and safety. The public concerns and incredulity about GM and transgenic plants results in regulatory challenges. Therefore, many countries have either banned GMOs or made the regulatory approval process more lengthy, expensive, and complex. For instance, countries like the United States and Brazil have reported significant gains in crop yield by embracing GM technology. On the contrary, many countries in the European Union have imposed strict bans on GMOs due to public concerns over the environmental impact and health risks [94].

4.4.2 Trade-offs in Trait Selection

In oilseed breeding, trade-offs in trait selection are crucial to develop cultivars adaptable to diverse environmental, agricultural, and market demands. The trade-offs are important to select the targeted traits for improvement as one trait might affect another. Higher seed yield in oilseed breeding is important but may be achieved at the expense of seed quality, and oil and fatty acid contents. To overcome trade-off challenges during trait selection, advanced breeding approaches at the molecular and genetic levels can help breeders to optimize the targeted traits [95, 96].

4.4.3 Maintaining Genetic Diversity

Resilience and long-term sustainability in oilseed crops can be achieved by maintaining genetic diversity. Genetic diversity is crucial for adaptation of crop plants to new environmental, biotic, and abiotic stresses. A narrow genetic base, reduced genetic diversity due to intensive selection pressure, genetic erosion, and climate change are the important challenges associated with maintaining genetic diversity in oilseed crops [79]. However, the use of advanced breeding approaches, preservation of landraces, and development of open-access genetic resources can help to maintain genetic diversity in oilseed crops, where genomic breeding and molecular/genetic engineering can contribute to the development of more resilient and adaptable oilseed varieties [97].

4.5 FUTURE DIRECTIONS AND EMERGING TECHNOLOGIES

4.5.1 Advancements in Breeding Techniques

Oilseed breeding has been revolutionized by modern breeding approaches. Modern phenotyping and genotyping tools enable breeders to study multi-trait associations through genome-wide association studies (GWAS) to analyze the accurate genetic makeup of targeted traits. QTL analysis and mapping of traits has helped to understand the relationship between plant's genotype and phenotype [98]. Similarly, next generation sequencing (NGS), gene cloning, molecular markers, allele mining, genome editing via CRISPR-Cas, base editing, and prime editing have emerged as valuable breeding tools for not only oilseed breeding but are also being used in other major crops. High-throughput genomics technologies, molecular breeding, and biotechnology to study the genome of *Brassica* species have been presented in detailed reviews [65, 99]. CRISPR-Cas-mediated genome editing aimed to improve the yield and oil content of oilseed crops under favorable and sub-optimal environments and molecular breeding tools are reviewed in detail [100, 101]. CRISPR-Cas9-mediated gene editing in rapeseed for plant type, yield and quality improvement, genetic improvement, and disease and stress resistance has been presented [102, 103]. Genome editing via CRISPR-Cas for improved oil content and fatty acid compositions in *Camelina sativa* and other oilseed crops has been reviewed comprehensively [64, 93]. Compared to traditional transgenic approaches, this technique offers greater accuracy and precision, ultimately reducing the risks of off-target effects, as it does not involve the introduction of foreign DNA. Furthermore, the CRISPR-Cas system also bypasses strict regulatory scrutiny and public acceptance as well, as it enables targeted modification to specific existing genes.

4.5.2 Predictive Modeling

The use of satellite and remote sensing technology has equipped researchers and farmers with access to detailed insights about their crops, such as forecasted yield outcomes under varying climate scenarios, and potential challenges such as pest outbreaks or drought. Satellite data can also be used to monitor the soil temperature, moisture level, and fertility status. Remote sensing can identify early signs of stresses, enabling farmers to make informed decisions and adopt proactive measures to optimize oilseed production. Furthermore, yield estimation algorithms may be useful for policymakers for understanding production trends at the macro level.

4.5.3 Innovative Farming Practices

Precision farming is a newly emerging concept, which incorporates advanced technology to predict the exact need of inputs by plants, enabling farmers to effectively manage their crops and resources, such as by using precise irrigation targeted fertilization, and careful pesticide application; hence it minimizes environmental hazards while improving crop production. Some examples include drones equipped with multispectral sensors that can provide high-resolution data allowing farmers to identify targeted areas in the field, where fertilizer and pesticide input is required, hence reducing waste and environmental impact. Similarly, the best time and amount of fertilizer application based on soil nutrient level and crop growth stage may be predicted by AI-based systems. Autonomous robots are being employed in various farm operations. Altogether, precision farming is an efficient and sustainable approach to address the prevailing challenges posed by climate change.

REFERENCES

1. Wittkop B, Snowdon RJ, Friedt W. Status and perspectives of breeding for enhanced yield and quality of oilseed crops for Europe. *Euphytica.* 2009 Nov;170:131–40.
2. Abiodun OA. The role of oilseed crops in human diet and industrial use. In *Oilseed crops: yield and adaptations under environmental stress.* John Wiley & Sons, 2017 Feb 17. pp. 249–63.
3. Murphy DJ. Designer oilseed crops. Genetic engineering of new oilseed crops for edible and non-edible applications. *Agro Food Industry Hi-Tech.* 1991;2(1):14–8.
4. Darr L, Cunicelli M, Bhandari H, Bilyeu K, Chen F, Hewezi T, Li Z, Sams C, Pantalone V. Field performance of high oleic soybeans with mutant FAD2-1A and FAD2-1B genes in Tennessee. *Journal of the American Oil Chemists' Society.* 2020 Jan;97(1):49–56.
5. Yuan X, Li S, Chen J, Yu H, Yang T, Wang C, Huang S, Chen H, Ao X. Impacts of global climate change on agricultural production: a comprehensive review. *Agronomy.* 2024;14(7):1360.
6. Gornall J, Betts R, Burke E, Clark R, Camp J, Willett K, Wiltshire A. Implications of climate change for agricultural productivity in the early twenty-first century. *Philosophical Transactions of the Royal Society B: Biological Sciences.* 2010 Sep 27;(1554):2973–89.
7. Mahato A. Climate change and its impact on agriculture. *International Journal of Scientific and Research Publications.* 2014 Apr 4;4(4):1–6.
8. Narasimham S, Subbarao DV. Relationship between climate change and agriculture–a review. *International Journal of Science Environmental Technology.* 2017;6(5):3011–25.
9. Ahmad M, Waraich EA, Skalicky M, Hussain S, Zulfiqar U, Anjum MZ, Habib ur Rahman M, Brestic M, Ratnasekera D, Lamilla-Tamayo L, Al-Ashkar I. Adaptation strategies to improve the resistance of oilseed crops to heat stress under a changing climate: an overview. *Frontiers in Plant Science.* 2021 Dec 15;12:767150.
10. Naqvi RZ, Siddiqui HA, Mahmood MA, Najeebullah S, Ehsan A, Azhar M, Farooq M, Amin I, Asad S, Mukhtar Z , Mansoor S. Smart breeding approaches in post-genomics era for developing climate-resilient food crops. *Frontiers in Plant Science.* 2022 Sep 16;13:972164.
11. Gaba Y, Pareek A, Singla-Pareek SL. Raising climate-resilient crops: journey from the conventional breeding to new breeding approaches. *Current Genomics.* 2021 Dec 12;22(6):450.
12. IPCC. *Fifth assessment report.* New York, NY: Cambridge University Press; 2018.

13. Sehgal A, Sita K, Siddique KHM, Kumar R, Bhogireddy S, Varshney RK, Hanumantha Rao B, Nair RM, Prasad PVV and Nayyar H. Drought or/and heat-stress effects on seed filling in food crops: impacts on functional biochemistry, seed yields, and nutritional quality. *Frontiers in Plant Sciences*. 2018. Nov 27:9:1705.
14. Rehman A, Atif RM, Azhar MT, Peng Z, Li H, Qin G, Jia Y, Pan Z, He S, Qayyum A, Du X. Genome wide identification, classification and functional .characterization of heat shock transcription factors in cultivated and ancestral cottons (*Gossypium* spp.). *International Journal of Biological Macromolecules*. 2021 Jul 1;182:1507–27.
15. Raza A, Razzaq A, Mehmood SS, Zou X, Zhang X, Lv Y, Xu J. Impact of climate change on crops adaptation and strategies to tackle its outcome: a review. *Plants*. 2019 Jan 30;8(2):34.
16. Drebenstedt I, Hart L, Poll C, Marhan S, Kandeler E, Böttcher C, Meiners T, Hartung J, Högy P. Do soil warming and changes in precipitation patterns affect seed yield and seed quality of field-grown winter oilseed rape. *Agronomy*. 2020 Apr 6;10(4):520.
17. Sünnemann M, Beugnon R, Breitkreuz C, Buscot F, Cesarz S, Jones A, Lehmann A, Lochner A, Orgiazzi A, Reitz T, Rillig MC. Climate change and cropland management compromise soil integrity and multifunctionality. *Communications Earth & Environment*. 2023 Oct 26;4(1):394.
18. Velásquez AC, Castroverde CD, He SY. Plant–pathogen warfare under changing climate conditions. *Current Biology*. 2018 May 21;28(10):619–34.
19. Singh BK, Delgado-Baquerizo M, Egidi E, Guirado E, Leach JE, Liu H, Trivedi P. Climate change impacts on plant pathogens, food security and paths forward. *Nature Reviews Microbiology*. 2023 May 2; *21*(8), 535–541Springer Nature. https://doi.org/10.1038/s41579-023-00897-0
20. Barzman M, Lamichhane JR, Booij K, Boonekamp P, Desneux N, Huber L, Kudsk P, Langrell SR, Ratnadass A, Ricci P, Sarah JL.Research and development priorities in the face of climate change and rapidly evolving pests. *Sustainable Agriculture Reviews*. 2015;17:1–27.
21. Hatfield JL, Boote KJ, Kimball BA, Ziska LH, Izaurralde RC, Ort D, Thomson AM, Wolfe D. Climate impacts on agriculture: implications for crop production. *Agronomy Journal*. 2011 Mar;103(2):351–70.
22. Tudi M, Daniel Ruan H, Wang L, Lyu J, Sadler R, Connell D, Chu C, Phung DT. Agriculture development, pesticide application and its impact on the environment. *International Journal of Environmental Research and Public Health*. 2021 Feb;18(3):1112.
23. Musser FR, Shelton AM. The influence of post-exposure temperature on the toxicity of insecticides to *Ostrinia nubilalis* (Lepidoptera: Crambidae). *Pest Management Science: Formerly Pesticide Science*. 2005 May;61(5):508–10.
24. Rosenzweig C, Iglesius A, Yang XB, Epstein PR, Chivian E. Climate change and extreme weather events–implications for food production, plant diseases, and pests. *Global Change and Human Health*;2(2):90–104.
25. Masters G, Norgrove L. *Climate change and invasive alien species*. UK: CABI Working Paper. 2010 Nov;1:30.
26. Wright C, Nyberg D, De Cock C, Whiteman G. Future imaginings: organizing in response to climate change. *Organization*. 2013 Sep;20(5):647–58.
27. Cagan L, Tancik J, Hassan S. Natural parasitism of the European corn borer eggs Ostrinia nubilalis Hbn.(Lep., Pyralidae) by Trichogramma in Slovakia—need for field releases of the natural enemy. *Journal of Applied Entomology*. 1998 Jan 12;122(1–5):315–8.

28. Burdon JJ, Zhan J. Climate change and disease in plant communities. *PLoS Biology*. 2020 Nov 24;18(11):e3000949.
29. Chaloner TM, Gurr SJ, Bebber DP. Plant pathogen infection risk tracks global crop yields under climate change. *Nature Climate Change*. 2021 Aug;11(8):710–5.
30. Ahuja I, Rohloff J, Bones AM. Defence mechanisms of Brassicaceae: implications for plant-insect interactions and potential for integrated pest management. *Sustainable Agriculture*. 2011;2:623–70.
31. Tixeront M, Dupuy F, Cortesero AM, Hervé MR. Understanding crop colonization of oilseed rape crops by the cabbage stem flea beetle (*Psylliodes chrysocephala* L. (Coleoptera: Chrysomelidae)). *Pest Management Science*. 2024 May;80(5):2260–6.
32. Nauen R, Denholm I. Resistance of insect pests to neonicotinoid insecticides: current status and future prospects. *Archives of Insect Biochemistry and Physiology: Published in Collaboration with the Entomological Society of America*. 2005 Apr;58(4):200–15.
33. Hervé MR, Cortesero AM. Potential for oilseed rape resistance in pollen beetle control. *Arthropod-Plant Interactions*. 2016 Dec;10:463–75.
34. Kolte SJ. *Diseases of annual edible oilseed crops: volume II: rapeseed-mustard and sesame diseases*. CRC Press; Boca Raton, Florida, USA, 2018 Jan 18. https://doi.org/10.1201/9781351073049
35. Narayanamma VL, Reddy KD, Reddy AV. Integrated pest and disease management practices in oilseeds–a critical review. *Indian Journal of Entomology*. 2013;75(1):34–56.
36. Jones RA, Sharman M, Trębicki P, Maina S, Congdon BS. Virus diseases of cereal and oilseed crops in Australia: current position and future challenges. *Viruses*. 2021 Oct 12;13(10):2051.
37. Philipsen H, Nielsen O. Host potential of insects from cruciferous crops to entomopathogenic nematodes and augmentation of nematodes through oil seed rape growing. *IOBC Bulletin*. 2003;26(1):141–6.
38. Pullens JW, Sharif B, Trnka M, Balek J, Semenov MA, Olesen JE. Risk factors for European winter oilseed rape production under climate change. *Agricultural and Forest Meteorology*. 2019 Jul 15;272:30–9.
39. Mahmuti MA, West JS, Watts JA, Gladders P, Hughes DJ, Fitt BD. Disease control on UK winter oilseed rape contributes to climate change mitigation. *Agriculture, Ecosystems & Environment*. 2011;295:106897.
40. Evans N, Butterworth MH, Baierl A, Semenov MA, West JS, Barnes A, Moran D, Fitt BD. The impact of climate change on disease constraints on production of oilseed rape. *Food Security*. 2010 Jun;2:143–56.
41. Noel K, Qi A, Gajula LH, Padley C, Rietz S, Huang YJ, Fitt BD, Stotz HU. Influence of elevated temperatures on resistance against phoma stem canker in oilseed rape. *Frontiers in Plant Science*. 2022 Mar 2;13:785804.
42. Jungers J, Runck B, Ewing PM, Maaz T, Carlson C, Neyhart J, Fumia N, Bajgain P, Subedi S, Sharma V, Senay S. Adapting perennial grain and oilseed crops for climate resiliency. *Crop Science*. 2023 Jul;63(4):1701–21.
43. Begna T. Conventional breeding methods widely used to improve self-pollinated crops. *International Journal of Research*. 2021;7(1):1–6.
44. Mehrabi Z, McDowell MJ, Ricciardi V, Levers C, Martinez JD, Mehrabi N, Wittman H, Ramankutty N, Jarvis A. The global divide in data-driven farming. *Nature Sustainability*. 2021 Feb;4(2):154–60.
45. Lamichhane S, Thapa S. Advances from conventional to modern plant breeding methodologies. *Plant Breeding and Biotechnology*. 2022 Mar 1;10(1):1–4.

46. Mwangangi IM, Muli JK, Neondo JO. Plant hybridization as an alternative technique in plant breeding improvement. *Asian Journal of Research in Crop Science*, 2019;4(1):1-11. https://doi.org/10.9734/ajrcs/2019/v4i130059
47. Schley RJ, Twyford AD, Pennington RT. Hybridization: a 'double-edged sword'for Neotropical plant diversity. *Botanical Journal of the Linnean Society*. 2022 May 1;199(1):331–56.
48. Kumari P, Singh KP, Bisht D, Kumar S. Somatic hybrids of Sinapis alba+ Brassica juncea: study of backcross progenies for morphological variations, chromosome constitution and reaction to Alternaria brassicae. *Euphytica*. 2020 Jun;216:1–4.
49. Liu CQ, Li GQ, Yao XQ, Huang L, Wu XY, Xie ZJ. Characterization of Ogura CMS fertility-restored interspecific hybrids and backcross progenies from crosses between broccoli and rapeseed. *Euphytica*. 2020 Dec;216 (9):1–12.
50. Singh KP, Kumari P, Rai PK. Phenotypic characterization and resistance response to *Sclerotinia sclerotiorum* of backcross lines developed from stable allohexaploids of *Sinapis alba+ Brassica juncea. Euphytica*. 2023 Mar;219(3):34.
51. Baydar HA, Marquard R, Turgut I. Pure line selection for improved yield, oil content and different fatty acid composition of sesame, *Sesamum indicum. Plant Breeding*. 1999 Oct;118(5):462–64.
52. Amaral LD, Bruzi AT, Resende PM, Silva KB. Pure line selection in a heterogeneous soybean cultivar. *Crop Breeding and Applied Biotechnology*. 2019 Oct 31;19:277–84.
53. Melchinger AE, Böhm J, Utz HF, Müller J, Munder S, Mauch FJ. High-throughput precision phenotyping of the oil content of single seeds of various oilseed crops. *Crop Science*. 2018 Mar;58(2):670–78.
54. Ramya P, Singh GP, Jain N, Singh PK, Pandey MK, Sharma K, Kumar A, Prabhu KV. Effect of recurrent selection on drought tolerance and related morpho-physiological traits in bread wheat. *PloS One*. 2016 Jun 14;11(6):e0156869.
55. Ribaut JM. The challenges of biotechnologies to improve plant breeding efficiency. *Current Opinion in Plant Biology*. 2007;13(2):174–181.
56. Cooper M, Eeuwijk FV, Chapman SC, Podlich DW, Löffler C. Genotype-by-environment interactions under water-limited conditions. *Drought Adaptation in Cereals*. Haworth Press; 2006:51–96.
57. Muthu V, Abbai R, Nallathambi J, Rahman H, Ramasamy S, Kambale R, Thulasinathan T, Ayyenar B, Muthurajan R. Pyramiding QTLs controlling tolerance against drought, salinity, and submergence in rice through marker assisted breeding. *PloS One*. 2020 Jan 7;15(1):e0227421.
58. Nogoy FM, Song JY, Ouk S, Rahimi S, Kwon SW , Kang KK, Cho YG. Current applicable DNA markers for marker assisted breeding in abiotic and biotic stress tolerance in rice (*Oryza sativa L.*). *Plant Breeding and Biotechnology*. 2016 Aug 31;4(3):271–84.
59. Yusoff MR, Harun AR, Ahmad F. Current applicable DNA markers for marker assisted breeding in rice (*Oryza sativa* L.). In: M.u.R. Ansari (Ed.), *Recent Advances in Rice Research*. IntechOpen; London, UK, 2020 Jul 15.
60. Dimitrijevic A, Horn R. Sunflower hybrid breeding: from markers to genomic selection. *Frontiers in Plant Science*. 2018 Jan 17;8: 2238.
61. Madhusudhana R. Marker-assisted breeding in sorghum. In: Aruna C., K. B. R. S. Visarada, B. Venkatesh Bhat & Vilas A. Tonapi (Eds.), *Breeding sorghum for diverse end uses*. Woodhead Publishing; Duxford, UK, 2019 Jan 1. pp. 93–114.
62. Hasan N, Choudhary S, Naaz N, Sharma N, Laskar RA. Recent advancements in molecular marker-assisted selection and applications in plant breeding programmes. *Journal of Genetic Engineering and Biotechnology*. 2021 Dec;19(1):1–26.

63. Obermeier C, Friedt W. Applied oilseed rape marker technology and genomics. In: P. Poltronieri & Y. Hong (Eds.), *Applied plant genomics and biotechnology*, 1st ed. Cambridge: Woodhead Publishing; Sawston, UK, 2015 Jan 1. pp. 253–295.
64. Jiang C, Shi J, Li R, Long Y, Wang H, Li D, Zhao J, Meng J. Quantitative trait loci that control the oil content variation of rapeseed (*Brassica napus L.*). *Theoretical and Applied Genetics*. 2014 Apr;127:957–68.
65. Snowdon RJ, Iniguez Luy FL. Potential to improve oilseed rape and canola breeding in the genomics era. *Plant Breeding*. 2012 Jun;131(3):351–60.
66. Spasibionek S, Mikołajczyk K, Ćwiek–Kupczyńska H, Piętka T, Krótka K, Matuszczak M, Nowakowska J, Michalski K, Bartkowiak-Broda I. Marker assisted selection of new high oleic and low linolenic winter oilseed rape (*Brassica napus L.*) inbred lines revealing good agricultural value. *PloS One*. 2020 Jun 4;15(6):e0233959.
67. Mehraj H, Akter A, Miyaji N, Miyazaki J, Shea DJ, Fujimoto R, Doullah MA. Genetics of clubroot and Fusarium wilt disease resistance in Brassica vegetables: the application of marker assisted breeding for disease resistance. *Plants*. 2020 Jun 9;9(6):726.
68. Kadirvel P, Veerraju C, Senthilvel S, Yadav P, Kiran BU, Shaik M, Shaw R, Selvaraj VM, Reddy YR, Mohanrao MD, Mukta N. Marker-assisted selection for fast-track breeding of high oleic lines in safflower (*Carthamus tinctorious L.*). *Industrial Crops and Products*. 2020 Dec 15;158:112983.
69. Goonde DB, Abady S. *Marker assisted selection in groundnut*. IntechOpen; **10** Lower Thames Street, London EC3R 6AF, United Kingdom, 2023 Feb 8. p. 108476.
70. Ahmed HG, Rizwan M, Naeem M, Khan MA, Baloch FS, Sun S, Chung G. Molecular characterization and validation of sunflower (*Helianthus annuus* L.) hybrids through SSR markers. *PloS One*. 2022 May 19;17(5):e0267383.
71. Altinkut Uncuoğlu, AH. SSR markers suitable for marker assisted selection in sunflower for Downy mildew resistance. *Biology Open*. 2018;7(11):bio035515.
72. Bhuiyan MS, Malek MA, Khana NA, Islam M, Rahman S, Alam MA. Validated molecular marker for downy mildew disease resistance breeding of sunflower: a short review. *Journal of Agrobiotechnology*. 2023 Aug 10;14(2):28–43.
73. Qi LL, Cai XW. Characterization and mapping of a downy mildew resistance gene, Pl 36, in sunflower (*Helianthus annuus L.*). *Molecular Breeding*. 2022 Feb;42(2):8.
74. Şahin EÇ, Kalenderoğlu A, Aydın Y, Evci G, Uncuoğlu AA. SSR markers suitable for marker assisted selection in sunflower for downy mildew resistance. *Open Life Sciences*. 2018 Sep 22;13(1):319–26.
75. Radanović A, Sprycha Y, Jocković M, Sundt M, Miladinović D, Jansen C, Horn R. KASP markers specific for the fertility restorer locus Rf1 and application for genetic purity testing in sunflowers (*Helianthus annuus L.*). *Genes*. 2022 Mar 5;13(3):465.
76. Hu J, Chen B, Zhao J, Zhang F, Xie T, Xu K, Gao G, Yan G, Li H, Li L, Ji G. Genomic selection and genetic architecture of agronomic traits during modern rapeseed breeding. *Nature Genetics*. 2022 May;54(5):694–704.
77. Langridge P, Braun H, Hulke B, Ober E, Prasanna BM. Breeding crops for climate resilience. *Theoretical and Applied Genetics*. 2021 Jun;134(6):1607–11.
78. Marsh JI, Hu H, Gill M, Batley J, Edwards D. Crop breeding for a changing climate: integrating phenomics and genomics with bioinformatics. *Theoretical and Applied Genetics*. 2021 Jun;134:1677–90.
79. Jaradat AA. Breeding oilseed crops for climate change. In: *Breeding oilseed crops for sustainable production*. Academic Press; London, United Kingdom, 2016 Jan 1. pp. 421–472.

80. Xiao Q, Bai X, Zhang C, He Y. Advanced high-throughput plant phenotyping techniques for genome-wide association studies: a review. *Journal of Advanced Research*. 2022 Jan 1;35:215–30.
81. Ashfaq H, Abubakar M, Ghulzar H, Farid M, Yaqoob S, Komal N, Azam Z, Hamza A, Ali S, Adrees M. Phytoremediation potential of oilseed crops for lead- and nickel-contaminated soil. *Plant Ecophysiology and Adaptation under Climate Change: Mechanisms and Perspectives II: Mechanisms of Adaptation and Stress Amelioration*. 2020:801–20.
82. Wang X, Xu Y, Hu Z, Xu C. Genomic selection methods for crop improvement: current status and prospects. *The Crop Journal*. 2018 Aug 1;6(4):330–40.
83. Sweeney DW, Sun J, Taagen E, Sorrells ME. Genomic selection in wheat. In: *Applications of genetic and genomic research in cereals*. Woodhead Publishing; 2019 Jan 1. pp. 273–302.
84. Robertsen CD, Hjortshøj RL, Janss LL. Genomic selection in cereal breeding. *Agronomy*. 2019 Feb 19;9(2):95.
85. Cowling WA, Balázs E. Prospects and challenges for genome-wide association and genomic selection in oilseed Brassica species. *Genome*. 2010 Nov;53(11):1024–8.
86. Sandhu KS, Merrick LF, Sankaran S, Zhang Z, Carter AH. Prospectus of genomic selection and phenomics in cereal, legume and oilseed breeding programs. *Frontiers in Genetics*. 2022 Jan 21;12:829131.
87. Werner CR, Voss-Fels KP, Miller CN, Qian W, Hua W, Guan CY, Snowdon RJ, Qian L. Effective genomic selection in a narrow-genepool crop with low-density markers: Asian rapeseed as an example. *The Plant Genome*. 2018 Jul;11(2):170084.
88. Cros D, Denis M, Sánchez L, Cochard B, Flori A, Durand-Gasselin T, Nouy B, Omoré A, Pomiès V, Riou V, Suryana E. Genomic selection prediction accuracy in a perennial crop: case study of oil palm (*Elaeis guineensis Jacq.*). *Theoretical and Applied Genetics*. 2015 Mar;128:397–410.
89. Mangin B, Bonnafous F, Blanchet N, Boniface MC, Bret-Mestries E, Carrère S, Cottret L, Legrand L, Marage G, Pegot-Espagnet P, Munos S. Genomic prediction of sunflower hybrids oil content. *Frontiers in Plant Science*. 2017 Sep 21;8:1633.
90. Miladinović D, Hladni N, Radanović A, Jocić S, Cvejić S. Sunflower and climate change: possibilities of adaptation through breeding and genomic selection. In C. Kole (Ed.), *Genomic Designing of Climate-Smart Oilseed Crops*. Springer Cham, 2019:173–238.
91. Patil R, Viswanatha KP, Upadhyaya HD, Lokesha R, Khan H, Gururaj S. Genetic diversity, association and principle component analyses for agronomical and quality traits in genomic selection training population of groundnut (*Arachis hypogaea L.*). *Indian Journal of Genetics and Plant Breeding*. 2020 Sep 25;80(03):282–90.
92. Zhao H, Lin Z, Khansefid M, Tibbits JF, Hayden MJ. Genomic prediction and selection response for grain yield in safflower. *Frontiers in Genetics*. 2023 Mar 27;14:1129433.
93. Subedi U, Jayawardhane KN, Pan X, Ozga J, Chen G, Foroud NA, Singer SD. The potential of genome editing for improving seed oil content and fatty acid composition in oilseed crops. *Lipids*. 2020 Sep;55(5):495–512.
94. Jha S, Thakur ND, Singh AA. Regulatory framework for GMOs-interplay between regulatory policies and socio-ethical concern. In: *Policy Issues in Genetically Modified Crops*. Academic Press; 2021 Jan 1. pp. 401–419.
95. Dwivedi SL, Reynolds MP, Ortiz R. Mitigating tradeoffs in plant breeding. *iScience*. 2021 Sep 24;24(9), 103087.

96. Peltonen-Sainio P, Jauhiainen L, Hyövelä M, Nissilä E. Trade-off between oil and protein in rapeseed at high latitudes: means to consolidate protein crop status. *Field Crops Research.* 2011 Mar 18;121(2):248–55.
97. Joshi BK, Ghimire KH, Neupane SP, Gauchan D, Mengistu DK. Approaches and advantages of increased crop genetic diversity in the fields. *Diversity.* 2023 Apr 28;15(5):603.
98. Razzaq A, Kaur P, Akhter N, Wani SH, Saleem F. Next-generation breeding strategies for climate-ready crops. *Frontiers in Plant Science.* 2021 Jul 21;12:620420.
99. Rai SK, Bawa V, Dar ZA, Sofi NR, Mahdi SS, Qureshi AM. Use of modern molecular biology and biotechnology tools to improve the quality value of oilseed Brassicas. In M. S. Kole (Ed.), *Quality Breeding in Field Crops.* Springer Nature, Cham, Switzerland, 2019:255–66.
100. He J, Zhang K, Tang M, Zhou W, Chen L, Chen Z, Li M. CRISPR-based genome editing technology and its applications in oil crops. *Oil Crop Science.* 2021 Jul 1;6(3):105–113.
101. Subedi U, Ozga JA, Chen G, Foroud NA, Singer SD. CRISPR/Cas-mediated genome editing for the improvement of oilseed crop productivity. *Critical Reviews in Plant Sciences.* 2020 May 3;39(3):195–221.
102. Chang T, Guan M, Zhou B, Peng Z, Xing M, Wang X, Guan C. Progress of CRISPR/Cas9 technology in breeding of Brassica napus. *Oil Crop Science.* 2021 May 1;6(2):53–7.
103. Tian Q, Li B, Feng Y, Zhao W, Huang J, Chao H. Application of CRISPR/Cas9 in rapeseed for gene function research and genetic improvement. *Agronomy.* 2022 Mar 28;12(4):824.

5 Modern Breeding Strategies for Heat Tolerance in Solanaceae Crops

Fatima Sajal, Muhammad Abu Bakar Saddique, Zulfiqar Ali, Sadia Hakeem, Muhammad Ali Sher, Nadia Ayub, and Mubashir Ali

5.1 INTRODUCTION

By the end of the 21st century (2081–2100), global temperature is expected to rise by an average of around 1–3.7°C over the 1986–2005 values, depending on various scenarios (Farnham and Bjorkman, 2011). Global average temperature is rising at the rate of 0.3°C per decade. Increased temperatures may benefit certain locations, but if adaptation measures are not taken, agricultural yields will severely decline. Each crop has a threshold temperature it can withstand at a certain growth stage. Plants may experience heat stress if the temperature is slightly raised or lowered above the threshold. Heat stress or heat shock is typically used to describe raising the temperature by 10–15°C above the optimum temperature. The effects of heat stress are complex and rely on both the rate and duration of temperature rise. The degree of heat stress in certain climatic zones is primarily determined by the frequency of heat shocks and the length of high day/night temperatures. Elevated temperatures have a negative impact on agricultural output as they not only hinder flowering and fruit set but also hinder fruit growth and maturity. Plant morphology, physiology, biochemistry, and agronomy are all impacted by temperature (Osei et al., 2018). Approximately 20–30% of plant species will become threatened when the average temperature rises by 1.5–2.5°C. Genetic advancements are vital in reducing the harmful effects of high temperatures (Collard and Mackill, 2008). The human population heavily relies on vegetable crops to meet its daily dietary requirements. The Solanaceae family is one of the most important families among plant species. It includes 3000–4000 species, categorized into 90 genera. The main genera of this family are *Solanum*, *Capsicum*, *Physalis*, and *Lycium*. The plant species belonging to the Solanaceae family are found across the world, but are most prevalent and prolific in the tropical parts of Latin America, where roughly 40 genera are endemic (Motti, 2021). Only over 50 species are found in the United States and Canada combined, and very few of them are found in

 DOI: 10.1201/9781032720142-6

temperate climates. Their habitat ranges from deserts to rainforests, and includes both annual herbs and trees (Rabara et al., 2015). They have significant economic importance and are used extensively as a source of food (potatoes, tomatoes, eggplant/aubergine, peppers, etc.), ornamental plants (petunias), and bio-chemical crops (tobacco, mandrake, belladonna, jimsonweed) (Motti, 2021). Some of these species are known to contain poisonous alkaloid compounds, which is why they go by the alternate name 'nightshades' (Shah et al., 2013). Others like potato, tomato, eggplant, and pepper are, however, important solanaceous vegetables comprising 39–66% of horticultural production globally and in Europe, respectively (Morris and Taylor, 2017). Members of the family are distinguished by having solitary or clustered flowers with five fused sepals and petals, five stamens, and a superior ovary made up of two fused carpels, or segments, that bear ovules, obliquely positioned within the flower on a basal disk of tissue (Gebhardt, 2016). The style, or upper end of the ovary, is simple and has two lobes on the stigma, which is the surface that accepts pollen.

Certain regions of the world have temperatures that are too high for solanaceous crops due to climate change effects (Katsoulas and Kittas, 2008). This rise in temperature drastically reduces fruit count, fruit weight, and the quantity of seeds per fruit (Arkoub-Djermoune et al., 2016) as the daytime temperature rises above 25°C (Xu et al., 2017). The complexity of heat stress, plant responses to it, and incomplete knowledge of the genetic mechanisms responsible for heat tolerance hinder efforts to breed heat-tolerant cultivars (La Pena and Hughes, 2007). This chapter highlights the impact of high temperature on crops, focusing on the Solanaceae family including mechanisms of adaptation in Solanaceae plants against heat stress and breeding strategies to develop heat-tolerant cultivars.

5.2 THE IMPACT OF HEAT STRESS ON CROP PRODUCTION

The global climatic changes have dominated discussions among environmentalists and decision-makers because these changes are adversely impacting agriculture and ultimately food security (Giordano et al., 2021). Crop yields are being affected by variations in meteorological parameters such as temperature, solar radiation, and precipitation. Heat stress affects all growth stages of a plant by altering its morpho-physiological and biochemical responses. It hinders plant root and shoot growth and canopy performance. Temperatures as high as 45°C for short periods of time can cause caspase-like enzyme synthesis, cytochrome c release, and programmed cell death (PCD) that lead toward stunted plants with low yields. Plants are more sensitive to high temperatures during their reproductive stage than during their vegetative stage. Heat stress also affects the photosynthetic machinery as it denatures the chloroplasts (Faiz et al., 2020). Moreover, once a plant experiences high temperature it closes its stomata to save water, which results in low CO_2 concentration in leaves, inhibition of photosynthetic enzymes and ATP synthases, reduced leaf expansion, and ultimately the senescence of leaves (Lai and He, 2016). The high temperature causes increased floral abscission and reduced fruit production in solanaceous crops. It also affects the pollen grain numbers and viability, particularly in the early stages.

It hinders the growth of the pollen tube and results in poor germination of the pollen and ultimately minimal fruit set (Hamilton Iii and Coleman, 2001).

The Intergovernmental Panel on Climate Change (IPCC) in its Fourth Assessment Report (AR4) stated that there has been a general decline in crop yields above threshold (Intergovernmental Panel on Climate Change, 1995) due to a 3°C rise in the global mean temperature (Abou-Hussein, 2012). A rise in CO_2 level positively impacts the photosynthesis rate and transpiration of plants, resulting in an enhanced yield in some crops due to the CO_2 fertilization effect. To predict these uncertainties about whether the change in climate pattern would be beneficial or detrimental for a specific crop, several yield prediction models have been developed. For instance, a new version of the global crop yield model PEGASUS (Predicting Ecosystem Goods And Services Using Scenarios) that forecasts heat stress sensitivity levels for the crop at anthesis and CO_2 fertilization effects for maize, spring wheat, and soybean, has been developed (Akhtar et al., 2015). Additionally, the CRUTS 2.10 model observed the climate dataset for the years 1971–2020 to efficiently use the PEGASUS model and produce a robust estimate of uncertainties related to future climate change (Hoshikawa et al., 2021). The model concluded that projected changes in the frequency and severity of extreme climatic events will have more serious consequences for food production and food insecurity than changes in the mean climate alone.

5.3 PLANT STRATEGIES TO ADAPT TO THE RISING TEMPERATURE

The internal dynamics of agricultural systems, particularly their capacity for adaptation to changes, will determine the potential effects of climate change on agricultural productivity in addition to the climate itself (Ifejika Speranza, 2010). Interactions between the environment and plant might affect the degree of impact of climate change or complicate the plant stress response. Plants may adapt to the rising temperature through different morphological or biochemical pathways (Capiati et al., 2006). The buildup of sugars, amino acids, and betaines is one tactic that helps plants survive in harsh environments. For better survival under heat stress, the metabolism of carbon absorption is changed to counteract the impacts of heat stress, remobilizing the starch reserve in chloroplasts of plants and allowing them to survive stress periods and avoid additional damage by releasing energy, sugars, and derived metabolites (Nemeskéri and Helyes, 2019). Plants have developed the ability to activate defense mechanisms to shield themselves from the detrimental effects of high temperatures (Guo et al., 2022). Acquired thermotolerance (ATT) is the ability of plants to withstand extremely high temperatures (HT) after being pre-exposed to mild HT (sublethal temperatures), whereas basal thermos tolerance (BTT) is the ability of plants to withstand high temperatures without having previously experienced heat stress. Plants' defensive mechanisms against high temperatures are intimately associated with rapid changes in the gene expression related to BTT and ATT phenomena (Dos Santos et al., 2022).

Plant stress physiology research is both developing new and evolving existing technologies, that may help lessen the effects of climate change on the production of vegetables (Yadav and Sharma 2016). Nonetheless, farmers in developing countries are typically smallholders with few alternatives, and are strongly dependent on the resources found on their farms or in their communities. Thus, to improve the resilience of agriculture for resource-constrained small-scale farmers, simple, inexpensive, and accessible technology must be employed. The World Vegetable Center (WVC) (previously called the Asian Vegetable Research and Development Centre – AVRDC) has been addressing how environmental stress affects the production of vegetables. Advanced breeding lines are being produced, and the germplasm of the key vegetable crops that can withstand extreme temperatures, flooding, and drought has been identified. The WVC has introduced modern ways to tackle the problems of flooding, drought, and salinity, and for the provision of optimum fertilizers (Hughes et al., 2007). They are practicing the use of soil additives and modifying the methods of fertilizer administration to boost soil fertility and improve plant uptake of nutrients, grafting to induce both biotic and abiotic stress tolerance and modern forms of drip irrigation for the effective use of water. Additionally, efforts are being made to find germplasm that efficiently uses nitrogen, as it is a key element that induces temperature tolerance in crops (Schafleitner, 2015). Research and development initiatives also include the invention of production methods intended to reduce the impacts of hot and dry weather on vegetable production systems and to increase water-use efficiency. Vegetables cultivated in the hot and humid lowland tropics may yield more when different management techniques are used.

5.4 UNDERSTANDING HEAT STRESS IMPACTS IN SOLANACEAE CROPS

Several significant vegetable crops, including tomato, pepper, beans, and sweet corn, are particularly sensitive to heat stress during reproductive development, and yield decline is likely to occur unless production is moved to colder months of the year (Hassan et al., 2021). Overproduction of reactive oxygen species and oxidative stress are also brought on by high temperatures, which also have an impact on photosynthesis, alter membrane fluidity, and undermine the overall stability of metabolic processes (Karapanos et al., 2008). Heat stress causes abortive male gametophyte, which lowers fruit set and overall plant growth. It reduces the size of vegetables, soluble solids in them, and their softness.

Tomatoes have dominated scientific research on how environmental pressures affect plants. They can thrive in a variety of environments but excessive temperatures can have a detrimental effect on fruit setting and productivity (Singh et al., 2017). High temperature of up to 120–125°F (49–52°C) scorches ripe tomato fruits, while green tomatoes get burnt if they are exposed for an hour or more to these temperatures. Dead patches that show up on immature leaves of various plants and hypocotyl leaves are signs of heat damage (Madani et al., 2019). Transpiration in plants lowers heating by around 15–25% (Madani et al., 2019). Depending on the species or tissue, a variety of plants can sustain heat damage. For instance, *Solanum chilense*, a species from the 'tomato' subgenus Lycopersicon, has evolved mechanisms to better withstand rising temperatures compared to commercial vegetables (Ayenan et al., 2019). This is especially true when the nighttime temperature rises above 26°C. Besides air temperature, increasing soil temperature also significantly damages plants. It affects microbial growth and development, soil organic matter degradation, seed germination, root development, and the root absorption of water and nutrients (Faiz et al., 2020). The high temperature of the soil has a significant impact on the form, size, and quality of storage organs, e.g., tubers, bulbs, rhizomes, corms, and stolons. Even minor heat stress can directly impair tuber induction and growth. Potato yields are especially vulnerable to high-temperature stress. Temperature and photoperiod interact in such a way that as temperature increases, the photoperiod needed for tuberization in a certain genotype becomes shorter (Snyder and Ewing, 1989). The early development of stolons from young tubers, known as 'heat sprouting,' is another way that high temperatures can negatively impact the quality of tubers. Further research is required about the impact of abiotic stressors on other vegetable crops by translating information from mechanisms of stress tolerance in tomato (Gerszberg and Hnatuszko-Konka, 2017).

5.5 EXPRESSION PATTERNS OF HEAT SHOCK PROTEINS AND HEAT SHOCK FACTORS IN VEGETABLES UNDER HEAT STRESS

Plant cells undergo cellular alterations when exposed to high temperature. To reduce damage and maintain cellular homeostasis, plants have developed a variety of physiological and molecular responses to stressors (Gong et al., 2021). Plants including vegetables produce a variety of heat stress-responsive proteins – heat shock proteins (HSPs) – and transcription factors (TFs) – heat stress-responsive TFs (HSFs) – in response to severe temperature shocks (Figure 5.1) (Kang et al., 2022). These proteins and transcription factors help plants to adapt to rising temperatures. Strong heat stress initiates molecular remodeling to retain equilibrium in plant cells (Aldubai et al., 2022). How plants detect HT and employ HT-induced signaling to change transcription downstream is one of the central topics that plant scientists have been pondering for a long time (Ding and Yang, 2022). Ca^{2+} is essential for plants to sense and respond under heat stress. In reaction to heat stress, the fluidity of plasma membranes (PM) alters and opens cyclic nucleotide-gated calcium channels (CNGCs) and allow Ca^{2+} to flow from the PM into the cytosol (Dong et al., 2015). The protein calmodulin 3 (CaM3) bind to Ca^{2+} ions to make Ca^{2+}-CaM3 complex. It regulates the phosphorylation and dephosphorylation of HSFA1 through the interactions between Ca^{2+}-CaM3-binding protein kinase 3 (CBK3) and phosphatase PP7. These interactions allow HSFA1 to transmit cytosolic heat-stress response (HSR) signals into the nucleus for the expression of responsive genes (Zhang et al., 2020). Furthermore, increased levels of inositol-1,4,5-triphosphate (IP3) through the phosphoinositide-signaling pathway cause intracellular Ca^{2+} pools such as the endoplasmic reticulum (ER) and vacuole to release Ca^{2+} into the cytoplasm to induce heat stress tolerance. According to Solankey et al. (2015), reactive oxygen species (ROS) produced by respiratory burst oxidase homolog (*RbohB*, *RbohD*), and NADPH oxidases are additional possible heat stress sensors. The formation of HSPs, a family of protective proteins that includes HSP100, HSP90, HSP70, HSP60, and certain mitogen-activated protein kinases (MAPKs), is most critical for plants under heat stress. A number of TFs control the expression of heat stress-responsive genes, including *NAC*, *MYB*, *WRKY*, *RAV*, *bZIP*, *AP2/ERF*, and *ZAT* (Guo et al., 2016). In the promoters of these stress-inducible genes, cis-elements

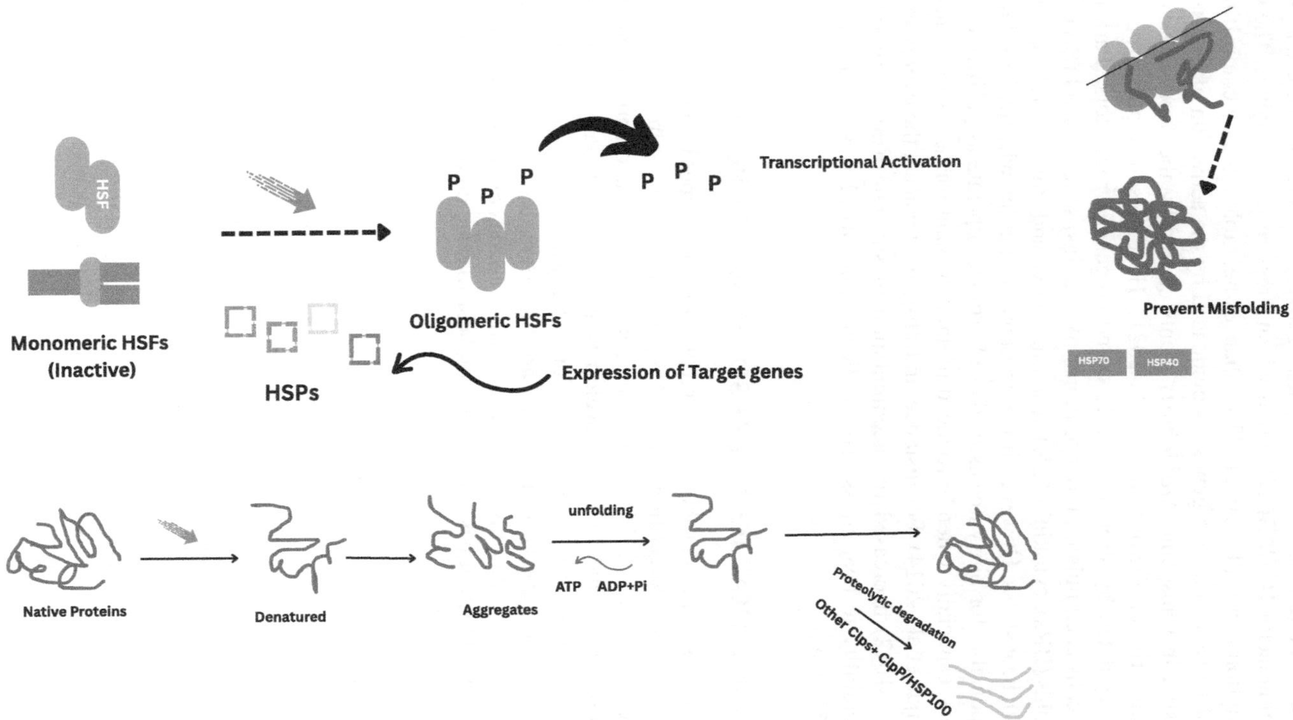

FIGURE 5.1 Transcriptional activation of Heat Shock Proteins (HSP) and Heat responsive transcription factors (HSFs). ADP: adenosine diphosphate, ATP: adenosine triphosphate, Caseinolytic proteases (Clps:, ClpP), Heat stress elements (HSE).

(ARE, CORE, W-box, GCC box, as-1 like) are bound to trigger transcription. Such HSFs including MAPK, NADPH oxidases (*Rboh*), dehydration-responsive element-binding protein 2A (*DREB2A*), heat shock (HsfA2, HsfA7a, HsfBs), and multi-protein bridging factor 1C (MBF1C), control heat stress-responsive genes. The wheat NAC transcription factor (TaNAC2L TF) enhances heat tolerance by controlling the expression of stress-responsive genes, as evidenced by the activation of heat-related gene expression in transgenic *Arabidopsis* plants caused by the overexpression of TaNAC2L transcription factor (Chen et al., 2021). Heat stress-responsive genes are additionally regulated by short RNAs, TFs, and epigenetic regulators (Park et al., 2013). Plant cells accumulate more hormones like auxins, salicylic acid (SA), ABA, brassinosteroids (BRs), cytokinin (CK), jasmonate (JA), and ethylene (ET) as a means of coping with heat stress. However, it is important to examine whether TFs that bind to genes responsible for biosynthesis of these hormones operate as positive or negative regulators, especially when introduced to another plant for heat stress tolerance (Hemantaranjan et al., 2014). For instance, in transgenic tobacco, the overexpression of NAC TF (SlJA2) decreased the accumulation of salicylic acid, increasing the plant's vulnerability to heat stress. This indicates that SlJA2 was negatively regulating the NAC genes.

5.6 BREEDING HEAT-TOLERANT SOLANACEAE CROPS

Heat stress causes several morphological and physiological types of damage to plants and has an impact on seed germination, and vegetative as well as reproductive development (Driedonks et al., 2016). To prepare for the present and anticipated rise in heat stress, efforts are underway to develop heat-tolerant cultivars. High-yielding cultivars that are resistant to biotic and abiotic stresses are the main goal of plant breeding (Babu et al., 2004). Gene banks are the source of genetic diversity that can help breed climate-smart high-yielding cultivars. In the coming sections, conventional and advanced techniques for developing heat-tolerant cultivars in solanaceous crops will be discussed.

5.7 TRADITIONAL BREEDING APPROACHES

High-throughput genotyping is being widely used due to rapid advances in genotyping and its falling cost, but accurate phenotyping has emerged as the main obstacle to determine the genetic basis of significant and complex traits, which is slowing down the advancement of breeding programs (Table 5.1). The development of phenotyping methods for the identification of essential characteristics that are easy to use, accurate, economical, and labor-efficient is crucial to improving the effectiveness of plant breeding programs (Ro et al., 2021). High-throughput phenotyping tools, such as thermal, multispectral, hyperspectral, and fluorescence sensors that may be used as imagers or stationary platforms, have been developed in recent years. Specifying the characteristics to be assessed and the screening parameters are just as crucial as implementing technology (trait-measuring tools and assessment platforms). Determining the ideal experimental procedure for inducing heat stress

TABLE 5.1
Wild sources of various crops having features for heat tolerance and other important traits

Crop	Wild species as source of heat tolerance	Characters responsible for heat tolerance	Reference
Tomato	*S. pimpinellifolium*	High pollen viability	Driedonks et al., 2018
	S. corneliomulleri	High pollen viability	
	S. pennellii Correll	High pollen viability, maintained photosynthesis rate, chlorophyll florescence	
Cucumber	*Cucumis hystris*	Improved stomatal conductance Increase in number of antioxidants	(Ali et al., 2019)
Egg plant	*Solanum insanum*	Enhanced membrane stability Higher expression of HSPs.	(Zhang et al., 2020)
Bell pepper	*Capsicum chacoense*	Increase in Water use Efficiency and antioxidant production	(Preet et al., 2020)
Sweet potato	*Ipomoea trifida*	High photosynthesis rate Increased membrane stability Increased osmotic adjustments	(Ponniah et al., 2017)

has been challenging (Chandler, 1983). Consequently, heat treatments vary widely, and there are no set procedures for determining heat tolerance in such cases.

The germplasm stored in gene banks is the best source to develop cultivars of solanaceous crops that can tolerate the present and future threats of high temperatures. Consequently, it's critical to identify, preserve, and make use of genetic sources. To identify heat-tolerant tomato genotypes, there is increasing interest in screening both farmed and wild tomato genetic resources (Poudyal et al., 2018). However, genetic variants are often released after years of breeding, and plants exhibiting diversity are selected from the local populations. This is crucial for selection since they have developed over many years in natural conditions (Collard and Mackill, 2008). Breeders may choose breeding materials and suitable breeding techniques, as well as carry out effective screening, by reviewing the knowledge that is currently available on these genetic and genomic resources for heat tolerance.

Breeding heat-resistant cultivars is one of the best practices to tackle the damages caused by heat stress. Vegetables have seen very few conventional breeding attempts toward the production of heat-tolerant cultivars. Most of the time, the process of developing genotypes resistant to heat is done in hotter climates, or artificially heated greenhouses or growth chambers. High-yielding farmed cultivars may be screened for heat tolerance using the indirect technique, which can also be used to identify genotypes that are heat tolerant (Priya et al., 2018). Through traditional tomato breeding, the WVC has developed lines like CL5195_93D4 and CLN-162IL. These lines have good fruit production under heat stress (Priya et al., 2019). A heat-tolerant tuber population of potatoes has been developed by three continuous rounds of recurrent selection. When compared to their mother population under heat stress, this population shows 37.8% genetic gain in yield (Ghuge and Mirza, 2021). By using the pedigree method, genotypes of broccoli with high-quality heads have been developed. Similarly, traditional breeding techniques are being used to develop advanced lines of heat-tolerant cauliflower, French beans, capsicum, and peas (Farnham and Bjorkman, 2011).

5.8 MOLECULAR APPROACHES

Recent progress in genetics and genomics has been accompanied by the development and deployment of novel tools, techniques, and approaches that could be used to enhance plant breeding programs. Molecular markers, genetic linkage maps, marker assays, and whole-genome sequences have been developed and published for many crop species, including several vegetables (Xu and Crouch, 2008). Marker-assisted selection (MAS) and genomic selection (GS) that can be used for precision breeding are in various stages of development for each vegetable crop, depending on available resources and the complexity of the species' genetics and breeding (Simko et al., 2021). Various approaches, such as linkage mapping, genome-wide association mapping (GWAS), nested association mapping (NAM), and multi-parent advanced generation inter-cross (MAGIC) populations have been developed for the detection and mapping of genes and QTLs (Lema, 2018). Furthermore, the genetic structure of some characteristics linked to heat tolerance, such as pollen viability, inflorescence count, pollen count, style protrusion and length, and fruit set, has been analyzed, and the corresponding quantitative trait loci (QTL) have been mapped (Chaudhary et al., 2020). The novel genetic and genomic tools and techniques were applied primarily and frequently in major cereal crops, such as maize, rice, and wheat, but they are gradually finding their way into vegetable genetics and breeding.

Given the complexity of abiotic stress tolerance, selecting plants based only on phenotypic characteristics influenced by environmental variables would have a very deceptive effect on breeding programs. One of the best methods for selecting tolerant plants is germplasm screening based on molecular markers (Barone, 2004). Plants with well-established genotypic traits that are unaffected by environmental factors should be used in breeding research. High polymorphic molecular markers can be utilized in breeding programs to select heat-tolerant best-performing parents in segregating populations. With the benefits of being quicker, less time-consuming, and more effective, DNA molecular marker technology offers strong tools for the selection of the desired cultivars in a variety of crops. The polymerase chain reaction (PCR) amplification products of genomic DNA fragments using short oligonucleotides (often 10-mers) of arbitrary sequence are used as primers that produce randomly amplified polymorphic DNA (RAPD) markers (Phan and Sim, 2017). The RAPD markers have been effectively used to fingerprint a large number of fruit species. Sequence-related amplified polymorphism (SRAP) is a PCR-based marker system (Moreno et al., 1995). SRAP can be used for gene tagging, map-based cloning, genomic and cDNA fingerprinting, and map creation. It is an effective and straightforward marker system (Bliss, 2009). It is superior to other systems in several ways. It targets open-reading frames (ORFs), reveals a large number of co-dominant markers, and makes it simple to isolate bands for sequencing.

In whole-genome sequencing of tomatoes and other solanaceous crops, the genetic linkage maps based on DNA marker technologies remain a crucial tool for mapping short genome sequences on chromosomes. Selectable markers for practical breeding have been developed with the use of biparental genetic analysis utilizing DNA marker linkage maps (He et al., 2014). This is especially helpful for features (such as vertical disease resistance) that are regulated by one or a few key genes. The development and practical implementation of next-generation breeding methodologies (like GWAS-based marker development and GS) based on comprehensive whole-genome polymorphism information for breeding lines and genetic resources in solanaceous crops will be fueled by next-generation technologies like massively parallel sequencing and array-based SNP typing platforms (Tiwari et al., 2022).

5.9 CRISPR/CAS SYSTEM FOR INDUCING HEAT TOLERANCE

The CRISPR-Cas system is being used for the improvement of many crops to induce stress tolerance and develop better quality, high-yielding crops (Hyun, 2020). The CRISPR-Cas targets noncoding RNAs as well as the open reading frame, untranslated region, and promoter region of a target coding gene, which is best to target stress-responsive genes in crops (Ahmad et al., 2021). Without causing DNA double-stranded breaks (DSBs), nickase Cas9 (nCas9) or catalytically inactive Cas9 (dead Cas9; dCas9) variants coupled with cytosine or adenine deaminases have also been employed to generate single-base alterations at genomic targets. To increase Proto spacer Adjuscent Motif (PAM) preferences, a great deal of work has gone into developing Cas9 proteins (Ma and Liang, 2021). Cas9 orthologs will increase the range of CRISPR-Cas9 genome editing in plants by including not only the conventional NGG PAM but also NG and additional PAMs. These CRISPR-based techniques are being used to develop climate-smart crops and could be further used for inducing heat tolerance in almost all solanaceous vegetables (Chaudhuri et al., 2022), as presented in Figure 5.2. The gene known as brassinazole-resistant 1 (*BZR1*) is involved in several development events mediated by brassinosteroid (BR). The activation of Respiratory Burst Oxidase Homolog1 (RBOH1) and the generation of H_2O_2 were hampered by the CRISPR-mediated mutation in *BZR1*. In the tomato *bzr1* mutant, exogenous H_2O_2 restored heat tolerance (Li et al., 2022). Furthermore, by editing specific genes, such as those regulating *CBF1* (C-repeat binding factor 1), which controls cold tolerance in plants, and MAPK3, which takes part in the drought stress response to shield tomato plant cell membranes from peroxidative damage, resistant genotypes to cold and drought can be produced (Ahmad, 2023).

Because tomatoes are diploid and have a high-quality genome sequence they are a prime candidate for genome editing (GE), which defines their appropriateness and application for efficient transformation (Younas et al., 2024). Furthermore, genomic architectural differences caused by environmental stress modalities have a tremendous impact on tomato plant growth. This is followed by changes in epigenetic memory and durable inheritance of modified epigenetic structures. Therefore, genuine stable

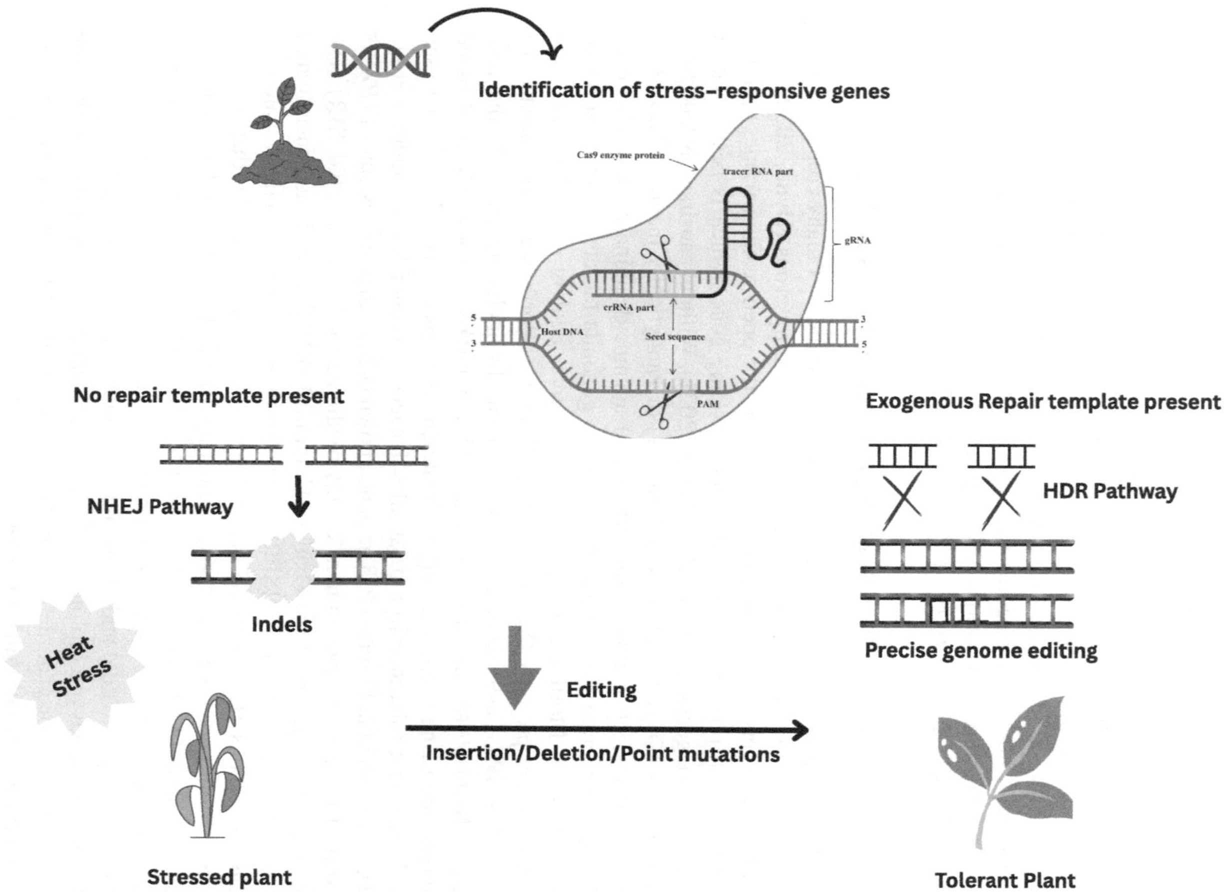

FIGURE 5.2 Genome editing through CRIPR/Cas9 and NHEJ process for the development of stress resistant varieties to confer heat stress. NHEJ (non-homologous end joining), HDR (homology-directed repair), Indels.

inheritance patterns for the generation of improved phenotypes envisioning yield optimization are made possible by precision genome alteration (Tofazzal Islam, 2019). A complex chemical process behind GE is predicated on the use of sequence-specific nucleases (SSNs) to induce chromosomal changes. These changes include the insertion, deletion, translocation, and transversion of genes at certain loci, which increase yield potentials through phenotypic changes (Bhat et al., 2021; Rahman et al., 2022). MicroRNAs (miRNAs) are a subclass of cross-kingdom RNAi (regulatory RNAs) that suppress genes to fight a variety of diseases that affect tomato plants. Breeding strategies for tomato plants can benefit greatly from TILLING (Targeting Induced Local Lesions IN Genomes), which enhances gene alterations through phenotypic variations through random mutagenesis. Likewise, the introduction of SSNs also suggests reduced specificity and sensitivity in well-targeted gene alteration (Salava et al., 2021).

5.10 CONCLUSION

The development of heat-tolerant cultivars is a need of the hour in the face of rising global temperatures. The World Vegetable Center has produced several heat-tolerant lines of vegetable crops through breeding approaches. High-yielding cultivars that are resistant to biotic and abiotic stresses are the main goal of plant breeding. For this purpose, advanced techniques like marker-assisted breeding and CRISPR-Cas technology have been utilized so far. However, to keep up with climatic uncertainties, breeders need to speed up their efforts by increasing the number of generations per year. Fast breeding strategies can be a revolutionary tool for the selection of heat-tolerant cultivars with rapidly increasing global warming and climate change.

REFERENCES

Abou-Hussein, S.D. 2012. Climate change and its impact on the productivity and quality of vegetable crops. *J. Appl. Sci. Res.* 8:4359–4383.

Ahmad, M. 2023. Plant breeding advancements with "CRISPR-Cas" genome editing technologies will assist future food security. *Front. Plant Sci.* 14:1133036.

Ahmad, S., Z. Sheng, R.S. Jalal, J. Tabassum, F.K. Ahmed, S. Hu, G. Shao, X. Wei, K.A. Abd-Elsalam and P. Hu 2021. CRISPR–Cas technology towards improvement of abiotic stress tolerance in plants. *CRISPR and RNAi Systems*. Elsevier.

Akhtar, S., P. Hazra and A. Naik. 2015. Harnessing heat stress in vegetable crops towards mitigating impacts of climate change. *Clim. Dyn. Hortic Sci.* 1:419.

Aldubai, A.A., A.A. Alsadon, H.H. Migdadi, S.S. Alghamdi, S.A. Al-Faifi and M. Afzal. 2022. Response of tomato (*Solanum lycopersicum* L.) genotypes to heat stress using morphological and expression study. *Plants* 11:615.

Ali, M., C.M. Ayyub, M. Amjad and R. Ahmad. 2019. Evaluation of thermo-tolerance potential in cucumber genotypes under heat stress. *Pak. J. Agric. Sci.* 56(1);53–61.

Arkoub-Djermoune, L., L. Boulekbache-Makhlouf, S. Zeghichi-Hamri, S. Bellili, F. Boukhalfa and K. Madani. 2016. Influence of the thermal processing on the physico-chemical properties and the antioxidant activity of a solanaceae vegetable: eggplant. *J. Food Qual.* 39:181–191.

Ayenan, M.A.T., A. Danquah, P. Hanson, C. Ampomah-Dwamena, F.A.K. Sodedji, I.K. Asante and E.Y. Danquah. 2019. Accelerating breeding for heat tolerance in tomato (Solanum lycopersicum L.): an integrated approach. *Agronomy* 9:720.

Babu, R., S.K. Nair, B. Prasanna and H. Gupta. 2004. Integrating marker-assisted selection in crop breeding–prospects and challenges.*Curr. Sci.* 87(5):607–619.

Barone, A. 2004. Molecular marker-assisted selection for potato breeding. *Am. Potato J.* 81:111–117.

Bhat, M.A., R.A. Mir, V. Kumar, A.A. Shah, S.M. Zargar, S. Rahman and A.T. Jan. 2021. Mechanistic insights of CRISPR/Cas-mediated genome editing towards enhancing abiotic stress tolerance in plants. *Physiol. Plant.* 172:1255–1268.

Bliss, F. 2009. Marker-assisted breeding in horticultural crops. *Int. Sympos. Mol. Markers Horticul.* 859:339–350.

Capiati, D. A., País, S. M., & Téllez-Iñón, M. T. (2006). Wounding increases salt tolerance in tomato plants: Evidence on the participation of calmodulin-like activities in cross-tolerance signalling. *Journal of Experimental Botany,* 57(10):2391–2400.

Chandler, R.F. 1983. *The potential for breeding heat tolerant vegetables for the tropics.*

Intergovernmental Panel on Climate Change. 1995. *Intergovernmental panel on climate change (IPCC).* CC BY. 4. Asian Vegetable Research and Development Center (AVRDC), Shanhua, Taiwan.

Chaudhary, S., P. Devi, A. Bhardwaj, U.C. Jha, K.D. Sharma, P. Prasad, K.H. Siddique, H. Bindumadhava, S. Kumar and H. Nayyar. 2020. Identification and characterization of contrasting genotypes/cultivars for developing heat tolerance in agricultural crops: current status and prospects. *Front. Plant Sci.* 11:587264.

Chaudhuri, A., K. Halder and A. Datta. 2022. Classification of CRISPR/Cas system and its application in tomato breeding.*Theor. Appl. Genet* . 135:367–387.

Chen, X., Z. Wang, R. Tang, L. Wang, C. Chen and Z. Ren. 2021. Genome-wide identification and expression analysis of Hsf and Hsp gene families in cucumber (*Cucumis sativus* L.). *J. Plant Growth Regul* . 95:223–239.

Collard, B.C. and D.J. Mackill. 2008. Marker-assisted selection: an approach for precision plant breeding in the twenty-first century. *Philosoph. Trans. R. Soc. B: J. Biol. Sci.* 363:557–572.

Ding, Y. and Yang, S., 2022. Surviving and thriving: How plants perceive and respond to temperature stress. *Developmental Cell,* 57(8), pp.947-958.

Dong, X., H. Yi, J. Lee, I.-S. Nou, C.-T. Han and Y. Hur. 2015. Global gene-expression analysis to identify differentially expressed genes critical for the heat stress response in Brassica rapa. *PLoS One.* 10:e0130451.

Dos Santos, T.B., A.F. Ribas, S.G.H. De Souza, I.G.F. Budzinski and D.S. Domingues. 2022. Physiological responses to drought, salinity, and heat stress in plants: a review. *Stresses* 2 :113–135.

Driedonks, N., I. Rieu and W.H. Vriezen. 2016. Breeding for plant heat tolerance at vegetative and reproductive stages. *Plant Reprod.* 29:67–79.

Faiz, H., C.M. Ayyub, R.W. Khan and R. Ahmad. 2020. Morphological, physiological and biochemical responses of eggplant (*Solanum melongena* L.) seedling to heat stress. *Pak. J. Agric. Sci.* 57(2);371-380.

Farnham, M.W. and T. Bjorkman. 2011. Breeding vegetables adapted to high temperatures: a case study with broccoli. *Hort Sci.* 46:1093–1097.

Gebhardt, C. 2016. The historical role of species from the Solanaceae plant family in genetic research. *Theor. Appl. Genet.* 129:2281–2294.

Gerszberg, A. and Hnatuszko-Konka, K., 2017. Tomato tolerance to abiotic stress: a review of most often engineered target sequences. *Plant growth regulation*, 83(2):175-198.

Ghuge, M.B. and A. Mirza. 2021. Vegetable breeding strategies. *Asian J. Agric. Res.* 8:28–37.

Giordano, M., S.A. Petropoulos and Y. Rouphael. 2021. Response and defence mechanisms of vegetable crops against drought, heat and salinity stress. *Agriculture.* 11:463.

Gong, C., Q. Pang, Z. Li, Z. Li, R. Chen, G. Sun and B. Sun. 2021. Genome-wide identification and characterization of Hsf and Hsp gene families and gene expression analysis under heat stress in eggplant (*Solanum melongema* L.). *Horticulturae.* 7:149.

Guo, M., J.-H. Liu, X. Ma, D.-X. Luo, Z.-H. Gong and M.-H. Lu. 2016. The plant heat stress transcription factors (HSFs): structure, regulation, and function in response to abiotic stresses. *Front. Plant Sci.* 7:180954.

Hamilton Iii, E. and J. Coleman. 2001. Heat-shock proteins are induced in unstressed leaves of Nicotiana attenuata (*Solanaceae*) when distant leaves are stressed. *Am. J. Bot.* 88:950–955.

Hassan, M.U., M.U. Chattha, I. Khan, M.B. Chattha, L. Barbanti, M. Aamer, M.M. Iqbal, M. Nawaz, A. Mahmood and A. Ali. 2021. Heat stress in cultivated plants: Nature, impact, mechanisms, and mitigation strategies—a review. *Plant Biosyst.* 155:211–234.

He, J., X. Zhao, A. Laroche, Z.-X. Lu, H. Liu and Z. Li. 2014. Genotyping-by-sequencing (GBS), an ultimate marker-assisted selection (MAS) tool to accelerate plant breeding. *Front. Plant Sci.* 5:107179.

Hemantaranjan, A., A.N. Bhanu, M. Singh, D. Yadav, P. Patel, R. Singh and D. Katiyar. 2014. Heat stress responses and thermotolerance. *Adv. Plants Agric. Res.* 1:1–10.

Hoshikawa, K., D. Pham, H. Ezura, R. Schafleitner and K. Nakashima. 2021. Genetic and molecular mechanisms conferring heat stress tolerance in tomato plants. *Front. Plant Sci.* 12:786688.

Hyun, T.K. 2020. CRISPR/Cas-based genome editing to improve abiotic stress tolerance in plants. *Bot. Serb.* 44:121–127.

Ifejika Speranza, C., 2010. *Resilient adaptation to climate change in African agriculture* (No. 54). Studies.

Kang, Y., K. Lee, K. Hoshikawa, M. Kang and S. Jang. 2022. Molecular bases of heat stress responses in vegetable crops with focusing on heat shock factors and heat shock proteins. *Front. Plant Sci.* 13:837152.

Karapanos, I.C., S. Mahmood and C. Thanopoulos. 2008. Fruit set in solanaceous vegetable crops as affected by floral and environmental factors. *Eur. J. Plant Sci. Biotechnol.* 2:88–105.

Katsoulas, N. and C. Kittas. 2008. Impact of greenhouse microclimate on plant growth and development with special reference to the Solanaceae. *Eur. J. Plant Sci. Biotechnol.* 2:31–44.

La Pena, R.D. and J. Hughes. 2007. Improving vegetable productivity in a variable and changing climate. *Journal of SAT Agricultural Research*, 4:1–22.

Lai, C.-H. and J. He. 2016. Physiological performances of temperate vegetables with response to chronic and acute heat stress. *Am. J. Plant Sci.* 7:2055–2071.

Lema, M. 2018. Marker assisted selection in comparison to conventional plant breeding. *Agric Res Technol.* 14:555914.

Li, X., S. Xu, M.B. Fuhrmann-Aoyagi, S. Yuan, T. Iwama, M. Kobayashi and K. Miura. 2022. CRISPR/Cas9 technique for temperature, drought, and salinity stress responses. *Curr. Issues Mol. Biol.*44:2664–2682.

Ma, L. and Z. Liang. 2021. CRISPR technology for abiotic stress resistant crop breeding. *Plant Growth Regul.* 94:115–129.

Madani, B., A. Mirshekari and Y. Imahori. 2019. In: Elhadi M. Yahia (Editor); Armando Carrillo-López (Associate Editor), Physiological responses to stress. *Postharvest physiology and biochemistry of fruits and vegetables.* Elsevier.

Moreno, S., Gogorcena, Y. and Ortiz, J.M., 1995. The use of RAPD markers for identification of cultivated grapevine (Vitis vinifera L.). *Scientia horticulturae*, 62(4):237-243.

Morris, W.L. and Taylor, M.A., 2017. The solanaceous vegetable crops: Potato, tomato, pepper, and eggplant.

Motti, R. 2021. The Solanaceae family: botanical features and diversity. In D. Carputo, R. Aversano & M. R. Ercolano (Eds.), *The Wild Solanums Genomes.* 1–9. Springer, Cham. https://doi.org/10.1007/978-3-030-30343-3_1

Nemeskéri, E. and L. Helyes. 2019. Physiological responses of selected vegetable crop species to water stress. *Agronomy.* 9:447.

Osei, M.K., R. Prempeh, J. Adjebeng-Danquah, J.A. Opoku, A. Danquah, E. Danquah, E. Blay and H. Adu-Dapaah. 2018. Marker-assisted selection (MAS): a fast-track tool in tomato breeding. In Seloame Tatu Nyaku; Agyemang Danquah, *Recent advances in tomato breeding and production.* 93–113, IntechOpen.

Park, H.J., W.Y. Jung, S.S. Lee, J.H. Song, S.-Y. Kwon, H. Kim, C. Kim, J.C. Ahn and H.S. Cho. 2013. Use of heat stress responsive gene expression levels for early selection of heat tolerant cabbage (*Brassica oleracea* L.). *Int. J. Mol. Sci.* 14:11871–11894.

Phan, N.T. and S.-C. Sim. 2017. Genomic tools and their implications for vegetable breeding. *Hortic. Sci. Technol.* 35:149–164.

Ponniah, S.K., J. Thimmapuram, K. Bhide, V. Kalavacharla and M. Manoharan. 2017. Comparative analysis of the root transcriptomes of cultivated sweetpotato (*Ipomoea batatas* [L.] Lam) and its wild ancestor (Ipomoea trifida [Kunth] G. Don). *BMC Plant Biol.* 17:1–14.

Poudyal, D., E. Rosenqvist and C.-O. Ottosen. 2018. Phenotyping from lab to field–tomato lines screened for heat stress using Fv/Fm maintain high fruit yield during thermal stress in the field. *Funct. Plant Biol.* 46:44–55.

Preet, T., N. Ghai and S. Jindal. 2020. Ameliorating thermo-tolerance in bell pepper (*Capsicum annuum* L. var. *grossum*) with plant growth regulators. *J. Veg. Sci.* 47:213–218.

Priya, M., K. Siddique, O. Dhankhar, P.V. Prasad, B. Hanumantha Rao, R.M. Nair and H. Nayyar. 2018. Molecular breeding approaches involving physiological and reproductive traits for heat tolerance in food crops. *Indian J. Plant Physiol.* 23:697–720.

Priya, M., Dhanker, O.P., Siddique, K.H., HanumanthaRao, B., Nair, R.M., Pandey, S., Singh, S., Varshney, R.K., Prasad, P.V. and Nayyar, H., 2019. Drought and heat stress-related proteins: an update about their functional relevance in imparting stress tolerance in agricultural crops. *Theoretical and Applied Genetics*, 132(6):1607–1638.

Rabara, R.C., P. Tripathi, R.N. Reese, D.L. Rushton, D. Alexander, M.P. Timko, Q.J. Shen and P.J. Rushton. 2015. Tobacco drought stress responses reveal new targets for Solanaceae crop improvement. *BMC Genomics.* 16:1–23.

Rahman, M.-U., S. Zulfiqar, M.A. Raza, N. Ahmad and B. Zhang. 2022. Engineering abiotic stress tolerance in crop plants through CRISPR genome editing. *Cells.* 11:3590.

Ro, S., L. Chea, S. Ngoun, Z.P. Stewart, S. Roeurn, P. Theam, S. Lim, R. Sor, M. Kosal and M. Roeun. 2021. Response of tomato genotypes under different high temperatures in field and greenhouse conditions. *Plants.* 10:449.

Salava, H., S. Thula, V. Mohan, R. Kumar and F. Maghuly. 2021. Application of genome editing in tomato breeding: Mechanisms, advances, and prospects. *Int. J. Mol. Sci.* 22:682.

Schafleitner, R., Nair, R.M., Rathore, A., Wang, Y.W., Lin, C.Y., Chu, S.H., Lin, P.Y., Chang, J.C. and Ebert, A.W., 2015. The AVRDC–The World Vegetable Center mungbean (Vigna radiata) core and mini core collections. *BMC genomics*, 16(1):344.

Shah, V.V., N.D. Shah and P.V. Patrekar. 2013. Medicinal plants from Solanaceae family. *RJPT.* 6:143–151.

Simko, I., M. Jia, J. Venkatesh, B.-C. Kang, Y. Weng, G. Barcaccia, S. Lanteri, G. Bhattarai and M.R. Foolad. 2021. Genomics and marker-assisted improvement of vegetable crops. *Crit. Rev. Plant Sci.* 40:303–365.

Singh, A.K., M.K. Singh, V. Singh, R. Singh, T. Raghuvanshi and C. Singh. 2017. Debilitation in tomato (*Solanum lycopersicum* L.) as result of heat stress. *RJPP.* 6:1917–1922.

Snyder, R.G. and Ewing, E.E., 1989. Interactive effects of temperature, photoperiod, and cultivar on tuberization of potato cuttings. *HortScience, 24*(2):336-338.

Solankey, S.S., Singh, R.K., Baranwal, D.K. and Singh, D.K., 2015. Genetic expression of tomato for heat and drought stress tolerance: An overview. *International Journal of Vegetable Science*, 21(5):496-515.

Tiwari, J.K., S.R. Yerasu, N. Rai, D.P. Singh, A.K. Singh, S.G. Karkute, P.M. Singh and T.K. Behera. 2022. Progress in marker-assisted selection to genomics-assisted breeding in tomato. *Crit. Rev. Plant Sci.* 41:321–350.

Tofazzal Islam, T.I. 2019. *CRISPR-Cas technology in modifying food crops.* CABI Reviews. 1–16.

Xu, J., M. Wolters-Arts, C. Mariani, H. Huber and I. Rieu. 2017. Heat stress affects vegetative and reproductive performance and trait correlations in tomato (*Solanum lycopersicum*). *Euphytica.* 213:1–12.

Xu, Y. and J.H. Crouch. 2008. Marker-assisted selection in plant breeding: From publications to practice. *Crop Sci.* 48:391–407.

Yadav, S. and Sharma, K.D., 2016. Molecular and morphophysiological analysis of drought stress in plants. In *Plant growth.* IntechOpen.

Younas, A., N. Riaz, M. Rashid, A. Tufail, S. Hyder and Z. Noreen. 2024. *CRISPR/Cas system for achieving abiotic stress tolerance.* OMICs-based Techniques for Global Food Security.

Zhang, A., Z. Zhu, J. Shang, S. Zhang, H. Shen, X. Wu and D. Zha. 2020. Transcriptome profiling and gene expression analyses of eggplant (*Solanum melongena* L.) under heat stress. *PloS One.* 15:e0236980.

Section II

Risk Assessment Strategies

6 Selection Criteria for Assessment of Drought Tolerance on Physio-morphic Attributes

Kashif Rashid, Husnain Rauf, and Muhammad Azam Khan

6.1 INTRODUCTION

Climate change is a term that describes the persistent changes in the earth's climate over time. These changes can have natural causes, such as variations in solar activity or volcanic eruptions, but they are predominantly caused by human activities that emit greenhouse gases into the atmosphere. Greenhouse gases, such as carbon dioxide and methane, capture heat from the sun and warm up the planet [1]. This warming has multiple impacts on the environment and human well-being, such as melting ice sheets, glaciers and sea ice, which raises sea levels and endangers coastal communities and ecosystems. Changes in rainfall patterns, which can result in droughts, floods, wildfires, and crop failures in different regions [2]. Shifts in the range and behavior of wildlife, which can affect biodiversity, food webs, and ecosystem services. Increased occurrence and intensity of extreme weather events, such as heat waves, storms, hurricanes, and cyclones, which can damage infrastructure, disrupt livelihoods and increase mortality and morbidity [3]. Increased threats of infectious diseases, malnutrition, mental health problems, and conflicts due to resource scarcity and displacement. The concentrations of the three major greenhouse gases – carbon dioxide, methane, and nitrous oxide – reached unprecedented levels in 2023, according to the World Meteorological Organization (WMO) 2024 report. Data from specific monitoring sites like Mauna Loa and Kennaook/Cape Grim indicate that these gas levels were expected to rise even further in 2024. Since 1750, the atmospheric concentration of carbon dioxide (CO_2) has surged from approximately 278 ppm to 420 ppm, marking a 51% increase. Over the past decade, the average annual growth rate of CO_2 has been 2.4 ppm. Fossil fuel emissions have been the primary source of human-induced emissions since the 1950s. Meanwhile, global average methane

DOI: 10.1201/9781032720142-8

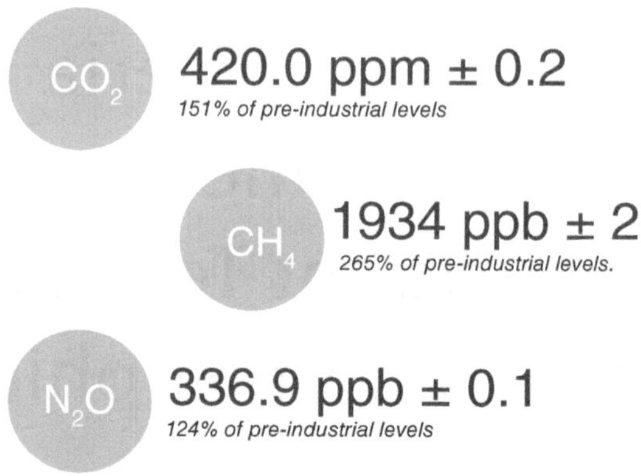

FIGURE 6.1 Concentration of three greenhouse gases (WMO COP29).

(CH_4) concentrations rose from 729 ppb in preindustrial times to 1934 ppb in 2023, reflecting a 165% increase. Nitrous oxide (N_2O) levels also increased, from 270 ppb in 1750 to 336.9 ppb in 2023, which is a 24% rise (Figure 6.1).

The global average temperature does not rise in a consistent manner from year to year. Instead, it exhibits significant fluctuations due to natural climate variability, which can be influenced by factors such as El Niño and La Niña events, volcanic eruptions, and shifts in ocean circulation. These natural variations occur alongside a long-term warming trend primarily driven by ongoing greenhouse gas emissions. This situation highlights the importance of focusing on long-term trends rather than short-term changes. As global temperatures continue to rise, there is an increasing necessity to establish clear indicators for measuring and monitoring temperature changes in relation to the targets set by the Paris Agreement. The latest assessment report from the Intergovernmental Panel on Climate Change (IPCC), known as AR6, defines global warming levels using 20-year averages compared to the baseline period of 1850–1900. According to this framework, the year in which a specific temperature threshold like 1.5°C or 2.0°C is exceeded is considered to be the midpoint of that 20-year average. For instance, a 1.5°C increase would only be confirmed once the average temperature rise has consistently reached that level over a 20-year span. This means that recognition of exceeding the 1.5°C threshold would occur a decade after the actual temperature surpasses it. Consequently, this approach could result in a significant delay of up to 10 years in acknowledging and responding to the long-term temperature goals established by the Paris Agreement. Even if a shorter averaging period of 10 years is used, as seen in IPCC AR6 and the First Global Stocktake, this still leads to a 5-year delay in recognizing temperature exceedances (Figure 6.2).

The World Meteorological Organization (WMO) and the international scientific community are exploring several alternative methods to provide more timely updates on when global temperature thresholds, such as 1.5°C or 2°C, are exceeded. These methods can be categorized into three main approaches:

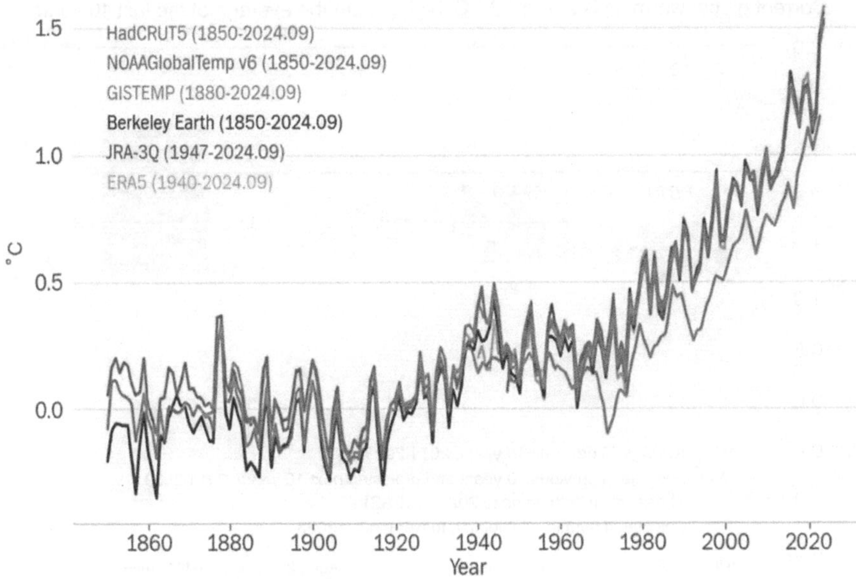

FIGURE 6.2 Annual global mean temperature from 1850 to 2024 (WMO COP29).

1. **Combining Historical Data with Climate Models**: This approach integrates observed historical warming data with projections from climate models to estimate current warming levels.
2. **Trend-Fitting Techniques**: Methods like Locally Estimated Statistical Smoothing (LOESS) are used to analyze historical data and better estimate the current state of long-term warming.
3. **Assessing Human Impact**: This method focuses on quantifying the human contributions to historical climate changes by evaluating the warming effects of key human activities, particularly greenhouse gas emissions.

According to these approaches, global warming as of 2023 was estimated to be around 1.3°C above the baseline average from 1850 to 1900. In comparison, using the IPCC AR6 method and the WMO's 2023 report, the average warming over the last decade (2014–2023) is about 1.2°C. For the average from 2011 to 2020, as used in the First Global Stocktake, the warming is approximately 1.1°C. It's crucial to understand that every fraction of a degree of warming is significant. Whether temperatures are below or above the 1.5°C threshold, each additional increment can lead to more severe climate extremes and risks. This underscores the urgency of addressing climate change and the need for effective monitoring and reporting mechanisms to inform global responses (Figure 6.3).

Climate change is a serious and urgent problem that demands collective and immediate action from all countries and sectors of society. Developing nations that have been most affected by climate change in the past 25 years are also the most vulnerable to the consequences of extreme weather events, sea level rise, droughts, floods, and

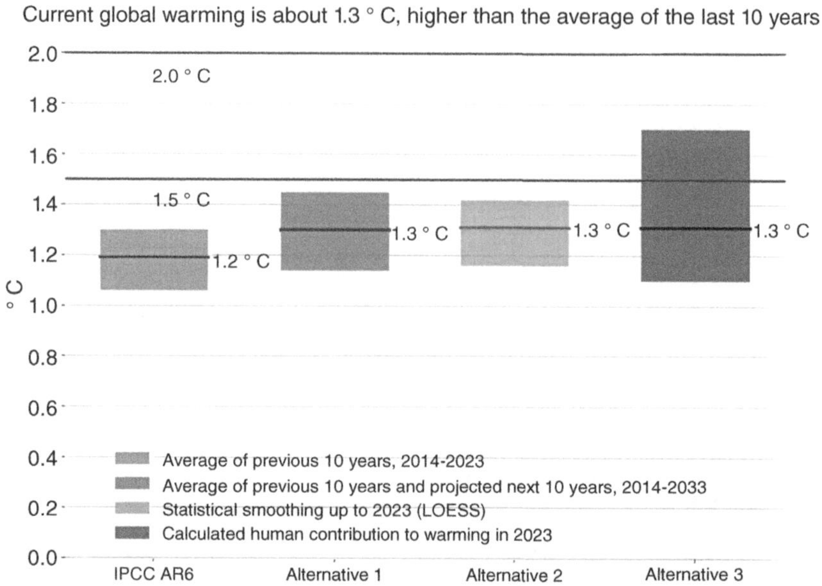

FIGURE 6.3 Different methods of assessment of global warming (WMO COP29).

food insecurity [4]. Climate change is a global problem that impacts every country, but some regions are more at risk than others. South Asian countries, including India and Pakistan, are among the countries that experience the worst effects of climate change, such as extreme heat, droughts, floods, and crop failures [3]. The World Meteorological Organization (WMO) reported that climate change increased the likelihood of the deadly heat wave that struck India and Pakistan in March and April 2022 by 30 times [5]. The heat wave caused widespread human suffering, with at least 90 people dying and millions more exposed to health risks. In March and April, the daily peak temperatures in the north-west of India and the south-east of Pakistan were 43–46°C, which is much higher than the normal range of 38–40°C [5]. The extreme heat was accompanied by a serious drought that affected crops and caused water scarcity. Heat waves are becoming more frequent and severe due to climate change, which also impacts other aspects of the weather and climate system in India and Pakistan [3]. For example, climate change is changing monsoon patterns, which are vital for the water and food security of the region (Figure 6.4).

Climate change is increasing the risk of glacial lake outburst floods in Pakistan's highlands, triggering ruinous flash floods below. Additionally, climate change is degrading air quality and heightening fire risks in both nations [1]. These climatic shifts pose significant threats to socio-economic progress, human safety, and regional equilibrium [3]. Some nations have initiated measures to acclimate to the evolving climate and curtail greenhouse gas emissions, including heat health initiatives, renewable energy investments, and improved disaster risk strategies. Nonetheless, more comprehensive and synchronized efforts are imperative at national, regional, and international tiers to cap warming at 1.5°C above pre-industrial figures, as stipulated

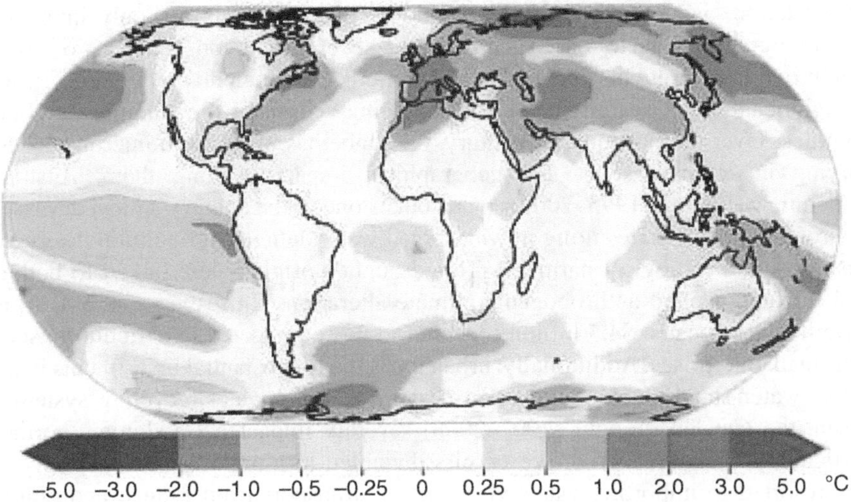

FIGURE 6.4 Near-surface temperature differences between 2022 and the 1991–2020 average (WMO 2022).

by the Paris Agreement. This demands swift and extensive decarbonization of the global economy and augmented support for adaptation and resilience in susceptible areas. Nations often cited as most impacted include Chad, Somalia, Syria, Haiti, Myanmar, and Puerto Rico [5].

Nations across the globe are grappling with the severe repercussions of civil unrest, mass displacement, dire poverty, and acute humanitarian emergencies, all of which are further aggravated by the relentless impact of climate change [6]. Despite these challenges, there is a commendable display of leadership and determination from these nations as they step up to confront the climate emergency. This is evident in their commitment to enhancing their national climate pledges (nationally determined contributions – NDCs) submitted to the UN Framework Convention on Climate Change (UNFCCC). A revealing report from the United Nations Development Programme (UNDP) highlights that an overwhelming 93% of least developed countries (LDCs) and small island developing states (SIDS) have either already submitted improved NDCs or intend to do so shortly. This contrasts sharply with some of the largest G20 emitters, who have been notably sluggish or inadequate in elevating their climate commitments. The UNDP report advocates for increased support and unity from the global community to bolster these vulnerable nations in their efforts to mitigate climate change effects and realize their developmental aspirations [4].

6.1.1 Impact of Water Shortage on Food Security

One significant consequence of climate change that these countries face is prolonged periods of drought, which stem from a complex interplay of factors including shifting weather patterns, diminished soil moisture levels, excessive water demands, rampant deforestation, and ongoing soil degradation. These elements not only interact but also fuel a vicious cycle that exacerbates the severity and consequences of drought conditions [5]. Droughts are prolonged periods of deficient precipitation leading to significant water scarcity, adversely affecting ecosystems, economies, and communities. Over the past quarter-century, the globe has observed numerous droughts varying in severity, span, and geographical impact. Among them, Australia's Millennium drought (1995–2009) stands out as one of the country's most devastating environmental crises, resulting in widespread water deficits, agricultural destruction, bushfires, and ecological harm [4]. This drought's origins were linked to both natural fluctuations and anthropogenic climate alterations. In response, the Australian government allocated $4.5 billion for relief to those impacted in agricultural sectors and small enterprises. Additionally, urban areas made substantial investments in alternative water supply technologies like desalination and water recycling systems. In Spain, the Catalan drought (2007–2008) severely impacted Catalonia's northeast, particularly Barcelona, where reservoirs dwindled to a mere 20% of capacity. The city resorted to importing water via maritime transport from France and enforced stringent water usage regulations on its populace [7]. Factors such as insufficient rainfall, elevated temperatures, water overconsumption, and suboptimal irrigation practices were held accountable for this drought. NASA's satellite observations disclosed that northern India experienced a staggering loss of 108 cubic kilometers of subterranean water from 2002 to 2008 due to intensive irrigation for crop cultivation. This alarming depletion of groundwater reserves poses imminent risks to agrarian stability, potable water availability, and public health in an area where groundwater serves as the primary resource for over 600 million inhabitants. These persistent drought conditions have led to widespread hunger, undernutrition, forced migration, conflicts, and increased incidence of diseases. Research conducted by the World Weather Attribution reveals that the likelihood of the 2023 drought occurring was magnified by a factor of 100 due to anthropogenic climate change, compared to a scenario where global temperatures were 1.2°C lower [8]. These instances represent only a fraction of the historical and ongoing droughts globally that have inflicted profound distress on humans and caused extensive damage to the environment. Future projections indicate an uptick in both the frequency and intensity of droughts, attributed to climatic shifts, demographic expansion, urban development, and inefficient water management practices. A recent analysis indicates a staggering 233% rise in flash droughts and rapid and intense water shortages in certain areas over the past 50 years. Such flash droughts tend to occur more in humid areas like northwest North America, Europe, and southern China, regions where climate models predict increasingly unpredictable and extreme precipitation patterns [3]. On the other hand, arid and semi-arid zones such as Somalia, Ethiopia, and Kenya are grappling with more consistent and extended periods of dry spells characterized by less-than-average rainfall.

The repercussions of arid shocks are far-reaching, undermining food stability, livelihoods, and health across various regions. The latest report from the IPCC cautions that the intensification of the global water cycle is a likely outcome of ongoing global warming. This phenomenon is expected to manifest as a reduction in the number of rainy days, coupled with heightened rainfall intensity over numerous terrestrial areas [7]. Consequently, extended periods of dryness during typically arid seasons are anticipated, elevating drought risks and water scarcity concerns. Research also indicates a correlation between enduring droughts in the United States over the past three centuries and prolonged La Niña episodes, marked by abnormally cool sea surface temperatures in the eastern Pacific Ocean. With the planet's temperature on an upward trajectory, leading to increased evapotranspiration and diminished precipitation, projections suggest a rise in the occurrence of sudden droughts, while enduring droughts may become more prevalent due to shifts in oceanic and atmospheric interactions. Agriculture stands as the sector most acutely impacted by water deficits. Drought ranks among the most catastrophic natural calamities for agriculture, inflicting substantial losses in yields and economic harm. Data from the Food and Agriculture Organization (FAO) reveal that from 2008 to 2018, drought accounted for 34% of crop and livestock production losses in least developed countries (LDCs) and low- to middle-income countries (LMICs), translating to an overall cost of USD 37 billion for the sector [9]. The influence of drought is predominantly felt in agriculture, which bears 82 percent of all drought-related consequences, a stark contrast to the 18 percent impact on other sectors. The degree to which drought affects agriculture hinges on various factors including its severity, frequency, duration, timing, crop type, irrigation infrastructure, and farmers' capacity to adapt. An analysis by the CMCC Foundation – Euro-Mediterranean Center on Climate Change (CMCC) delved into these intricate patterns of droughts and their associations. Recent research indicates that crop yields are notably impacted by suboptimal moisture levels, with rice being the exception. This correlation is particularly pronounced for maize and wheat compared to soybean [10]. Additionally, the research highlights years with diminished yields also experienced a higher frequency of suboptimal moisture conditions across all crops. These findings underscore the significant influence of droughts on global agricultural output, particularly affecting maize and wheat [11]. The looming challenges of climate change, population increase, and water shortages are expected to exacerbate the effects of drought on farming. The United Nations has issued a stark warning, noting a nearly one-third rise in drought occurrences over the past 10 years, signaling a critical juncture for the world. The repercussions of droughts extend beyond crop yields, potentially instigating food shortages, malnutrition, displacement, conflicts, and poverty. Droughts represent periods of water scarcity that can have detrimental effects on plant life and ecosystems, undermining growth, productivity, and survival, while also impacting the carbon cycle and ecosystem services delivery [4]. The reduction in plant species diversity and abundance due to droughts can compromise ecosystem functionality and resilience [12]. Furthermore, droughts can degrade the quality and volume of water resources that plants typically help maintain through pollutant filtration, water storage, and streamflow regulation. The broader implications of droughts include significant alterations to the carbon cycle and climate system. Plants play a crucial role as carbon sinks by absorbing carbon dioxide during photosynthesis and sequestering it within their biomass and soil. However, drought conditions can diminish this carbon sequestration by curtailing plant growth and elevating respiration rates.

Drought conditions pose significant threats to both plant life and human livelihoods by exacerbating carbon losses through increased soil decomposition and plant mortality. This release of carbon dioxide into the atmosphere can intensify global warming, leading to a detrimental cycle of more droughts and higher emissions. The impact on crop yields and agricultural productivity is profound, jeopardizing food security and economic stability.

6.2 PHYSIOLOGICAL MECHANISMS OF DROUGHT TOLERANCE

Drought represents a multifaceted challenge characterized by insufficient water supply, elevated temperatures, and low humidity that disrupts the complete life cycle of crops or persists for an extended period [13]. Occasionally, water scarcity may not coincide with these factors and may only affect a specific growth stage of the plant, resulting in a water-deficit condition. Understanding plant responses to drought is crucial for devising strategies to mitigate and adapt to its effects. Different species exhibit varied reactions based on their species-specific traits, physiology, and environmental context. For instance, C3 plants like rice (*Oryza sativa*) and C4 plants like maize (*Zea mays*) exhibit distinct responses to drought stress due to differences in their photosynthetic pathways, anatomical structures, and physiological characteristics. These differences are critical for their adaptation to water-deficient environments (Figure 6.5).

Research involving meta-analysis and network analysis has shed light on the transcriptional response of these plants to drought stress. Several genes are differentially expressed under such conditions in both species, many of which are linked to photosynthesis, molecular metabolic processes, and stress response mechanisms. Understanding these genetic expressions is vital for improving plant resilience to drought and ensuring sustainable agricultural practices in the face of climate change. Within the regions identified as quantitative trait loci, certain genes that are expressed have been linked to various plant traits such as carbon isotope signature (d13C), photosynthetic gas exchange, and root characteristics. Studies have highlighted the significant role of the mitogen-activated protein kinases (MAPK) signaling pathway in the divergent traits of C3 and C4 species [14]. When comparing physiological responses to nanomaterials (NMs) in soil, C4 plants like amaranth and maize show a greater resilience to NM-induced stress compared to C3 plants such as wheat and rice [15]. This resilience is evident through variations in chlorophyll content, Fv/Fm levels, NM uptake, and metal concentration. In rice, a C3 plant, CuO nanoparticles have been observed to alter chloroplast thylakoid structures and accumulate in vascular sheath cells, whereas maize, a C4 plant, showed no significant changes or accumulation. Furthermore, an RNA-seq meta-analysis aimed at comparing drought stress responses at the gene expression level in C3 and C4 plants revealed that drought stress triggers the activation of more genes in C4 plants than in C3 plants [4]. These genes are predominantly associated with photosynthesis, carbon fixation, oxidative stress response, hormone signaling, and transcription factors. The analysis also identified both common and specific genes and pathways that are regulated by drought stress

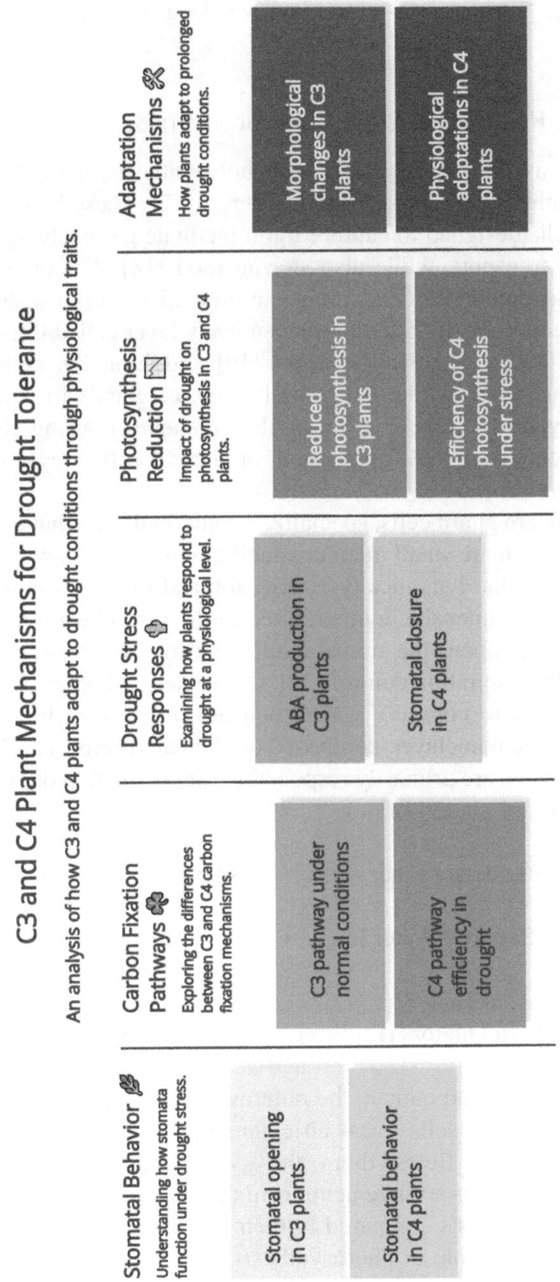

FIGURE 6.5 An analysis of C3 and C4 plant mechanisms for drought tolerance through physiological traits.

across these plant types. A comprehensive examination of transcriptomic data comparing the responses of C3 and C4 plants to drought stress and other environmental factors uncovered unique and shared mechanisms that underpin their adaptation to water scarcity [16]. These insights could be pivotal for developing drought-resistant crop varieties.

6.2.1 Stomatal Regulation Under Drought Stress

Plant leaves serve as the primary sites for photosynthesis, a vital process where plants transform light energy into chemical energy [17]. These leaves are equipped with specialized cells designed to capture light, facilitate gas exchange, and manage water and nutrient transport, while also storing food [18]. The outermost layer of leaf cells, known as epidermal cells, creates a protective barrier against water loss and insect damage. These cells often produce a waxy layer called the cuticle to minimize evaporation and reflect surplus light [19]. Additionally, certain epidermal cells develop unique features like hairs (trichomes) or glands for extra defense or secretion. Typically, devoid of chloroplasts, the organelles housing chlorophyll, the green pigment essential for light absorption, epidermal cells play a crucial role in safeguarding the leaf [20].

Adjacent to these are guard cells, specialized epidermal cells that encircle stomata (singular: stoma), which are small openings facilitating gas exchange with the environment [21]. Guard cells dynamically adjust stomatal openings based on external conditions such as light intensity, temperature, humidity, and carbon dioxide levels. Swelling of guard cells opens the stomata, allowing for gas exchange necessary for photosynthesis and transpiration control [22]. Conversely, when guard cells contract, they close the stomata to conserve water. Beneath the surface lies the mesophyll tissue's palisade parenchyma layer, composed of elongated cells densely packed with chloroplasts. These cells are primarily responsible for conducting photosynthesis due to their high chloroplast content [23].

Drought Stress → Roots detect water stress → ABA production → Transport to leaves
↓
Stomatal Closure → Reduced Water Loss → Altered Metabolism (reduced photosynthesis)
↓
Activation of Osmotic Regulators (Proline) → Long-term Adaptations (root growth)

Beneath the epidermis and cuticle, the outermost leaf layers, lie the palisade parenchyma cells [24]. These cells boast an extensive surface area, optimizing light capture for photosynthesis. Below them, the spongy parenchyma cells, with their irregular shapes and spacious arrangement, form the mesophyll's lower tier. Although they house fewer chloroplasts compared to their palisade counterparts, spongy parenchyma cells still play a role in photosynthesis [24]. Their design includes ample air gaps that enhance gas exchange within this mesophyll segment. Meanwhile, vascular bundle cells constitute the leaf's intricate transport network. Comprising xylem and phloem tissues, vascular bundles perform distinct functions: xylem channels water and minerals upward from roots using its hollow, deceased cellular tubes, while phloem distributes organic nutrients like sugars throughout the plant via living conduits [25]. Additionally, these bundles lend crucial support to the leaf's structure.

During periods of water scarcity, stomata play a crucial role in managing the equilibrium between water conservation and carbon acquisition in vegetation. These minute openings on the leaf surface, flanked by a pair of guard cells, adjust their dimensions to modulate the stomatal gap, responding to external cues and plant necessities [26]. In response to soil water deficits, stomata constrict to curtail transpiration and avert excessive moisture depletion that could cause plant desiccation or drought-induced stress. Nonetheless, this action also constrains carbon dioxide absorption vital for photosynthesis and vegetative growth [1]. Consequently, flora must fine-tune stomatal activity to sustain adequate water use efficiency (WUE), defined as the proportion of carbon assimilation relative to water expenditure [27]. Various flora exhibit distinct adaptive tactics under hydric stress conditions, including modifications in stomatal count, dimensions, placement, or reactivity to diverse stimuli. Certain species may evolve specialized adaptations like augmented root depth, water-retentive tissues, or crassulacean acid metabolism (CAM) to bolster water absorption or retention [27]. Additionally, plants can diminish water loss by partially sealing their stomata minuscule leaf openings that permit carbon dioxide ingress and water vapor egress through a process known as stomatal conductance, which is affected by atmospheric carbon dioxide levels [26]. Elevated carbon dioxide levels may boost plant photosynthesis and water efficiency, leading to increased biomass production with reduced water consumption. Yet, this benefit varies among plant species and is constrained by other elements like nutrients, temperature, and light availability. Stomatal regulation plays a pivotal role in managing water loss through transpiration by adjusting stomatal openings. This regulation is affected by factors including leaf water potential, abscisic acid (ABA) presence, CO_2 levels, light intensity, and ambient temperature. ABA, a crucial drought-response hormone, induces stomatal closure by binding to guard cell receptors, initiating a sequence of events that expel potassium ions and diminish cell turgor. Such closure limits CO_2 entry, impacting photosynthesis and carbohydrate metabolism [26]. Wheat plants can mitigate this effect by enhancing photosynthetic efficiency through mechanisms like increasing Rubisco activity which is key in carbon fixation. Rubisco, essential for photosynthesis, enables plants to convert light energy into organic compounds using CO_2 and water. It primarily facilitates carboxylation, the initial step of carbon fixation in the Calvin cycle, while also performing oxygenation [28]. Carbon fixation is a critical biochemical process where inorganic carbon dioxide (CO_2) is converted into organic compounds usable by plants and other organisms [29]. During this process, the enzyme RuBisCO catalyzes the attachment of CO_2 to ribulose-1,5-bisphosphate (RuBP), a five-carbon sugar, resulting in two molecules of 3-phosphoglycerate (3-PGA). However, RuBisCO can also catalyze an undesired reaction known as oxygenation, where O_2 is attached to RuBP instead of CO_2, leading to the production of one molecule of 3-PGA and one molecule of 2-phosphoglycolate [30]. This latter molecule undergoes photorespiration, a complex process involving multiple organelles and enzymes that ultimately causes a loss of fixed carbon and energy. RuBisCO is not only the most abundant protein on Earth making up to 50% of soluble leaf protein in C3 plants and 30% in C4 plants but is also among the most ancient enzymes, having emerged over three billion years ago in an oxygen-poor atmosphere [22]. The enzyme consists of large (L) and small (S) subunits; the large subunits house the active site for substrate binding, while the small subunits play a role in regulating enzyme activity and stability [21]. The large subunits of RuBisCO are produced from a single rbcL gene found within chloroplast DNA, whereas the small subunits originate from several rbcS genes within nuclear DNA. These smaller units are created in the cytoplasm and then transported to the chloroplasts to join with the large units, forming the active enzyme [31]. Despite being vital for life on Earth by contributing to most organic carbon in the biosphere, RuBisCO's effectiveness is hindered by its slow catalytic rate, processing only a handful of substrate molecules each second, and its low specificity, as it binds CO_2 and O_2 similarly [32]. This results in RuBisCO being a significant limiting factor for photosynthesis and plant growth. To overcome this, various adaptations have emerged to boost RuBisCO's carboxylation efficiency and decrease its oxygenation rate. C4 plants, for instance, achieve this through spatial separation of RuBisCO from the initial CO_2 fixation site, while CAM plants do so temporally [31]. Both adaptations elevate CO_2 levels around RuBisCO and curb photorespiration. With its pivotal role in carbon fixation, RuBisCO is also a prime candidate for genetic modification aimed at improving plant productivity and tackling climate change by increasing carbon capture. Yet, altering RuBisCO presents considerable challenges due to its intricate structure, regulatory mechanisms, and evolutionary history [33].

6.2.2 Role of Abscisic Acid

Abscisic acid (ABA), a crucial plant hormone, orchestrates various physiological processes including seed dormancy, leaf aging, fruit maturation, and stomatal regulation. ABA's role is particularly crucial under abiotic stress conditions like drought, salinity, and extreme temperatures, aiding plants by adjusting gene expression, antioxidant activity, osmotic balance, and ion management [34]. Synthesized in plastids from carotenoids, ABA travels through xylem and phloem to various tissues. Its perception and signaling cascade involve receptors, kinases, phosphatases, transcription factors, and other proteins that facilitate ABA-induced physiological and developmental alterations in plants. Moreover, ABA interplays with other phytohormones, gibberellin, auxin, cytokinin, ethylene, jasmonic acid, salicylic acid, brassinosteroids, and strigolactone, to modulate the environmental response. Remarkably conserved across all plant forms including mosses, ABA signifies an ancient adaptation to terrestrial life [1]. Nonetheless, plants exhibit resilience to drought by maintaining photosynthesis and translocation processes, enabling them to utilize reserves even under severe stress [9]. The sustained translocation of assimilates underpins plant resistance to drought across various physiological aspects. ABA plays a pivotal role in the plant response to drought, particularly within mesophyll cells. As water content diminishes, ABA accumulates in chloroplasts. This increase in ABA production prompts its movement toward the guard cells of stomata, leading to stomatal closure.

This process is influenced not only by signals from the roots but also by soil water availability, which affects stomatal conductance more than leaf water status [34]. Roots, being directly impacted by dehydration, initiate stomatal closure via ABA, pH changes, and redistribution of inorganic ion key elements in root-to-shoot communication. Consequently, photosynthesis rates decline due to reduced transpiration and CO_2 concentration (Figure 6.6).

ABA is synthesized via two distinct pathways: the direct pathway in phytopathogenic fungi and the indirect pathway in plants. The indirect pathway uses the methylerythritol phosphate (MEP) pathway as a source of isopentenyl pyrophosphate (IPP). The ABA catabolic pathway involves the breakdown of ABA, which is crucial for regulating its levels within the plant. Enzymes such as ABA 8'-hydroxylase play a significant role in this process. The ABA signaling pathway involves the perception of ABA by PYR/PYL/RCAR receptors, which then inhibit PP2C phosphatases, leading to the activation of SnRK2 kinases. These kinases phosphorylate downstream targets, including transcription factors like ABI5, which regulate stress-responsive genes. ABI5 (ABA-Insensitive 5) is a key transcription factor in the ABA signaling pathway that regulates genes involved in seed germination and early seedling growth under stress conditions. LEA (Late Embryogenesis Abundant) proteins are involved in protecting cells from damage during dehydration and are regulated by ABA. RD29A Responsive to Desiccation 29A is a gene that is highly induced by ABA and drought stress, playing a role in osmotic stress tolerance. CRISPR-Cas9 genome editing technology has been used to modify genes involved in the ABA signaling pathway to enhance drought tolerance in crops. For example, editing the genes encoding ABA receptors or signaling components can improve a plant's response to drought. Introducing genes that enhance ABA signaling or biosynthesis into crops can improve their drought tolerance. For instance, overexpressing genes like ABI5 or RD29A has been shown to enhance drought resistance. Utilizing plant growth-promoting rhizobacteria (PGPR) can enhance ABA production or signaling in plants, thereby improving their drought tolerance. These advancements highlight the potential of biotechnological interventions in improving crop resilience to drought through the manipulation of ABA signaling pathways.

6.2.3 Maintenance of Cell Turgor and Osmotic Adjustment

Osmotic adjustment refers to the process where plants accumulate solutes in their cytoplasm, reducing osmotic potential and preserving cell turgor. This mechanism is crucial for plants to continue growing, conducting photosynthesis, and transpiring even when water is scarce. Solutes like proline, glycine betaine, sugars, amino acids, and ions such as sodium and potassium play a significant role in this adjustment [35]. Osmotic adjustment is a vital process that enables plants to absorb water that is tightly bound to soil particles, beyond what is available in soil spaces. This phenomenon predominantly occurs in the root meristems, where it plays a crucial role in enhancing turgor pressure and sustaining root growth, especially as water becomes scarce in the soil. As an integral aspect of the root growth pattern, osmotic adjustment ensures that plants can continue to develop even under water-limited conditions [9]. Droughts

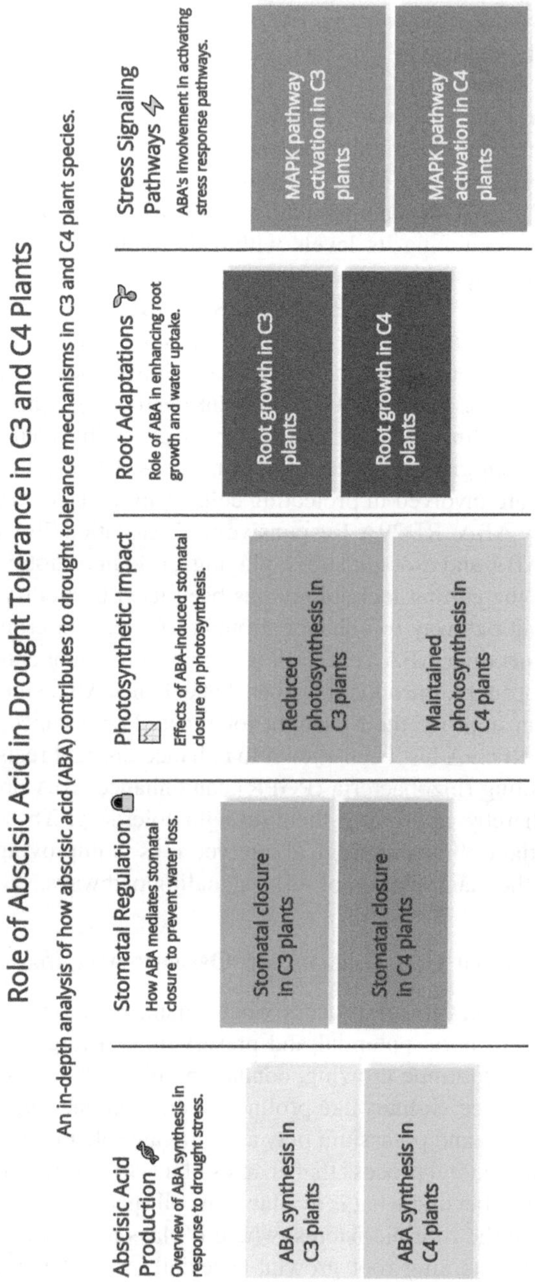

FIGURE 6.6 Role of abscisic acid in drought tolerance in C3 and C4 plants.

impact plant growth by constraining water availability in the soil, hindering water and nutrient transport from roots to leaves. This leads to a loss of turgor pressure in plant cells, causing wilting, diminished photosynthetic ability, and heightened mortality risk. Plants initiate various physiological and morphological adaptations to counteract this, striving for osmotic pressure equilibrium where solvent movement is neutral [36]. Osmoregulation actively adjusts osmotic pressure in plant fluids, ensuring water content homeostasis, neither overly diluted nor concentrated. Additionally, drought prompts biochemical and molecular shifts in plants, enabling them to endure water scarcity through the production of osmolytes, antioxidants, and stress hormones [20]. These adaptive responses depend on the drought's severity and duration, as well as the plant's acclimation and adaptation capabilities. Some plants employ drought escape strategies, completing their life cycle before drought onset by accelerating vegetative growth early in the season [19]. Upon detecting drought signals like low moisture or high temperatures, these drought escapers hasten through the reproductive stage. Conversely, drought-tolerant plants cope with limited water by conserving usage and maintaining function despite dehydration. Plants adept at enduring arid conditions possess various defense strategies to combat water scarcity [36]. A primary response to diminishing water levels within a plant is a reduction of leaf surface area. This adaptation involves a decrease in turgor pressure and an increase in cellular solute concentration as water content drops [37]. Consequently, the plasma membrane condenses, enveloping a diminished area compared to its prior state. Some flora may shed their foliage altogether, minimizing leaf area to preserve moisture. Additionally, a reduction in turgor can decelerate or even halt growth temporarily [38].

6.2.4 Role of Compatible Solutes

Proline, often found in high levels in drought-stressed plants, serves multiple functions: it stabilizes proteins, membranes, and enzymes; scavenges reactive oxygen species; and influences gene expression. Sodium and potassium are vital for sustaining membrane potential, enzyme functions, and overall cellular balance [39]. During drought conditions, plants may increase sodium and potassium levels by triggering specific transporters and channels within their plasma membrane and tonoplast. Nonetheless, plants must regulate sodium intake to prevent toxicity and maintain an optimal sodium–potassium ratio to avoid salinity stress. Osmotic adjustment ultimately involves a reduction in growth rate coupled with an accumulation of various solutes such as sugars, organic acids, amino acids, and particularly potassium ions within the vacuole. Other compounds like sugars, alcohols, amino acids including proline, and the quaternary amine glycine betaine are stored in the cytoplasm without disrupting enzymatic activities [9].

6.2.5 Proline Contents

Proline is an amino acid pivotal in plant abiotic stress tolerance [40]. Synthesized from glutamate or ornithine, proline's degradation is mediated by enzymes such as proline dehydrogenase (PDH) or proline oxidase (POX). Its accumulation is observed across various plant species as a defensive response to environmental stressors like salinity, drought, temperature fluctuations, heavy metal exposure, nutrient scarcity, and pathogen invasion. Proline biosynthesis and degradation take place within distinct cellular compartments, including cytosol, chloroplasts, and mitochondria, adding layers to its functional significance. It aids stress tolerance by modulating cellular osmotic pressure and maintaining redox balance. Engaging in redox reactions, proline contributes to cellular homeostasis [32]. It also serves as an antioxidant, scavenging free radicals to shield cells from oxidative damage, thus preserving membrane integrity and stabilizing enzymes and proteins. It's chelating action on heavy metals, modulation of cellular functions, and gene expression triggering further underscore its protective role [20]. The stability imparted by proline to proteins, membranes, and subcellular structures is crucial for sustaining cellular integrity under duress. It also functions as a stress signal, fostering cross-tolerance to varied stress conditions [2]. Proline metabolism intersects with several metabolic pathways including pentose phosphate, tricarboxylic acid, or urea cycles, influencing purine synthesis and the phenylpropanoid pathway. This interplay extends to gene expression regulation and signal transduction pathways. Intriguingly, proline's ability to confer cross-tolerance underscores its multifaceted role in plant stress response [1].

When plants endure drought, they often accumulate proline, which may inadvertently boost their resilience to other adversities such as salinity or extreme temperatures. Despite proline's critical functions, its molecular regulatory mechanisms are still largely enigmatic. Nonetheless, manipulating proline levels through genetic engineering could pave new pathways for augmenting plant stress endurance [41]. Applying proline externally at minimal concentrations has shown promise in fortifying plants against stress, although excessive amounts might be detrimental. Beyond the stress response, proline influences pivotal developmental stages, including seed sprouting and bloom timing [34]. Thus, proline stands as a pivotal element in both plant stress resistance and growth phases.

6.2.6 Ethylene Production

Another pivotal hormone in combating abiotic stress is ethylene. This gaseous plant hormone is integral to plant maturation and coping mechanisms against various biotic and abiotic challenges [15]. Plants synthesize ethylene by transforming methionine into KMBA and subsequently into ACC, which ACC oxidase then converts into ethylene. A sophisticated array of receptors, transcription factors, and genes orchestrate ethylene perception and signaling, steering a multitude of physiological and molecular plant activities. Ethylene's regulatory scope encompasses diverse facets of plant development and stress adaptation, notably influencing root dynamics and drought acclimatization [42]. It can modulate root architecture through interactions with other hormones like auxin, cytokinin, abscisic acid, and jasmonic acid, impacting root hair growth, lateral root emergence, elongation, and gravitropism. Ethylene plays a pivotal role in regulating root growth, which significantly impacts plant water absorption and drought resilience. This hormone aids plants in enduring drought by prompting stomatal closure, thus diminishing water loss and bolstering antioxidant defenses. Additionally, ethylene triggers adaptive mechanisms like osmotic adjustment, accumulation of compatible solutes, and production of stress-related proteins. It also enhances root tolerance to hypoxia by suppressing growth and mitigating oxidative harm during episodes of flooding and subsequent re-oxygenation. Consequently, ethylene is a crucial regulator that harmonizes plant development with responses to abiotic stress. Salinity stress is another major abiotic factor that impairs plant yield and quality through osmotic stress, ion toxicity, oxidative stress, and hormonal disequilibrium. Ethylene's role as a modulator of salinity stress responses is well-documented; it bolsters salt tolerance by managing ion transport, water relations, antioxidant defense, cell wall adjustments, and gene regulation. Ethylene's interaction with other hormones like ABA, JA, and SA further refines plant adaptation to salinity stress. However, ethylene's influence on plant stress tolerance is not universally beneficial; overproduction can adversely affect plant growth and development under stressful conditions. Thus, controlling ethylene biosynthesis and signaling pathways emerges as a viable approach to enhance plant stress tolerance. For instance, certain beneficial microbes produce KMBA that modulates ethylene signaling, fostering plant salt stress tolerance [43]. Alternatively, genetic engineering can alter ethylene receptors or transcription factors to modulate ethylene responses in stressed plants [44].

6.2.7 Jasmonic Acid

Jasmonic acid is a plant hormone that plays a crucial role in regulating plant stress responses, including drought tolerance. It plays a crucial role in the drought tolerance of wheat plants. It is a plant hormone that modulates various physiological processes, including growth, reproductive development, and defense responses [45]. During drought conditions, jasmonic acid levels increase, which helps in the regulation of stomatal closure to reduce water loss through transpiration [46]. Additionally, it enhances root growth, allowing the plant to access deeper soil moisture. Jasmonic acid also upregulates the expression of certain genes associated with osmoprotectant synthesis, antioxidant defense, and stress-responsive proteins, contributing to cellular stability and recovery during and after stress events [47]. Its role is vital in signaling pathways that activate plant responses to drought stress, making it an essential component for wheat plants to survive and adapt to water-deficient environments [48]. These physiological traits are indicators of drought tolerance in plants under drought stress [49].

6.2.8 Relative Water Content (RWC)

Relative water content is an important measure of plant tissue water status, reflecting the amount of water present relative to the tissue's full turgor capacity. RWC is a straightforward and commonly employed method for assessing plant water status, taking into account cellular hydration influenced by both leaf water potential and osmotic adjustment [26]. It is calculated as the ratio of current water content to saturation water content, expressed as a percentage. RWC values can vary widely, from 98% in fully hydrated leaves to 30–40% in severely wilted leaves, depending on the species. This metric is invaluable for evaluating the impact of drought stress and osmotic adjustment on plant hydration and physiological processes. Additionally, RWC correlates with other vital parameters such as water potential, leaf area, photosynthesis rates, and overall growth [50]. RWC is determined by weighing fresh, saturated, and dry samples of leaf tissue and applying the following formula:

$$\text{RWC} = [(\text{Fresh weight} - \text{Dry weight}) / (\text{Saturated weight} - \text{Dry weight})] \times 100$$

6.2.9 Excised-leaf Water Loss

Plants adept at maintaining their water content during periods of scarce water supply exhibit drought tolerance. The metric known as excised-leaf water loss quantifies the volume of water an isolated leaf dispenses via transpiration, reflecting the leaf's capacity for water conservation and drought resistance [22]. Influences on excised-leaf water loss encompass leaf dimensions, configuration, external area, stomatal frequency, cuticle robustness, and prevailing environmental elements like temperature, moisture levels, illumination, and more. Transpiration serves dual purposes for plants: thermal regulation and establishing a water uptake gradient from the soil. Nonetheless, excessive transpiration can precipitate leaf wilting, impede photosynthesis, and diminish crop yields [37]. Consequently, plants have evolved assorted strategies to modulate their hydration equilibrium and acclimate to varying hydration conditions. Employing excised-leaf water loss as a criterion in plant breeding can aid in identifying varieties with superior drought resilience. Such plants typically exhibit reduced excised-leaf water loss, augmented relative water content, enhanced tissue pliability, and improved grain production under hydric stress [38]. Additionally, excised-leaf water loss can signal wheat leaves' thermal tolerance since elevated temperatures can escalate transpiration rates and harm leaf structures. ELWL of the leaf is calculated using the formula:

$$\text{ELWL} = (\text{Fresh weight} - \text{wilted weight}) / \text{Dry weight}$$

6.2.10 Cell Membrane Thermostability (CMT)

Cell membrane thermostability gauges the resilience of plant cell membranes against high temperatures and limited water availability [51]. It's a key indicator of a plant's ability to withstand drought and heat, which is pivotal for maintaining cellular balance and metabolic activities. CMt's effectiveness hinges on factors like lipid bilayer composition, membrane proteins, antioxidants, osmolytes, and secondary metabolites. Evaluating CMT involves subjecting leaf samples to thermal and osmotic stress, followed by tracking electrolyte leakage or chlorophyll fluorescence [9]. This metric has been instrumental in pinpointing robust crop varieties, including wheat, rice, maize, and sorghum, and is linked to specific genomic regions and genes that govern solute transport, ROS neutralization, hormonal pathways, and stress-related biochemical functions. Given its complexity, CMT is best understood through omics studies which dissect the intricate molecular and regulatory networks at play.

Cell membrane thermostability can be calculated using the formula:

$$CMT = 1 - \%\ injury = 1 - (C1/C2) \times 100$$

where C1 and C2 are the first and second readings of conductance, respectively.

6.3 MORPHOLOGICAL AND STRUCTURAL ADAPTATIONS

6.3.1 Root System Architecture

Roots, the subterranean extensions of vascular plants, secure them within the earth while facilitating water and mineral absorption. Drought conditions can impede root functionality by curtailing water uptake and nutrient assimilation, modifying cellular water permeability, and affecting root growth and structure. Roots comprise various cell types, each fulfilling distinct roles. Root hair cells, specialized epidermal cells with elongated projections known as root hairs, augment the root's absorptive surface area for soil nutrients and moisture [18]. These cells also contribute to plant stability within the soil. Notably, root hair cells lack chloroplasts since they do not engage in photosynthesis. Parenchyma cells, prevalent and adaptable within plant anatomy, feature thin walls and sizable vacuoles. They serve as reservoirs for water, starches, sugars, and oils, among other substances, and can partake in photosynthesis or respiration given the presence of chloroplasts or mitochondria, respectively [17]. These cells are also capable of division and differentiation into other cell types as necessary. Collenchyma cells are characterized by their elongated shape and corner-thickened cell walls. Collenchyma cells offer versatile reinforcement to the plant's youthful segments like stems and leaves, aiding in the transport of water and nutrients. These cells might contain chloroplasts to carry out photosynthesis [15]. Sclerenchyma cells, on the other hand, are rigid, with thickened, lignified walls that grant mechanical robustness and support to mature plant parts such as stems, roots, and seeds, alongside aiding in water conduction. At maturity, sclerenchyma cells are typically deceased, lacking cytoplasm or organelles. They come in two varieties: fibers are elongated cells forming bundles, while sclereids are compact, irregular cells constituting hard structures like nutshells and seed coats [16]. Xylem cells, specialized for vascular functions, shuttle water and minerals from roots throughout the plant and contribute structural support. These cells are also dead at maturity, with hollow walls that create unbroken tubes or vessels. Xylem includes tracheids, long, narrow cells with wall pits for water transit and vessel elements, shorter, broader cells with wall perforations for water flow. Phloem cells specialize in transporting organic materials such as sugars and amino acids from leaves to other plant parts [16]. They also lend support to the plant structure. Unlike xylem, phloem cells remain alive at maturity, complete with cytoplasm and organelles. Phloem cells are categorized into sieve elements and companion cells. Sieve elements are specialized, elongated cells featuring sieve plates on their walls, facilitating substance movement. Companion cells, smaller in size, are closely associated with sieve elements, aiding in metabolic functions and transport processes [52]. Meristematic cells represent the undifferentiated stem cells located in active growth zones like the root and shoot tips. These cells possess thin walls, minimal vacuoles, and numerous nuclei, enabling them to rapidly divide through mitosis and produce new cells that differentiate into diverse plant cell types. Meristematic cells are divided into apical meristem, lateral meristem, and intercalary meristem. The apical meristem drives primary growth, contributing to plant lengthening [53]. The lateral meristem facilitates secondary growth, enhancing plant thickness. Intercalary meristem promotes internode growth – the segments between nodes where leaves are attached. To combat drought stress, plant roots employ various strategies by modifying root depth, density, angle, length, diameter, and surface area – traits crucial for efficient water and nutrient uptake from soil in arid conditions. Roots are categorized into taproot, fibrous, and adventitious based on their origin and structure [54]. The taproot system features a primary root descending vertically from the stem with numerous lateral root hairs and is typical of dicotyledonous plants like mustard, carrot, and beetroot. This system allows deep soil penetration and food storage, enabling access to deeper water reserves during droughts. Taproots serve multiple functions, including water and carbohydrate storage, which prove vital during droughts [55]. However, their hydraulic conductivity is often less than that of fibrous roots, hindering efficient water transport to shoots. Additionally, taproots typically exhibit a lower specific root length (SRL) compared to fibrous roots. SRL, the root length to mass ratio, reflects a root's foraging ability; a higher SRL indicates a greater surface area per biomass unit. Conversely, the fibrous root system, prevalent in monocots like rice and maize, consists of numerous slender roots branching out from the stem [56]. These roots stay near the soil's surface, contributing to erosion prevention but lack food storage capabilities. They excel in exploiting nutrient-rich shallow soil layers and enhance soil water retention while minimizing evaporation through a dense surface network. However, fibrous roots are prone to drying and have less water storage capacity than taproots, making them less effective in buffering water fluctuations [57]. Adventitious roots emerge from non-radicle plant parts, offering structural support and additional absorption capabilities. The adventitious root system, a hallmark of angiosperms, serves diverse functions ranging from respiration and support to food storage. This versatile system may manifest as either aerial or subterranean structures. Notable examples include the sprawling roots of the banyan tree, the water-thriving mangrove, and the sturdy corn [58]. When considering drought resilience, the utility of a root system is contingent upon various factors: drought severity and duration, soil properties, water and nutrient availability at different soil strata, alongside plant species, and genetic makeup. Research indicates that taproot systems might outperform fibrous ones under harsh, extended droughts, whereas fibrous systems could have an edge in less severe, sporadic drought scenarios [40]. Nonetheless, these findings are not absolute and can fluctuate with specific environmental contexts and plant traits. It's crucial to comprehend the interplay between root system types and environmental elements, as well as potential enhancements through breeding or cultivation techniques to bolster plant drought resistance [6]. The balance between a plant's root and shoot growth is pivotal, hinging on the root's water absorption and the shoot's photosynthetic activity [50]. A shoot expands until water scarcity curtails further growth, while roots proliferate until their demand for photosynthates aligns with the shoot's provision. Typically, abundant water leads to shorter roots, but scarcity prompts roots to delve deeper in search of moisture, marking a plant's secondary defense against drought. The tertiary defense involves stomatal closure, with hydropassive closure occurring under low humidity, and hydroactive closure when leaves or roots are dehydrated [27].

6.3.2 Root–shoot Ratio

Confronting the task of selecting drought-resistant elite plant varieties poses a formidable challenge for breeders aiming to cultivate new, high-yield strains. Nowadays, a comprehensive grasp of plant reactions to drought stress is imperative for accurate drought tolerance assessment. The root–shoot ratio stands as a critical metric, delineating the balance between supportive (roots) and growth-promoting (shoots) plant tissues [1]. This ratio not only mirrors the relative availability of aerial (light and CO_2) versus subterranean (water and nutrients) resources but also significantly sways plant growth dynamics and morphological progression [26]. A higher root proportion endows plants with a competitive edge in nutrient absorption from the soil, whereas a greater shoot proportion amplifies light energy capture [22]. Environmental stressors, both chemical and physical, also impinge upon this ratio, eliciting a biphasic dose–response curve emblematic of hormesis. Furthermore, variations in the root–shoot ratio are indicative of plant maturity, size, ontogeny, species diversity, and ecological succession stages [59]. As an index, it's pivotal for gauging plant vitality, carbon sequestration capacity, and overall productivity. Traditional selection methods favor plants with elongated roots and an elevated root–shoot ratio for their optimized performance and resilience under drought-induced duress. Such traits confer a survival advantage to plants during water scarcity [33]. When a plant increases its root system, it utilizes all available resources to enhance both root and shoot length. However, this can lead to a compromise where the plant's height remains short. Interestingly, plant height has a positive correlation with grain yield, meaning that shorter plants may ultimately yield less. Selecting plants with a root–shoot ratio close to 1, especially under drought stress conditions, serves as a morphological indicator of drought tolerance [41].

6.3.3 Wax Deposition

Additionally, plants employ other strategies to conserve water, such as producing a thicker cuticle layer on their epidermis to minimize water loss. This cuticle layer is fortified with waxes on its surface and within its inner layer, with the latter being particularly significant in regulating the rate of water loss. In response to high temperatures, plants typically transpire to cool down their leaves. However, when faced with a shortage of water, transpiration ceases, and plants must dissipate heat through other means to maintain an optimal temperature.

6.4 MOLECULAR RESPONSE OF PLANTS TO DROUGHT STRESS

6.4.1 Production of Reactive Oxygen Species

During periods of water deficit, plants may experience an increase in reactive oxygen species (ROS) such as oxides, peroxides, and hydroxyl radicals. These ROS can activate certain genes and lead to oxidative stress. To counteract this, plants have developed an antioxidant defense mechanism that protects cellular components from damage caused by ROS. These reactive molecules or ions are usually produced during normal metabolic processes like photosynthesis and respiration but increase under drought stress due to an energy production and consumption imbalance. ROS have the potential to harm lipids, proteins, nucleic acids, and other biomolecules, which can result in cell death and tissue damage [8]. To prevent or repair this oxidative damage, plants activate antioxidant enzymes such as superoxide dismutase (SOD), catalase (CAT), ascorbate peroxidase (APX), and glutathione reductase (GR). These enzymes help scavenge ROS and convert them into less harmful substances. Moreover, plants can produce antioxidant compounds like phenolic acids, flavonoids, and anthocyanins. These compounds not only aid in modulating plant growth and development but also play a role in responding to stress conditions [8].

6.4.2 Transcriptome Analyses

Gene products are divided into two distinct categories. The first category encompasses those that are pivotal in the regulation of compatible solute accumulation, such as proline, betaine, and sugars. These solutes are instrumental in shielding the cell from desiccation and mitigating ROS damage through detoxification enzymes like superoxide dismutase and glutathione S-transferase. The second category comprises a variety of transcription factors, including MYC, MYB, leucine zipper proteins, and dehydration response element-binding proteins (DREBs), as well as MAP kinase pathways that are ABA-independent. The MYC genes are further subdivided into four subfamilies – I, II, III, and IV – based on phylogenetic affiliations with other plant species. Subfamily I includes five genes: TaMyc-A1, TaMyc-B1, TaMyc-D1, TaMyc-A2, and TaMyc-D2. These genes are allocated across chromosomes 2A, 2B, 2D, 5A, and 5D, respectively. Notably, TaMyc-D1 is classified as a pseudogene due to a frameshift mutation. The remaining four genes exhibit diverse expression patterns across different tissues and under various stress conditions. For instance, TaMyc-B1 is predominantly expressed in the coleoptile and functions as a co-regulator of anthocyanin synthesis alongside TaC1-A1 (an R2R3-MYB gene) within the MBW complex. The genes TaMyc-A1, TaMyc-A2, and TaMyc-D2 are osmotically inducible and are implicated in enhancing drought and salinity tolerance. In contrast to MYC genes, MYB genes are more prevalent and exhibit greater diversity [27]. They are categorized into four types: R2R3-MYB, R1R2R3-MYB, MYB-related, and Atypical MYB. The R2R3-MYB type is the most extensively studied and is characterized by two contiguous MYB domain repeats (R2 and R3) at the N-terminus. The R1R2R3-MYB type bears a resemblance to R2R3-MYB but includes an additional repeat (R1) at the N-terminus [2]. MYB-related type has only one incomplete repeat of the MYB domain, while Atypical MYB has no typical MYB domain but other conserved motifs. Wheat MYB genes are involved in various biological processes such as anthocyanin biosynthesis, phenylpropanoid metabolism, cell wall formation, seed development, hormone signaling, and stress responses [24]. MYC and MYB are two important transcription factor families that regulate a wide range of plant functions. They have evolved from a common ancestor but have diversified in their structures and functions during the course of evolution. They interact with each other and with other proteins to form complex regulatory networks that modulate gene expression in response to internal and external cues. MYC and MYB transcription factors interact with Dehydration-Responsive Element Binding (DREB) transcription factors to regulate plant responses to abiotic stresses such as drought, salinity, and extreme temperatures. MYC proteins are involved in various stress responses and developmental processes in plants. They can interact with DREB transcription factors to modulate the expression of stress-responsive genes. MYC transcription factors can form complexes with DREB proteins, enhancing their ability to bind to dehydration-responsive elements (DREs) in the promoters of target genes. This interaction can lead to the activation of genes involved in stress tolerance. MYB proteins play crucial roles in regulating plant development, metabolism, and stress responses. They can interact with DREB transcription factors to fine-tune the expression of stress-responsive genes. MYB transcription factors can physically interact with DREB proteins, influencing their DNA-binding activity and transcriptional regulation. This interaction can enhance the plant's ability to cope with abiotic stresses by regulating the expression of key stress-responsive genes. There are two major types of DREBs (dehydration-responsive element-binding proteins): DREB1 and DREB2, which differ in their expression patterns and functions. DREB1 genes are mainly induced by low temperature and osmotic stress, while DREB2 genes are mainly induced by dehydration and high-temperature stress [26]. Both types of DREBs can regulate the expression of genes involved in various stress responses, such as cold acclimation, drought adaptation, salt tolerance, and heat shock. However, some DREB2 genes also have a role in regulating plant growth and development under normal conditions. In wheat, which is one of the most important cereal crops in the world, several DREB genes have been identified and characterized [9]. Wheat DREB1 genes, such as TaDREB1, TaDREB3, and TaDREB4, have been shown to confer enhanced tolerance to cold, drought, and salt stress in transgenic plants. Wheat DREB2 genes, such as TaDREB2A, TaDREB2B, and TaDREB2C, have been shown to confer enhanced tolerance to drought and heat stress in transgenic plants. Moreover, some wheat DREB genes have been found to be involved in biotic stress responses, such as resistance to fungal pathogens and insect pests [60]. DREBs, or dehydration-responsive element-binding proteins, serve as crucial transcription factors that enhance plant resilience to both abiotic and biotic stresses. They achieve this by orchestrating the regulation of numerous genes responsive to stress, thereby equipping plants with robust mechanisms to withstand adverse conditions. Delving into the molecular intricacies and functionalities of DREBs unlocks opportunities for pioneering crop enhancement techniques, especially pertinent in the face of fluctuating environmental scenarios [24]. By overexpressing or modifying MYC, MYB, and DREB transcription factors, researchers can develop transgenic plants with improved tolerance to abiotic stresses. This approach has been used to enhance drought and salinity tolerance in various crops. CRISPR-Cas9 Genome Editing technology allows precise editing of MYC, MYB, and DREB genes to improve their interactions and enhance stress tolerance in plants. For example, editing the promoter regions of these genes can increase their expression under stress conditions.

The selection criteria for an ideal drought-tolerant plant on the basis of physio-morphic attributes is a root–shoot ratio near to 1, having high relative water contents, low excise leaf water loss, and high cell membrane thermostability [61] and high proline contents under drought stress. After carefully observing the role of ethylene, under drought stress ethylene contents should also be monitored.

REFERENCES

1. Ahmed, H. G. M.-D., Sajjad, M., Li, M., Azmat, M. A., Rizwan, M., Maqsood, R. H., et al. (2019). Selection criteria for drought-tolerant bread wheat genotypes at seedling stage. *Sustainability* 11, 2584. doi: 10.3390/su11092584
2. Gibbes, C., Southworth, J., Waylen, P., Child, B. J. A. G. Gibbes, C., Southworth, J., Waylen, P., and Child, B. J. A. G. 2014. Climate variability as a dominant driver of post-disturbance savanna dynamics. *Appl. Geogr.* 53, 389–401.
3. GOP (2021). Pakistan Economic Survey Finance Division. Islamabad: Economic Advisor's Wing.
4. Uzair, M., Ali, Z., Mahmood, T., Karim, I., Akram, U., Mahmood, N., et al. (2016). Genetic basis of some yield related traits in wheat (*Triticum aestivum* L.) under drought conditions. *Imp. J. Interdiscip. Res.* 2, 444–449.
5. UN Enviroment Programme. www.unep.org/explore-topics/climate-action/what-we-do/climate-action-note/state-of-climate.html
6. Abbas, M., Sheikh, A., Sabir, H. M., and Nighat, S. (2005). Factors responsible for low wheat productivity in Central Punjab. *Pak. J. Agri. Sci.* 42, 3–4.
7. Global Climate Change Impact Studies Centre. www.gcisc.org.pk/PCCCV.php
8. Wang, W., Vinocur, B., and Altman, A. (2003). Plant responses to drought, salinity and extreme temperatures: towards genetic engineering for stress tolerance. *Planta* 218, 1–14. doi: 10.1007/s00425-003-1105-5
9. Taiz, L., Zeiger, E., Møller, I. M., and Murphy, A. (2015). Plant Physiology and Development. Sunderland, MA: Sinauer Associates Incorporated.
10. FAO (2021). Crop Prospects and Food Situation. Quarterly Global Report No. 4, December 2021. Rome: FAO.
11. Khan, M. H., Bukhari, A., Dar, Z. A., and Rizvi, S. M. (2013). Status and strategies in breeding for rust resistance in wheat. *Agri. Sci.* 4, 292. doi: 10.4236/as.2013.46042.
12. Ahmed, H. G. M.-D., Zeng, Y., Iqbal, M., Rashid, M. A. R., Raza, H., Ullah, A., et al. (2022b). Genome-wide association mapping of bread wheat genotypes for sustainable food security and yield potential under limited water conditions. *PLoS ONE* 17, e0263263. doi: 10.1371/journal.pone.0263263.
13. Ruan, S., Xue, Q., and Tylkowska, K. (2002). The influence of priming on germination of rice (*Oryza sativa* L.) seeds and seedling emergence and performance in flooded soil. *Seed Sci. Technol.* 30, 61–67.

14. Qayyum, A., Al Ayoubi, S., Sher, A., Bibi, Y., Ahmad, S., Shen, Z., et al. (2021). Improvement in drought tolerance in bread wheat is related to an improvement in osmolyte production, antioxidant enzyme activities, and gaseous exchange. *Saudi J. Biol. Sci.* 28, 5238–5249. doi: 10.1016/j.sjbs.2021.05.040
15. Kramer, P. J., and Boyer, J. S. (1995). Water Relations of Plants and Soils. Cambridge, MA: Academic Press. doi: 10.1016/B978-012425060-4/50003-6
16. Mahpara, S., Zainab, A., Ullah, R., Kausar, S., Bilal, M., Latif, M. I., et al. (2022). The impact of PEG-induced drought stress on seed germination and seedling growth of different bread wheat (*Triticum aestivum* L.) genotypes. *PLoS ONE* 7, e0262937. doi: 10.1371/journal.pone.0262937.
17. Khalilzadeh, R., Seyed Sharifi, R., and Jalilian, J. (2016). Antioxidant status and physiological responses of wheat (*Triticum aestivum* L.) to cycocel application and bio fertilizers under water limitation condition. *J. Plant Interact.* 1, 130–137. doi: 10.1080/17429145.2016.1221150
18. Kapoor, D., Bhardwaj, S., Landi, M., Sharma, A., Ramakrishnan, M., and Sharma, A. (2020). The impact of drought in plant metabolism: how to exploit tolerance mechanisms to increase crop production. *Appl. Sci.* 10, 5692. doi: 10.3390/app10165692
19. Fathi, M., Mohebbi, M., and Koocheki, A. (2016). Introducing Prunus cerasus gum exudates: chemical structure, molecular weight, and rheological properties. *Food Hydrocoll.* 61, 946–955. doi: 10.1016/j.foodhyd.2016.07.004.
20. Fang, Y., and Xiong, L. (2015). General mechanisms of drought response and their application in drought resistance improvement in plants. *Cell. Mol. Life Sci.* 72, 673–689. doi: 10.1007/s00018-014-1767-0.
21. Datta, J., Mondal, T., Banerjee, A., and Mondal, N. (2011). Assessment of drought tolerance of selected wheat cultivars under laboratory condition. *J. Agri. Technol.* 7, 383–393.
22. Cochard, H., Coll, L., Le Roux, X., and Améglio, T. (2002). Unraveling the effects of plant hydraulics on stomatal closure during water stress in walnut. *Plant Physiol.* 128, 282–290. doi: 10.1104/pp.010400.
23. Bayoumi, T. Y., Eid, M. H., and Metwali, E. M. (2008). Application of physiological and biochemical indices as a screening technique for drought tolerance in wheat genotypes. *Afr J. Biotech.* 7, 2341–2352.
24. Bala, P., and Sikder, S. (2017). Evaluation of heat tolerance of wheat genotypes through membrane thermostability test. *MAYFEB J. Agri. Sci.* 2, 1–6. doi: 10.22161/ijeab/2.4.50
25. Ali, M., Zulkiffal, M., Anwar, J., Hussain, M., Farooq, J., and Khan,S. (2015). Morpho-physiological diversity in advanced lines of bread wheat under drought conditions at post-anthesis stage. *JAPS J. Anim. Plant Sci.* 25, 431–441.
26. Ahmed, H. G. M.-D., Zeng, Y., Yang, X., Anwaar, H. A., Mansha, M. Z., Hanif, C. M. S., et al. (2020). Conferring drought-tolerant wheat genotypes through morpho-physiological and chlorophyll indices at seedling stage. *Saudi J. Biol. Sci.* 27, 2116–2123. doi: 10.1016/j.sjbs.2020.06.019.
27. Ahmed, H. G. M.-D., Naeem,M., Zeng,Y., Rashid, M. A. R., Ullah, A., Saeed, A., et al. (2022a). Genome-wide association mapping for high temperature tolerance in wheat through 90k SNP array using physiological and yield traits. *PLoS ONE* 17, e0262569. doi: 10.1371/journal.pone.0262569
28. Asl, R. G., Arbat, H. K., Yarnia, M., Aminzade, G., Asl, L. G., and Ghanifathi, T. (2011). Evaluation of drought tolerance indices and grain yield in wheat genotypes using principal components analysis. *Adv. Environ. Biol.* 2, 1230-1232.

29. Bayoumi, T. Y., Eid, M. H., and Metwali, E. M. (2008). Application of physiological and biochemical indices as a screening technique for drought tolerance in wheat genotypes. *Afr J Biotech.* 7, 2341–2352.
30. Gibbes, C., Southworth, J., Waylen, P., & Child, B. (2014). Climate variability as a dominant driver of post-disturbance savanna dynamics. *Appl. Geogr.* 53, 389–401.
31. Faisal, S., Mujtaba, S., Khan, M., and Mahboob, W. (2017). Morpho physiological assessment of wheat (*Triticum aestivum* L.) genotypes for drought stress tolerance at seedling stage. *Pak. J. Bot.* 49, 445–452.
32. Djanaguiraman, M., Prasad, P. V., and Seppanen, M. (2010). Selenium protects sorghum leaves from oxidative damage under high temperature stress by enhancing antioxidant defense system. *Plant Physiol. Biochem.* 48, 999–1007. doi: 10.1016/j.plaphy.2010.09.009
33. Fan, Y., Shabala, S., Ma, Y., Xu, R., and Zhou, M. (2015). Using QTL mapping to investigate the relationships between abiotic stress tolerance (drought and salinity) and agronomic and physiological traits. *BMC Genom.* 6, 1–11. doi: 10.1186/s12864-015-1243-8.
34. Sallam, A., Alqudah, A. M., Dawood, M. F., Baenziger, P. S., and Börner, A. (2019). Drought stress tolerance in wheat and barley: advances in physiology, breeding and genetics research. *Int. J. Mol. Sci.* 20, 3137. doi: 10.3390/ijms20133137
35. Shewry, P. (2019). What is gluten—Why is it special? *Front. Nutr.* 6, 101. doi: 10.3389/fnut.2019.00101
36. Hussain, M., Khan, A. S., Khaliq, I., and Maqsood, M. (2012). Correlation studies of some qualitative and quantitative traits with grain yield in spring wheat across two environments. *Pak. J. Agri. Sci.* 49, 1–4.
37. Jaynes, D., Kaspar, T., Colvin, T., and James, D. (2003). Cluster analysis of spatiotemporal corn yield patterns in an Iowa field. *Agro. J.* 95, 574–586. doi: 10.2134/agronj2003.5740
38. Kalaji, H. M., Jajoo, A., Oukarroum, A., Brestic, M., Zivcak, M., Samborska, I. A., et al. (2016). Chlorophyll a fluorescence as a tool to monitor physiological status of plants under abiotic stress conditions. *Acta Physiologiae Plantarum* 38, 1–11. doi: 10.1007/s11738-016-2113-y
39. Bashir, A., Mahmood, M. A., Sheikh, A., and Kashif, M. (2006). Causes of wheat yield decline in the irrigated Punjab. *J. Agric. Res.* 44, 71–83.
40. Ristic, Z., Bukovnik, U., and Prasad, P. V. (2007). Correlation between heat stability of thylakoid membranes and loss of chlorophyll in winter wheat under heat stress. *Crop Sci.* 47, 2067–2073. doi: 10.2135/cropsci2006.10.0674
41. Gugino, B. K., Abawi, G. S., Idowu, O. J., Schindelbeck, R. R., Smith, L. L., Thies, J. E., et al. (2009). *Cornell Soil Health Assessment Training Manual.* Ithaca, NY: Cornell University College of Agriculture and Life Sciences.
42. Pour-Aboughadareh, A., Mehrvar, M. R., Sanjani, S., Amini, A., Nikkhah-Chamanabad, H., and Asadi, A. (2021). Effects of salinity stress on seedling biomass, physiochemical properties, and grain yield in different breeding wheat genotypes. *Acta Physiologiae Plantarum* 43, 1–14. doi: 10.1007/s11738-021-03265-7
43. Moebius-Clune, B. N. (2016). *Comprehensive Assessment of Soil Health: The Cornell Framework Manual.* Ithaca, NY: Cornell University.
44. Marˇ cek, T., Kova,ˇ c, T., Juki,´ c, K., Lonˇ cari,´ c, A., and IŽakovi,´ c, M. (2021). Application of high voltage electrical discharge treatment to improve wheat germination and early growth under drought and salinity conditions. *Plants* 10, 2137. doi: 10.3390/plants10102137

45. Comas, L., Becker, S., Cruz, V. M. V., Byrne, P. F., and Dierig, D. A. (2013). Root traits contributing to plant productivity under drought. *Front. Plant Sci.* 4, 442. doi: 10.3389/fpls.2013.00442.
46. Upadhyaya, N. M., Mago, R., Panwar, V., Hewitt, T., Luo, M., Chen, J., et al. (2021). Genomics accelerated isolation of a new stem rust avirulence gene–wheat resistance gene pair. *Nature Plants* 7, 1220–1228. doi: 10.1038/s41477-021-00971-5
47. Hussain, H. A., Men, S ., Hussain, S., Chen, Y., Ali, S., Zhang, S., et al. (2019). Interactive effects of drought and heat stresses on morpho-physiological attributes, yield, nutrient uptake and oxidative status in maize hybrids. *Sci. Rep.* 9, 1–12. doi: 10.1038/s41598-019-40362-7
48. Kamel, E., Arminian, A., and Houshmand, S. (2009). Karyomorphological and morphometric studies of ploidy levels in some wheat (*Triticum* aestivum L.) genotypes. *Biol. Sci. PJSIR* 2, 200–207.
49. Qayyum, A., Al Ayoubi, S., Sher, A., Bibi, Y., Ahmad, S., Shen, Z., et al. (2021). Improvement in drought tolerance in bread wheat is related to an improvement in osmolyte production, antioxidant enzyme activities, and gaseous exchange. *Saudi J. Biol. Sci.* 28, 5238–5249. doi: 10.1016/j.sjbs.2021.05.040
50. Ahmad, A., Aslam, Z., Javed, T., Hussain, S., Raza, A., Shabbir, R., et al. (2022). Screening of wheat (*Triticum aestivum* L.) genotypes for drought tolerance through agronomic and physiological response. *Agronomy* 12, 287. doi: 10.3390/agronomy12020287
51. Sisodia, B., and Rai, V. (2017). An application of principal component analysis for pre-harvest forecast model for wheat crop based on biometrical characters. *Int. Res. J. Agri. Econ. Stat.* 8, 83–87. doi: 10.15740/HAS/IRJAES/8.1/ 83-87
52. Mickky, B. M., and Aldesuquy, H. S. (2017). Impact of osmotic stress on seedling growth observations, membrane characteristics and antioxidant defense system of different wheat genotypes. *Egypt. J. Basic Appl. Sci.* 4, 47–54. doi: 10.1016/j.ejbas.2016.10.001
53. Mohi-Ud-Din, M., Hossain, M., Rohman, M., Uddin, M., Haque, M., Ahmed, J. U., et al. (2021). Multivariate analysis of morpho-physiological traits reveals differential drought tolerance potential of bread wheat genotypes at the seedling stage. *Plants* 10, 879. doi: 10.3390/plants10050879
54. Mujtaba, S., Faisal, S., Khan, M., Mumtaz, S., and Khanzada, B. (2016). Physiological studies on six wheat (Triticum aestivum L.) genotypes for drought stress tolerance at seedling stage. *Agric. Res. Technol. Open Access J.* 1, 1–5. doi: 10.19080/ARTOAJ.2016.01.555559
55. Paul, S., Aggarwal, C., Manjunatha, B., and Rathi, M. S. (2018). Characterization of osmotolerant rhizobacteria for plant growth promoting activities in vitro and during plant-microbe association under osmotic stress. *Indian J. Exp. Biol.* 56, 582–589.
56. Nezhadahmadi, A., Prodhan, Z. H., and Faruq, G. (2013). Drought tolerance in wheat. *Sci. World J.* 2013, 610721. doi: 10.1155/2013/610721
57. Pour-Aboughadareh, A., Mehrvar, M. R., Sanjani, S., Amini, A., Nikkhah-Chamanabad, H., and Asadi, A. (2021). Effects of salinity stress on seedling biomass, physiochemical properties, and grain yield in different breeding wheat genotypes. *Acta Physiologiae Plantarum* 43, 1–14. doi: 10.1007/s11738-021-03265-7
58. Hoagland, D., and Arnon, D. (1950). *The Water-Culture Method for Growing Plants Without Soil. Circular.* 2nd ed. Davis, CA: California Agricultural Experiment Station. p. 347.
59. Ellis, R., and Roberts, E. (1981). The quantification of ageing and survival in orthodox seeds. *Seed Sci. Technol.* 9, 373–409.

60. Pour-Aboughadareh, A., Omidi, M., Naghavi, M. R., Etminan, A., Mehrabi, A. A., Poczai, P., et al. (2019). Effect of water deficit stress on seedling biomass and physio-chemical characteristics in different species of wheat possessing the D genome. *Agronomy* 9, 522. doi: 10.3390/agronomy9090522
61. Sallam, A., Alqudah, A. M., Dawood, M. F., Baenziger, P. S., and Börner, A. (2019). Drought stress tolerance in wheat and barley: advances in physiology, breeding and genetics research. *Int J Mol. Sci.* 20, 3137. doi: 10.3390/ijms20133137
62. The World Meteorological Organization (WMO). (2022). *State of the Global Climate.*
63. The World Meteorological Organization WMO. (2024). *State of the Climate.* COP29.

7 Climate Change and Stress Mitigation Strategies in Plants

Hamna Maqsood, Rabia Nasir Khan, Fiaz Rasul, Muhammad Hussnain Siddique, Habibullah Nadeem, Muhammad Bilal, and Ijaz Rasul

7.1 INTRODUCTION

Environmental changes have drastic effects on agriculture production, natural systems, and human health (1). Due to the rising population and competition for food, there are worries and concerns about the global economy and food stability. Water availability and soil fertility have affected agriculture. Abiotic stressors can affect a plant's physiology, metabolism, and productivity directly or indirectly. These effects worsen when the environment undergoes sudden changes (2). Plants frequently face various stresses like salinity, drought, cold, water logging, heat, ions and heavy metals (3). Plants have adapted themselves to survive in harsh environments. The plant can survive by achieving tolerance to that particular stress. Plants avoid drought by reducing the duration of their life in order to finish their growth before experiencing extreme stress (4). Drought tolerance occurs in osmotic management and the scavenging of reactive oxygen species (ROS). High-temperature ranges are the cause of wilting and yellowing of leaves. The most affected phase in plants by extreme temperatures is pollination. Recent studies indicate that frost and excessive heat lead to the abortion of proximal grains in wheat spikes. Different fertilizers have been used to fight against all types of stresses. Genetic engineering plays a crucial role in producing stress-tolerant varieties of plants. The growth and development of plants are significantly influenced by both macro- and micronutrients. Mineral deficiency is also a great problem for plants. A shortage of certain nutrients can affect a plant's ability to withstand biotic and abiotic stresses. The major deficiencies which have gained scientists' attention are iron and zinc deficiency. These are particularly common in developing countries where the main foods are legumes and grains which are the major sources of these elements. The impact of drought has significantly decreased the production of legumes and grains. Drought has reduced mash bean (*Vigna mungo*) production to 31–57% during the flowering phase and during the reproductive phase to 26%. Drought stress has been shown to diminish soybean production by 42% at the grain-filling stage. Maize yield improves at a temperature of 29°C, but the yield decreases with a further increase in temperature. Recent studies have shown that developing nations are more susceptible to climate change than developed nations (5).

DOI: 10.1201/9781032720142-9

The interactions between crops, diseases, pests, and weeds are also being disrupted by climate change. It is also responsible for decreasing numbers of pollinator insects, increasing water deficiency and ozone concentration. The fishery industry has also suffered from climatic changes.

Climate change influences not only crops, but also the geographical distribution, quantity, population dynamics, and severity of several major pests. Continuous greenhouse gas emissions have raised the atmospheric concentration of CO_2, which is another factor contributing to climate change. The high amount of CO_2 in the atmosphere has caused further warming and climate changes which are damaging ecosystems and human health. Since pre-industrialization, global warming has risen by 1°C, and if the present pattern continues, it is predicted to reach 1.5°C by 2050 (6). Recent studies project that up to two-thirds of the global population (around 65%) may face clean water scarcity by the end of this century. The variability in the frequency and severity of typical precipitation patterns is also turning into a significant worldwide issue for agriculture, as it directly affects the soil's capacity to provide plant nutrients (7). Because soil water acts as a medium for plants to absorb and transport nutrients, it can have a considerable impact on nutrient allocation. Additionally, disruption of the metabolic pathways may result in a sharp shift in soil moisture, which is important for nutrient acquisition.

Plants have developed appropriate phytochemical defense systems to handle abiotic and biotic stresses. They develop a defense system that contributes to their resilience. Phytochemicals are secreted and accumulated in plants' bodies for different metabolic functions as well as for defense (8). Stress perception triggers the activation of conserved signal transduction cascades, such as certain ion channels, kinase cascades, ROS, and phytochemical signaling pathways. These cascades result in molecular alterations that trigger the unique phytochemical biosynthetic machinery. An abiotic stress response is controlled by the phytohormone abscisic acid; on the other hand, resistance to biotic stress is primarily regulated by salicylic acid (SA) and jasmonic acid (JA) signaling pathways, while ethylene signaling can also be analyzed for its complementary role. Recent research indicates that abscisic acid is critical in adjusting observed responses to concurrent biotic and abiotic stresses (9). The defense ability is reliant on the plant species throughout all its developmental growth stages. It is also dependent on the nature, timing, and intensity of the stress stimulus. Below we will discuss some risks which can affect plants in various ways.

7.2 DROUGHT AND WATER STRESS

Climate change has led to water scarcity and temperature extremes that are influencing plant reproduction. It has been demonstrated that water stress has an unfavorable effect on cereal inflorescence and flower initiation. Similarly, poor yield and sterility in cereals might result from temperature increases during floret growth. A 35–75% decline in grain set occurred during the meiotic period, which a critical phase for gamete formation in cereals, including rice and wheat, which are important cash crops in many nations. Meiosis ensures the creation of the haploid gametes required for sexual reproduction, and disturbances during this period can impair grain development. Drought stress significantly disrupts the fertilization and anthesis processes in rice. The main factor causing yield instability in agricultural production across a variety of crop habitats is drought stress. Plants react to drought stress in many physiological and biochemical ways. Abscisic acid (ABA), a phytohormone that is increased in cells under water stress, triggers the activation of many stress-responsive genes. Exogenous ABA therapy has no effect on a number of genes that are affected by water stress (10). Salinity and drought cause high ABA accumulation in plants, as well as significant alterations in gene expression. The hydration rate of plants has an important impact on the genes NCED3 and CYP07A3 (11). ABA is typically preserved as an ABA glycosyl ester that is dehydrated and liberated from the ester generated by β-glycosidase. Several signaling components involved in ABA signal transduction have been identified using newly generated genetic and biochemical devices. Resistance to pyrabatin plant stress is perceived by PYR1/pyrabactin-1-like proteins/AB regulatory components as an accumulation of ABA (12). Many mechanisms linked to stress detection, gene expression, signal transduction, and the production of various proteins and other chemicals are engaged in plants' drought stress response. There are two types of water stress-induced gene products. The first category includes components that play a direct role in stress tolerance. It consists of proteins that protect macromolecules and membranes, like osmotin, anti-freeze proteins, chaperones, mRNA binding proteins, etc., as well as water channel proteins that control the water flow across membranes and enzymes required for the production of several osmoprotectant compounds, such as proline, sugar, betaine, glycine, etc. The detoxifying enzymes like superoxide dismutase (SOD) ascorbate peroxidase are also included. The second group of protein factors engages in the control of stress-responsive gene expression and signal transmission. These include phospholipases, transcription factors, and kinases. It has been demonstrated that during a drought, the transcripts of certain proteins accumulate. The specific type of plant proteins known as LEA proteins are those that build up in plants' vegetative tissues when they are forming seeds and when they are stressed by heat, drought, cold, or salt (13). These hydrophilic proteins are high in glycine. It is believed that the innate disturbed structure and elevated hydrophilicity help to replenish or make up for water lost during desiccation. A plant's ability to withstand abiotic stressors is correlated with its LEA overproduction or underproduction.

7.3 TEMPERATURE

A slight temperature shift may boost pasture production in moist and temperate grassland zones. These advancements must be made to address the accelerated rate of climate change, such as intense heat and drought in the Mediterranean region, or heavy rainstorms that raise the danger of floods in temperate zones. However, in arid and semiarid areas, it may reduce livestock growth and increase mortality rates. Physiological processes in plants occur at the temperature range of 0–40°C. We can distinguish the direct physiological and morphological changes in plant bodies affected by temperature. The root bleeding activity of tomato stems was used to quantify how temperature affected sugar translocation in the stems (14). It was revealed that translocation decreases linearly as temperature increases from 2°C. At 2°C the radioactive sugar translocation was faster than at higher temperatures. However, this is not what happens in all cases. In many circumstances, plants growing at cold temperatures have higher sugar contents. A high sugar content is connected with frost resistance, as shown in the potato. It was discovered that starch hydrolysis at low temperature was associated with high phosphorylase activity. An inhibitor was discovered at higher temperatures. Temperatures fluctuate between day and night in a regular manner, known as thermo-periodicity, where the physiological reaction of plants and some other organisms alternates between day and night temperatures. This temperature change impacts growth, development, flowering, and other biological processes since some species need different temperatures between day and night in order to flourish. Deciduous trees, shrubs, and most plants in temperate regions with basal storage organs exhibit annual thermos-periodicity (15). In these plants, development is possible when high-temperature periods alternate with low-temperature periods. Although these cycles typically last about 1 year, variations in temperature patterns can shorten or extend their duration by approximately 50%.

7.4 SALT STRESS

Soil water that contains salts inhibits plant development for two reasons. Salt present in the soil reduces water uptake, and thus causes a reduction in the growth rate. This phenomenon is described as the osmotic effect of salinity or the water deficit effect of salinity (16). Secondly, if a large amount of salt enters the plant in the transpiration stream, it will cause injuries to the tissues and hence growth will be retarded. This is called the ion-specific effect of salinity. Seed germination is the key and critical stage in the growth of a plant during its life cycle and determines the yield. Salinity drastically influences the activity of germination in several plants such as *Oryza sativa*, *Tritium astivum*, and *Zea mays* (17).

Salinity influences the germination phase by several means. It inhibits the absorption of water by seeds due to low osmotic potential. Osmotic potential, or solute potential, is the component of water potential that depends on solutes in the solution. It sets the course for water travel across cell membranes. Water flows from areas of lower (more negative) osmotic potential to areas of higher (less negative) osmotic potential, which has important implications for processes such as nutrient uptake, maintenance of turgor, and growth. For example, less osmotic potential in root cells allows water to flow from the soil into the plant. It also causes toxicity in plants and changes the enzyme activation in nucleic acid metabolism. It also modifies the protein metabolism. It disturbs the hormonal balance and causes a reduction in seed reserve utilization. The excess amount of soluble salts in soil hinders the absorption and metabolic processes of essential mineral nutrients in plants. The optimum ion ratio serves as a tool for a plant's physiological response in relation to growth and development. High salt intake, on the other hand, triggers specific ion toxicities, for example caused by excessive sodium, calcium, potassium, nitrogen, or phosphorus (16). Most plants acquire significant concentrations of both Cl^- and Na^+ ions in their shoot tissues when growing. These ions can disrupt cellular functions by interfering with enzyme activities, nutrient uptake, and membrane stability, ultimately leading to reduced growth and productivity. Cl^- toxicity is a key factor in the reduction of growth in saline soils.

7.5 SOIL HEALTH AND ROS

The main role of soil is to sustain crop productivity while maintaining environmental and human health. There is an increasing awareness of the relationship between human health and soil in terms of health-boosting substances like phosphorus (P) and nitrogen (N), and other elements that are harmful to human health and agriculture, such as arsenic (As) and cadmium (Cd). Soil health is important for plant growth and yield. If the soil is affected by salinity, heavy metals, or excessive use of fertilizers then it affects the health of plants. Prolonged usage of chemical fertilizers has degraded agricultural soil quality by reducing soil organic matter (SOM) content and even increasing pollution and soil acidity. Chemical fertilizers are harmful to agriculture since salt, one of their main components, is harmful to plants and soil. Chemical fertilizers reduce the amount of critical soil minerals and nutrients. They are ineffective at restoring soil fertility and nutrients. ROS can induce protein oxidation, lipid peroxidation, enzyme deactivation, DNA damage, and interact with other important plant cell constituents (Figure 7.1) (18).

Stomata closure brought on by salt stress can lower the amount of carbon dioxide available in leaves and prevent carbon fixing. This results in excess excitation energy being absorbed by chloroplasts, potentially causing photodamage (19). It has been demonstrated that the formation of ROS during environmental conditions such as high salt presence is one of the prime reasons for crop yield declines (20). As a result, regulating ROS is a vital procedure to prevent oxidative damage and cellular cytotoxicity (21). Many plant investigations have found that ROS generation increases salt-induced cellular toxicity in different kinds of agricultural plants, for example rice, tomato, pea, citrus, and mustard, and has been linked to ROS-mediated membrane damage in saline environments (22).

7.6 IMPACTS ON PLANTS' MORPHOLOGICAL, PHYSIOLOGICAL, AND BIOCHEMICAL PROCESSES

The capability of plants to fight against several environmental factors is directly or indirectly connected with their photosynthetic ability. The rate of photosynthesis affects the physiological and biochemical progress of plants, and eventually their growth and yield. Abiotic stresses have a drastic effect on photosynthesis. Abiotic stresses disrupt the stomatal control of CO_2 supply, carbon cycle, and thylakoid electron transport – all of the key elements of photosynthesis. These together increase carbohydrate build-up, per-oxidative degradation of lipids, and disruption of water balance (23). Growth and yield are the morphological factors that are most affected by drought, cold, and salt in many plants. Drought has the greatest immediate impact on germination and stand establishment. Because of the lack of water, cell development is regarded as one of the most drought-sensitive physiological processes (24).

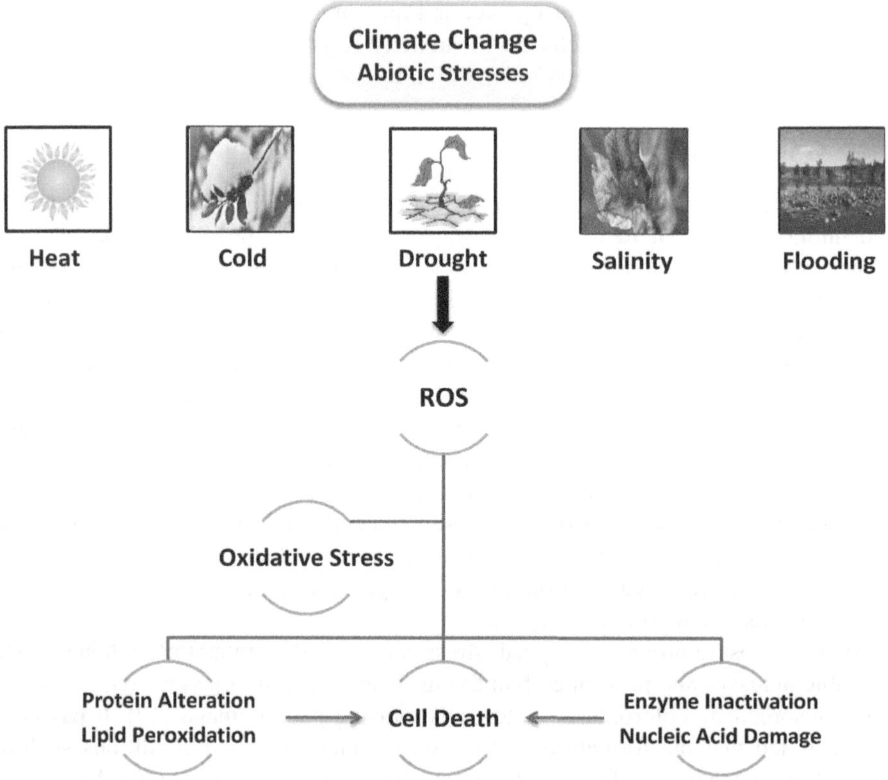

FIGURE 7.1 Plants' responses to abiotic stress and molecular adaptability.

Growth is the outcome of meristem cell divisions producing daughter cells, and the successive tremendous growth of young cells. In acute water scarcity, water flow from the xylem the neighboring elongating cells can be cut off to limit greater plant cell elongation (25).

Chlorophyll is the major component of relative content and photosynthetic pathways, and has a positive relation to photosynthetic rate (26). Under stress, a decrease in chlorophyll has been identified as a typical sign of oxidative stress, pigment photo-oxidation, and chlorophyll degradation (27). A decreased amount of chlorophyll causes yellowing of leaves and eventually death. Chloroplast membrane loss, severe swelling, lamellae vesiculation distortion, and the production of lipid droplets have all been linked to water-deficit-induced chlorophyll content reduction. Relative water content (RWC), temperature of the leaves, temperature of the canopy, stomatal resistance, transpiration rate, and leaf water potential are all key factors in plant–water relations (28). Plants that have undergone drought and salt stress have significantly lower leaf water potential, transpiration rate, and relative water content, with a concurrent rise in leaf temperature. One of the earliest metabolic responses of eukaryotic cells to biotic and abiotic stressors is the formation of ROS.

The generation of the oxidative burst, or ROS, is an initial event in the plant's defense reaction to a shortage of water and functions as a secondary massager to stimulate the plant's subsequent defense reaction. ROS comprises free radicals, peroxides, and oxygen ions that arise naturally as a by-product of metabolism. ROS levels rise rapidly during environmental stress, resulting in oxidative damage to proteins, lipids, and DNA (Figure 7.2) (29).

Plants have a delicate and complex internal defensive enzyme-catalyzed clean-up system that prevents active oxygen damage and ensures appropriate cellular function. The balance between ROS generation and antioxidative enzyme activity affects whether oxidative signaling and/or damage happen to plants with a cutting-edge enzymatic and non-enzymatic antioxidant system, comprising ROS-scavenging enzymes such as ascorbate peroxidase (APX), superoxide dismutase (SOD), and low-molecular-mass antioxidants like ascorbate, glutathione, and carotenoids (30). Preserving an increased anti-oxidative enzyme level activity may help to reduce stress by boosting resistance to oxidative damage.

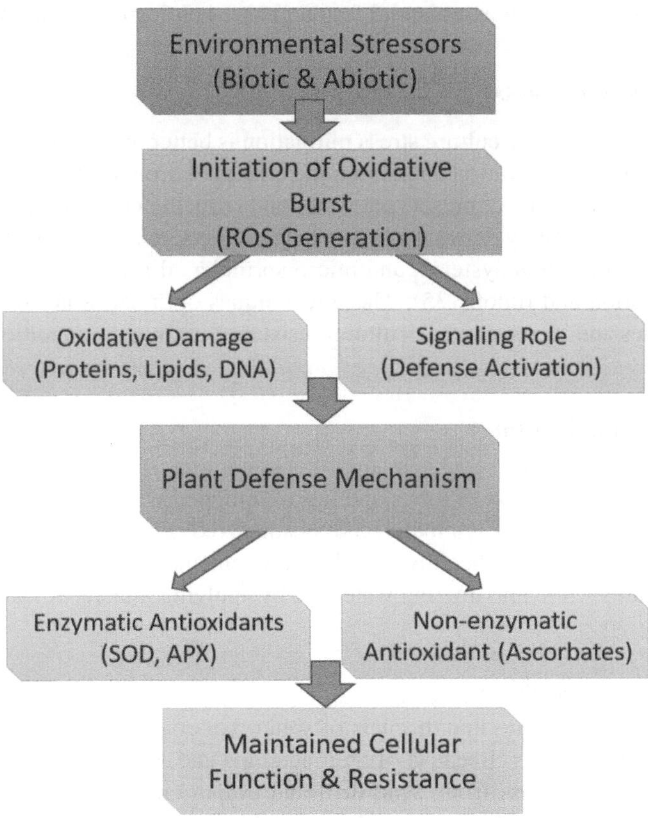

FIGURE 7.2 Flowchart depicting the role of ROS in plant defense: Balancing oxidative damage and signaling to activate antioxidants for stress tolerance.

7.7 STRESS MITIGATION STRATEGIES

7.7.1 Culturing Methods

In recent studies, scientists have investigated the strategies farmers employ to address environmental variability for plant adaptation. Farmers use a variety of effective strategies, including abiotic ones like adjusting planting and harvesting periods, selecting crops with fast growing seasons, rotating their crops, using irrigation methods, and changing cropping plans (31). Each of these strategies is extremely helpful for agricultural resilience under conditions of climate stress. Modifications in the time of seeding, the utilization of drought-tolerant cultivars, and the creation of new crops are some essential approaches for reducing the risks associated with climate variability and enhancing plants' ability to respond (32). Crop management strategies can also help plants adapt to their environment by promoting crop development under a variety of conditions. Selecting the proper sowing timing, planting density, and watering techniques is crucial to addressing weather pressures (33). Fertilizers are also necessary to aid in the plant's adaptation to the effects of global warming. They provide plants with a lot of energy and also maintain and increase soil fertility and productivity. Thus, it is evident how essential fertilizers are helping to feed the world (34).

7.7.2 Improved Irrigation Techniques

A vital component of agriculture stress mitigation is better irrigation practices, which emphasize water conservation and lessen the effects of drought. Innovative irrigation techniques are essential for preserving water and reducing crop stress during periods of drought. By accurately delivering water to plant roots, technologies like drip irrigation, precision irrigation systems, and micro-sprinklers drastically reduce water loss from evaporation and runoff (35). These techniques optimize water use, preserving water supplies and increasing agricultural resistance to drought conditions and water scarcity.

7.7.2.1 Drip Irrigation

This method involves the regulated supply of water directly to the root region of the plant through a connection of tubes and hoses equipped with emitters (36). It makes sure that water is applied gradually and steadily, reducing runoff and evaporation. Drip irrigation encourages healthy root development and facilitates effective nutrient uptake by plants while maximizing water use by applying moisture exactly where it's needed.

7.7.2.2 Micro-sprinklers

Micro-sprinklers generate fine droplets of water that cover a smaller area than traditional sprinkler systems. Because of their accurate and consistent delivery of water, they minimize water loss from wind drift and evaporation (37). These systems are flexible and adjustable, which makes them suitable for a range of crop varieties.

The incorporation of smart irrigation systems is an important turning point in modern agriculture. These systems use automation, data analytics, and sensor technologies to track crop requirements, weather patterns, and soil moisture levels in real time (38). Crops can receive the appropriate amount of water at the appropriate time thanks to smart irrigation, which allows for a precise, on-demand water supply (39). This integration reduces water waste as well as maximizes water use efficiency, which lessens the pressure that water shortage places on agricultural output. Agriculture becomes more flexible and efficient when modern irrigation techniques which prioritize water conservation are combined with the integration of smart technologies. In addition to reducing the effects of drought, these strategies also support resource conservation, sustainable practices, and increased crop resilience, all of which increase the long-term viability and productivity of agricultural systems.

7.7.3 Breeding Resilient Varieties

Modern biotechnology techniques are used with conventional breeding techniques to create crops that are tolerant to climate change. The introduction of characteristics like heat resistance, insect resistance, and drought tolerance is largely facilitated by genetic changes. Selection criteria for stress-tolerant plants include water use efficiency, the ability to survive temperature extremes, and pest and disease resistance (40). Breeders can use genomics to discover and include specific genes responsible for stress tolerance, resulting in crop varieties that are more suited to changing climates. Plant breeding demonstrates dynamic crop development and improvement approaches under a variety of environmental conditions (41). By developing stress-resistant cultivars, it provides a technique to enable plants to avoid different stressors at a critical period of growth as well as providing assurance of food security and safety during severe weather changes (42). Genetic divergence analysis is a key tool for identifying successful inbreeding. It is applied to inbreeding, evaluation, polymorphism, assortment, and recombination to achieve plant perfection. Genetic divergence studies are believed to provide a major benefit in the production of novel cultivars built on genetic similarity and distance (43).

7.7.4 Genetically Engineered Plants for Stress Tolerance

Global agriculture is facing serious challenges due to climate change and increasing environmental stresses like drought, salt, and extreme temperatures, which endanger crop productivity and food security. These issues demand novel approaches, such as genetic engineering, to create plants with increased stress tolerance. The intentional modification of DNA to change an organism's properties (phenotype) in certain ways is called genetic engineering. GM crops are agricultural plants whose DNA has been altered through the application of genetic engineering techniques (44). Genetics has gathered promising evidence that can be used to considerably alleviate a variety of biotic and abiotic challenges, comprising salt, drought, high temperature, and cold. Agricultural cultivars that are resistant to stress can be produced using information related to stress-responsive transcription factors (TFs), which are proteins that regulate the expression of specific genes involved in plant responses to environmental stresses such as drought, salinity, and extreme temperatures. In GM crops, these TFs can regulate the expression of genes linked to various stressors (45). To combat biotic and abiotic stressors, genetic engineering has created a wide variety of transgenic plants. Compared to natural plants, these GM plants display remarkable climate resistance (46). The AP2/ERFBP group is one of several TFs that are recognized as being plant-specific TFs (47). This family has a role in both abiotic and biotic stress responses in plants, as well as in numerous plant growth pathways (48). On the basis of resemblance and number, AP2/EREBP TFs are divided into four subgroups. The subfamilies are the RAV (related with ABI3/VP1), AP2 (Apetala 2), DREB (dehydration-responsive element-binding protein), and ERF TFs. Two prominent subfamilies that have attracted a lot of interest are DREB and ERF, owing to their functions in abiotic and biotic responses in plants (49). Under various conditions of cold stress and shortage of water, these DREB TFs exhibit robust regulatory capabilities. Moreover, DREB1 genes have been introduced into tobacco, rice, soybeans, and tomatoes to enhance their ability to withstand cold stress (50). Many DREB1 genes that are resistant to different abiotic stresses have been detected in wheat, corn, oilseed rape; rice and rye are utilized to make transgenic crops. ZmDREB2A has been isolated from maize and overexpressed in *Arabidopsis* to produce transgenic plants that are more resistant to drought (31). Transgenic rice with enhanced resistance to drought and salinity conditions was produced by effectively overexpressing the OsDREB2A gene (51).

Studies show that the implementation of TaPIE1 into wheat increased its resilience to disease and its ability to withstand chilling stress (52). In an effort to develop transgenic tobacco, the GmERF3 gene was overexpressed. This transgenic tobacco gene exhibits boosted resistance to dehydration, TMV, and salt stress tolerance (53). The myeloblastosis oncogene, or MYB TFs family, is an enormous family of TFs that are widely expressed in plants and have been identified in eukaryotes (54). Several MYB TFs control various physiological and biochemical pathways, such as hormone production, cell cycle, and primary and secondary metabolism. Furthermore, it is known that these TFs participate in biotic and abiotic stress responses (55). It has been discovered that a number of MYB TFs, including AtMYB61, AtMYB60, and AtMYB44, enhance drought tolerance in transgenic *Arabidopsis* by controlling stomata movement (56). The goal of the creation of transgenic rice was to make it more resistant to cold, salt, and dehydration by overexpressing the OsMYB2 gene (57). The ZmMYB30 gene was introduced into *Arabidopsis* to improve salt tolerance (58).

The WRKY family is a prominent TF family that is extensively distributed in response to biotic and abiotic stresses on plants (59). To increase the transgenic plants' resistance to abiotic stress, the WRKY genes were, for instance, overexpressed in transgenic rice. Its tolerance to the effects of heat and drought was increased by the addition of the OsWRKY11 gene (60). The overexpression of the TaWRKY19 gene was examined in order to investigate the impacts of cold, drought, and salinity stressors on transgenic wheat. In a way similar to this, transgenic *Arabidopsis* with resilience to heat and drought was created using TaWRKY1 and TaWRKY33, which were derived from wheat (61).

7.7.5 Biochemical and Molecular Approaches

Plant stress tolerance is studied genetically and physiologically using biochemical and molecular methods. Researchers examine the metabolic processes that allow specific plants to flourish in adverse environments, pinpointing essential components and pathways. These stress-tolerant characteristics can be specifically introduced into crops by genetic engineering (62). These may include genes related to hormone signaling, antioxidant defense, and osmotic control. Molecular approaches enable precise manipulation of plant genomes, enhancing their ability to withstand environmental stressors and ensuring a more resilient agricultural landscape.

7.7.5.1 Omics-led Breeding

Omics aims to characterize and quantify biological molecule pools collectively, which translates into the dynamics, structure, and function of organisms (63). Omics methods offer helpful tools to clarify the biological purposes of any genetic data for crop development and improvement (64). In population genomics, a variety of biological markers are examined across diverse populations involving a large number of individuals in order to find new patterns of variation and help determine whether the genes are involved in important ecological aspects (65). The breeding program is combined with genomic techniques in several crops in order to advance molecular breeding and screen elite germplasms using multi-trait assembly (66). Genetics and transcriptome analysis are utilized to identify phenotypes under various environmental variations (67). Recognition of the genes and pathways accountable for the desired features should result from the combination of genetic and phenotypic data with functional and genome-wide omics data (68). The potential of molecular genomics in conjunction with conventional quantitative genetics to integrate and evaluate phenotypic, pedigree, and omics data for important traits will be investigated further. Furthermore, genomics makes it feasible to look into the molecular mechanisms behind the resistance to abiotic stress. These methods support the creation of climate-smart crops for increased output and yield in response to changing climatic conditions (69). High-throughput sequencing and phenotyping have made it possible for genomic-led breeding to discover a variety of stresses that are anticipated to negatively affect crop yields (70). Researchers such as Haley and colleagues have utilized quantitative trait loci (QTL) mapping to create the drought-tolerant wheat variety known as 'Ripper.' This innovative approach pinpointed specific genetic regions tied to drought resistance, enabling the development of a wheat variety capable of sustaining both grain yield and quality in Colorado's dry conditions. In addition to wheat, QTL mapping has played a critical role in breeding maize, barley, and rice varieties with enhanced resilience to environmental stresses, showing the remarkable impact of omics-based methods on advancing agricultural progress (31).

7.7.5.2 Genome-wide Association Studies (GWAS)

Genome-wide association studies (GWAS) aim to recognize correlations between genotypes and phenotypes by comparing the allele frequencies of genetic variants in people with similar ancestry but diverse behaviors. It can account for sequence variations or copy number variations in the human genome, however the most commonly studied genetic variable in GWAS is single nucleotide polymorphisms (SNPs) (71). It is an effective technique for comprehending the full range of genetic variants present in various crop cultivars and for identifying allelic variants associated with any given feature (72). It emphasizes the relationship between SNPs and characteristics and is built around the design of the study, genotyping instruments, statistical models for analysis, and result interpretation (73). The genetic process that underlies genetic resilience to environmental change has been used in a number of crops (74). Plants use it extensively to respond to both biotic and abiotic stresses. GWAS has been utilized to investigate drought, salt, and heat tolerance (75).

Scientists have used reverse genetic techniques in an *Arabidopsis thaliana* GWAS analysis to identify distinct genes that store proline during drought stress. Specific areas influencing proline accumulation were identified by looking at the relationship between SNPs from both phenotypic and genotypic data. This study provided insights into proline accumulation during drought stress (76). A total of 7185 SNPs were selected to analyze the phenotypic behavior for the GWAS, and the relationships between SNPs and phenotypic traits were identified using mixed linear models and general linear models (77). GWAS was used to investigate QTLs involved in plant salinity resistance. Researchers have used Infinium high-throughput SNP arrays to identify several genes that regulate salt tolerance in rice. For genes related to stress, 6000 genotype-based SNPs were created, and linkage analyses between SNPs and phenotypic data were performed. It was discovered that the 'Saltol' new QTL, found on chromosome 1, is connected to seedling salt tolerance (78). In another study, researchers discovered numerous genes in *Sorghum bicolour* that exhibited resistance to heat and cold stress. They conducted genetic and phenotypic analyses utilizing GWAS methods. Thirty SNPs were discovered for genes associated with the synthesis of anthocyanins and the metabolism of carbohydrates, both of which are closely linked with cold stress throughout the sorghum seedling growth stage. A similar set of studies revealed that 12 SNPs regulated by genes involved in sugar metabolism and ion transport mechanism were linked to heat stress in seedlings (79). Thus, GWAS for complex trait loci in crops can identify important genes associated with stress and their favorable alleles (80).

GWAS offer valuable insights but are not without challenges, particularly in studying complex traits. A key difficulty lies in phenotyping limitations, where accurately assessing traits under varied environmental conditions is often problematic. This inconsistency can weaken the connections drawn between genetic and phenotypic data. Another challenge involves the complexity of trait dissection, as many traits are governed by multiple genes and are influenced by interactions between genes and the environment. Such traits often display low heritability, making it difficult to pinpoint specific genetic contributors. Moreover, detecting subtle genetic effects typically demands large and diverse populations, but the high costs associated with genotyping and phenotyping can restrict sample sizes. To overcome these hurdles, integrating GWAS with advanced technologies such as multi-omics approaches, high-throughput phenotyping, and improved statistical methods can enhance the identification of key genetic loci.

7.7.5.3 Marker-assisted Selection (MAS)

MAS involves the selection of plants that carry genomic areas implicated in the production of desired traits through the usage of molecular markers. It is feasible for qualities to be controlled by both main genes and quantitative trait loci (QTLs) (81). Genetic regions known as quantitative trait loci are often the product of genetic interactions with one another and the environment. They have an impact on a complicated characteristic's phenotypic variance. These are frequently found using a statistical genetic analysis technique called QTL mapping (82). The process of QTL dissection for yield-related attributes in stressed crops facilitates the creation of innovative cultivars that exhibit increased resilience to abiotic stress (83). Molecular breeding of plants is now an important method for increasing agricultural yield and productivity in the presence of biotic and abiotic stressors (83). MAS perform a key function in improving crop characteristics and production when breeding progresses quickly. Recent progress in crop genomics has led to the discovery of DNA markers helpful for marker-assisted breeding (84).

QTL mapping was carried out in 2009 to identify superior maize genotypes with higher production in drought stress (85). A rice variety that can resist salt conditions without harming its yield and quality has been created using QTL mapping for the molecular design of salt-tolerant crops (86). MAS was used to classify drought stress QTLs in wheat varieties (*Triticum aestivum* and *T. turgidum*) (87). The success rate of this method is likely to rise as a result of advancements in the creation of gene-based markers, improved QTL mapping techniques, and more affordable genotyping tools. Furthermore, the creation of high-throughput phenotyping technologies for QTL mapping, greater knowledge of genotypes by environment interaction as well as epistasis, and the creation of publicly available computational techniques are required to fully realize the potential of MAS in public service breeding programs.

MAS faces several challenges, particularly with phenotyping limitations, as accurately measuring complex traits across varying environments is often difficult. Additionally, dissecting complex traits is challenging due to the involvement of multiple genes and gene–environment interactions. QTL mapping can also be complicated by environmental factors and epistasis (gene interactions), which may hinder the identification of consistent markers. To overcome these issues, advancements in high-throughput phenotyping, better understanding of genotype–environment interactions, and improved computational methods are needed to maximize MAS's effectiveness in crop breeding.

7.7.6 Genome-editing Strategies

The term 'gene editing,' or 'genome editing,' describes a set of scientific methods that make it possible to alter an organism's DNA. These technologies enable the addition, modification, or removal of particular genomic areas (88). Gene-editing (GE) crop enhancement has the potential to significantly reduce food insecurity and build a weather-smart agriculture system (89). Due to widespread agricultural exploitation, which has been linked in a number of ways to increased vulnerability to a number of abiotic and biotic stressors (90), the genetic diversity of many elite cultivars has been significantly reduced. GE technology has significantly influenced plant breeding approaches, and novel methods for accurate, quick alterations of crop genomes to boost crop output are being investigated. There are three significant developments in genome-editing technology:

- ZFNs (zinc finger nucleases)
- TALENs (transcription activators-like effector nucleases)
- CRISPR-Cas9

The CRISPR/Cas9 system has been evolving as the most effective GE strategy in comparison to the ZFNs and TALENs genome-editing techniques because it is affordable, quick, specific, and allows for multiple-location precise editing in the genome (91).

7.7.6.1 The CRISPR/Cas9 System for Crop Advancement

CRISPR/Cas9 is a contemporary genome-editing method established on a prokaryotic defense mechanism begun by type II RNA organization that protects prokaryotes from virus invasion (92). The CRISPR/Cas-9 genome-editing process can be roughly divided into the following steps:

- A virus enters a bacterial cell.
- A new spacer is added to the CRISPR sequence that is obtained from a virus (adaptation).
- CRISPR RNA is created during the production process.
- CRISPR RNA directs the molecular apparatus to locate and destroy the viral genome (targeting).

CRISPR-Cas9 assembly has revolutionized genome editing by generating specific mutations in genes and by removing individual nucleotides from a genome (Figure 7.3) (93).

Additionally, this technique has enormous potential to help crop breeding in creating high-yielding, stress-resilient types (94). In order to combat environmental extremes and ensure food security, the CRISPR/Cas9 tool is most importantly evolving into a clear, ecologically favorable method for the creation of non-transgenic plants with altered genomes (95). As the field of genome editing continues to evolve, newer techniques such as prime editing and base editing are refining the precision and effectiveness of CRISPR-based technologies. Prime editing, for example, enables highly accurate changes by directly switching one DNA base for another without causing double-strand breaks. On the other hand, base editing allows for the precise alteration of individual nucleotides with minimal unintended changes, creating new opportunities for improving crops. These advancements make it possible to develop non-transgenic plants with targeted genetic modifications, offering a more environmentally sustainable approach to crop engineering. Plant genome editing using CRISPR/Cas9 has been widely exploited to adapt to biotic and abiotic stressors (96). In cassava, 21 KUP genes that were hyper-expressed in response to abiotic stressors

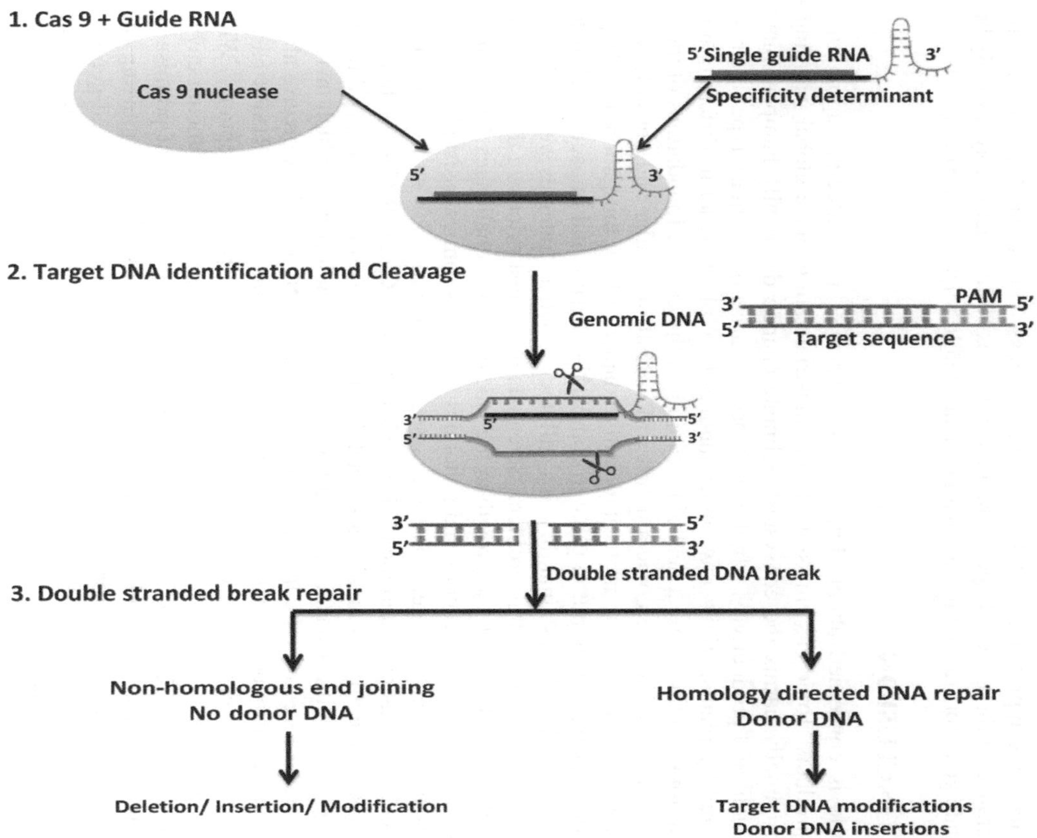

FIGURE 7.3 CRISPR-Cas9-based targeted genome editing. (1) CRISPR-Cas9 = Cas9 + guide RNA(s) that determines the DNA target. (2) Guide RNA + Cas9 cleave DNA to create double-stranded break. (3) NHEJ (non-homologous end joining) and HDR (homology-directed repair) repair double-stranded breaks, causing mutations or gene insertion.

have been discovered. KUP genes caused drought resistance, according to an investigation of their differential expression (97). MAPKKK genes were studied using genome-wide analysis for research into drought tolerance (98). Researchers employed this system to induce parthenocarpy in response to elevated temperatures, leading to suppression of the SlAGL6 gene. The resultant mutant tomato lines were identical to normal plants in terms of weight and form (99). Scientists investigated japonica rice using the Osann3 gene knockout. The application of CRISPR/Cas9 technology to mutant rice lines improved their tolerance to cold stress. This research established the cold tolerance of the OsANN3 gene and raised the possibility that it may possibly be a candidate gene for transgenic rice with enhanced cold tolerance (100).

7.8 CONCLUSION

The world is concerned about climate change because it is affecting agriculture and its products. Toxic gas emissions from industrialization cause global warming, which eventually harms the ecosystem. Climate change drastically hampers plant development and productivity. Abiotic stressors comprise the primary type of stress encountered by plants. Right now, understanding how plants react to different abiotic conditions requires a great deal of research into the genetic foundations of these systems. Plants need to overcome a number of crucial molecular and physiological obstacles in order to become more abiotically adapted. Many mitigation and adaptation strategies have been put in place to combat the detrimental impacts of climate change on agriculture's sustainability. It's still unclear what strategies we should use and how we should tackle this problem. Scientists now must prioritize their work to enhance plant development and growth in the face of abiotic challenges. These mitigation strategies include genetics and genomics strategies, omics-assisted breeding, marker-assisted selection, genome-wide association studies for stress tolerance, genome-editing strategies, and the CRISPR/Cas9 system. CRISPR/Cas9 technology has emerged as a highly sustainable approach for improving crops, especially through non-transgenic modifications. Unlike traditional genetic engineering, which involves the insertion of foreign DNA into a plant's genome, CRISPR enables precise alterations without introducing external genetic material. This makes CRISPR-based modifications more eco-friendly, which can facilitate greater public acceptance. With its ability to create crops resilient to climate-induced stresses while reducing environmental impacts, CRISPR/Cas9-mediated genome editing holds great promise in developing crops that can withstand climate-related stresses while minimizing the environmental impact. Therefore, we need to use CRISPR/Cas9-mediated genome editing to create environmentally friendly genome-modified crops in order to combat the detrimental effects of climate change in the future.

REFERENCES

1. Whitmee S, Haines A, Beyrer C, Boltz F, Capon AG, de Souza Dias BF, et al. Safeguarding human health in the Anthropocene epoch: report of The Rockefeller Foundation–Lancet Commission on planetary health. *The Lancet*. 2015;386(10007):1973–2028.

2. Vaughan MM, Block A, Christensen SA, Allen LH, Schmelz EA. The effects of climate change associated abiotic stresses on maize phytochemical defenses. *Phytochemistry Reviews*. 2018;17(1):37–49.
3. Raza A, Ashraf F, Zou X, Zhang X, Tosif H. Plant adaptation and tolerance to environmental stresses: mechanisms and perspectives. In M. Hasanuzzaman (ed), *Plant ecophysiology and adaptation under climate change: mechanisms and perspectives I*. Singapore: Springer; 2020. p. 117–45.
4. Shavrukov Y, Kurishbayev A, Jatayev S, Shvidchenko V, Zotova L, Koekemoer F, et al. Early flowering as a drought escape mechanism in plants: how can it aid wheat production? *Frontiers in Plant Science*. 2017;8:1950.
5. Yu S, Xia J, Yan Z, Zhang A, Xia Y, Guan D, et al. Loss of work productivity in a warming world: Differences between developed and developing countries. *Journal of Cleaner Production*. 2019;208:1219–25.
6. Mozaffari GA. Climate change and its consequences in agriculture. *The Nature, Causes, Effects and Mitigation of Climate Change on the Environment*. 2022;83:83–100.
7. Soares JC, Santos CS, Carvalho SM, Pintado MM, Vasconcelos MW. Preserving the nutritional quality of crop plants under a changing climate: importance and strategies. *Plant and Soil*. 2019;443:1–26.
8. Stiller A, Garrison K, Gurdyumov K, Kenner J, Yasmin F, Yates P, et al. From fighting critters to saving lives: Polyphenols in plant defense and human health. *International Journal of Molecular Sciences*. 2021;22(16):8995.
9. Sharma A, Shahzad B, Kumar V, Kohli SK, Sidhu GPS, Bali AS, et al. Phytohormones regulate accumulation of osmolytes under abiotic stress. *Biomolecules*. 2019;9(7):285.
10. Lata C, Muthamilarasan M, Prasad M. Drought stress responses and signal transduction in plants. *Elucidation of Abiotic Stress Signaling in Plants: Functional Genomics Perspectives*. 2015;2 :195–225.
11. Lee KH, Piao HL, Kim H-Y, Choi SM, Jiang F, Hartung W, et al. Activation of glucosidase via stress-induced polymerization rapidly increases active pools of abscisic acid. *Cell*. 2006;126(6):1109–20.
12. Hirayama T, Shinozaki K. Research on plant abiotic stress responses in the post-genome era: past, present and future. *The Plant Journal*. 2010;61(6):1041–52.
13. Ganguly M, Datta K, Roychoudhury A, Gayen D, Sengupta DN, Datta SK. Overexpression of Rab16A gene in indica rice variety for generating enhanced salt tolerance. *Plant Signaling & Behavior*. 2012;7(4):502–9.
14. Arai-Sanoh Y, Ishimaru T, Ohsumi A, Kondo M. Effects of soil temperature on growth and root function in rice. *Plant Production Science*. 2010;13(3):235–42.
15. Sabogal A, Sabogal A. Botany for landscapists. Urban ecology: a case study of Lima City, *Perú*. 2021:27–76.
16. Parihar P, Singh S, Singh R, Singh VP, Prasad SM. Effect of salinity stress on plants and its tolerance strategies: a review. *Environmental Science and Pollution Research*. 2015;22:4056–75.
17. Srivastava P, Wu Q-S, Giri B. Salinity: an overview. In B. Giri & A. Varma (Eds.), *Microorganisms in saline environments: Strategies and functions*. 2019:3–18. Cham: Springer Nature Switzerland AG. https://doi.org/10.1007/978-3-030-18975-4_1
18. Czarnocka W, Karpiński S. Friend or foe? Reactive oxygen species production, scavenging and signaling in plant response to environmental stresses. *Free Radical Biology and Medicine*. 2018;122:4–20.
19. Gupta DK, Palma JM, Corpas FJ. *Reactive oxygen species and oxidative damage in plants under stress*: Springer; 2015.

20. Kumar A, Singh S, Gaurav AK, Srivastava S, Verma JP. Plant growth-promoting bacteria: biological tools for the mitigation of salinity stress in plants. *Frontiers in Microbiology*. 2020;11:1216.
21. Farooq MA, Niazi AK, Akhtar J, Farooq M, Souri Z , Karimi N, et al. Acquiring control: the evolution of ROS-induced oxidative stress and redox signaling pathways in plant stress responses. *Plant Physiology and Biochemistry*. 2019;141:353–69.
22. Polash MAS, Sakil MA, Hossain MA. Plants responses and their physiological and biochemical defense mechanisms against salinity: a review. *Tropical Plant Research*. 2019;6:250–74.
23. Deka D, Yadav B, Yadav P, Narayan OP. Background level, occurrence, speciation, bioavailability, uptake detoxification mechanisms and management of Si-polluted soils. *Appraisal of metal (Loids) in the ecosystem*: Elsevier; 2022. p. 15–32.
24. Hussain MI, Lyra D-A, Farooq M, Nikoloudakis N, Khalid N. Salt and drought stresses in safflower: a review. *Agronomy for Sustainable Development*. 2016;36:1–31.
25. Saud S, Fahad S, Yajun C, Ihsan MZ, Hammad HM, Nasim W, et al. Effects of nitrogen supply on water stress and recovery mechanisms in Kentucky bluegrass plants. *Frontiers in Plant Science*. 2017;8:983.
26. Maghsoudi K, Emam Y, Pessarakli M. Effect of silicon on photosynthetic gas exchange, photosynthetic pigments, cell membrane stability and relative water content of different wheat cultivars under drought stress conditions. *Journal of Plant Nutrition*. 2016;39(7):1001–15.
27. Nahakpam S. Chlorophyll stability: a better trait for grain yield in rice under drought. *Indian Journal of Ecology*. 2017;44(4):77–82.
28. Salehi-Lisar SY, Bakhshayeshan-Agdam H. Drought stress in plants: causes, consequences, and tolerance. *Drought stress tolerance in plants*, Vol 1: Physiology and Biochemistry; 2016. p. 1–16.
29. Juan CA, Pérez de la Lastra JM, Plou FJ, Pérez-Lebeña E. The chemistry of reactive oxygen species (ROS) revisited: outlining their role in biological macromolecules (DNA, lipids and proteins) and induced pathologies. *International Journal of Molecular Sciences*. 2021;22(9):4642.
30. Shen Y, Li J, Gu R, Yue L, Wang H, Zhan X, et al. Carotenoid and superoxide dismutase are the most effective antioxidants participating in ROS scavenging in phenanthrene accumulated wheat leaf. *Chemosphere*. 2018;197:513–25.
31. Raza A, Razzaq A, Mehmood SS, Zou X, Zhang X, Lv Y, et al. Impact of climate change on crops adaptation and strategies to tackle its outcome: a review. *Plants*. 2019;8(2):34.
32. Ali A, Erenstein O. Assessing farmer use of climate change adaptation practices and impacts on food security and poverty in Pakistan. *Climate Risk Management*. 2017;16:183–94.
33. Battisti R, Sentelhas PC, Parker PS, Nendel C, Gil MDS, Farias JR, et al. Assessment of crop-management strategies to improve soybean resilience to climate change in Southern Brazil. *Crop and Pasture Science*. 2018;69(2):154–62.
34. Henderson B, Cacho O, Thornton P, van Wijk M, Herrero M. The economic potential of residue management and fertilizer use to address climate change impacts on mixed smallholder farmers in Burkina Faso. *Agricultural Systems*. 2018;167:195–205.
35. Agarwal R, Shekhar S, Thakur A, Olakkangal DJ, Singh YP. *Microirrigation: The water saving technology. Pharma Innovation Journal*, 2023;12(6), 2386–93.
36. Shareef TME, Ma Z, Zhao B. Essentials of drip irrigation system for saving water and nutrients to plant roots: as a guide for growers. *Journal of Water Resource and Protection*. 2019;11(9):1129–45.

37. Kaur P, Kaur K, Singh H. Role of micro-irrigation in vegetable crops. *International Journal of Agriculture Innovations and Research.* 2020;9:89–96.
38. Obaideen K, Yousef BA, AlMallahi MN, Tan YC, Mahmoud M, Jaber H, et al. An overview of smart irrigation systems using IoT. *Energy Nexus.* 2022:7:100124.
39. Shahdany SH, Taghvaeian S, Maestre JM, Firoozfar A. Developing a centralized automatic control system to increase flexibility of water delivery within predictable and unpredictable irrigation water demands. *Computers and Electronics in Agriculture.* 2019;163:104862.
40. Tripathi L, Ntui VO, Tripathi JN. Application of genetic modification and genome editing for developing climate-smart banana. *Food and Energy Security.* 2019;8(4):e00168.
41. Abdelrahman M, Burritt DJ, Tran L-SP, editors. The use of metabolomic quantitative trait locus mapping and osmotic adjustment traits for the improvement of crop yields under environmental stresses. *Seminars in cell & developmental biology*: Elsevier; 2018.
42. Sharma M, Gupta BB. Role of wild relatives for development of climate-resilient varieties. In Sharma, Manmohan, Punya, and Bharat Bhushan Gupta (Eds.), *Rediscovery of genetic and genomic resources for future food security:* Springer; 2020. p. 303–14.
43. Govindaraj M, Vetriventhan M, Srinivasan M. Importance of genetic diversity assessment in crop plants and its recent advances: an overview of its analytical perspectives. *Genetics Research International.* 2015;2015:431487.
44. Phillips T. Genetically modified organisms (GMOs): transgenic crops and recombinant DNA technology. *Nature Education.* 2008;1(1):213.
45. Baillo EH, Kimotho RN, Zhang Z, Xu P. Transcription factors associated with abiotic and biotic stress tolerance and their potential for crops improvement. *Genes.* 2019;10(10):771.
46. Babar U, Nawaz MA, Arshad U, Azhar MT, Atif RM, Golokhvast KS, et al. Transgenic crops for the agricultural improvement in Pakistan: a perspective of environmental stresses and the current status of genetically modified crops. *GM Crops & Food.* 2020;11(1):1–29.
47. Kumar J, Singh S, Singh M, Srivastava PK, Mishra RK, Singh VP, et al. Transcriptional regulation of salinity stress in plants: a short review. *Plant Gene.* 2017;11:160–9.
48. Agarwal PK, Agarwal P, Reddy M, Sopory SK. Role of DREB transcription factors in abiotic and biotic stress tolerance in plants. *Plant Cell Reports.* 2006;25(12):1263–74.
49. Xu ZS, Chen M, Li LC, Ma YZ. Functions and application of the AP2/ERF transcription factor family in crop improvement F. *Journal of Integrative Plant Biology.* 2011;53(7):570–85.
50. Sanghera GS, Wani SH, Hussain W, Singh N. Engineering cold stress tolerance in crop plants. *Current Genomics.* 2011;12(1):30.
51. Mallikarjuna G, Mallikarjuna K, Reddy MK, Kaul T. Expression of OsDREB2A transcription factor confers enhanced dehydration and salt stress tolerance in rice (Oryza sativa L.). *Biotechnology Letters.* 2011;33(8):1689–97.
52. Zhu X, Qi L, Liu X, Cai S, Xu H, Huang R, et al. The wheat ethylene response factor transcription factor pathogen-induced ERF1 mediates host responses to both the necrotrophic pathogen Rhizoctonia cerealis and freezing stresses. *Plant Physiology.* 2014;164(3):1499–514.
53. Zhai Y, Wang Y, Li Y, Lei T, Yan F, Su L, et al. Isolation and molecular characterization of GmERF7, a soybean ethylene-response factor that increases salt stress tolerance in tobacco. *Gene.* 2013;513(1):174–83.
54. Cao Y, Li K, Li Y, Zhao X, Wang L. MYB transcription factors as regulators of secondary metabolism in plants. *Biology.* 2020;9(3):61.

55. De Geyter N, Gholami A, Goormachtig S, Goossens A. Transcriptional machineries in jasmonate-elicited plant secondary metabolism. *Trends in Plant Science*. 2012;17(6):349–59.
56. Shin D, Moon S-J, Han S, Kim B-G, Park SR, Lee S-K, et al. Expression of StMYB1R-1, a novel potato single MYB-like domain transcription factor, increases drought tolerance. *Plant Physiology*. 2011;155(1):421–32.
57. Shakiba E, Edwards JD, Jodari F, Duke SE, Baldo AM, Korniliev P, et al. Genetic architecture of cold tolerance in rice (Oryza sativa) determined through high resolution genome-wide analysis. *PloS One*. 2017;12(3):e0172133.
58. Chen YH, Cao YY, Wang LJ, Li LM, Yang J, Zou MX. Identification of MYB transcription factor genes and their expression during abiotic stresses in maize. *Biologia Plantarum*. 2018;62(2):222–30.
59. Wu M, Liu H, Han G, Cai R, Pan F, Xiang Y. A moso bamboo WRKY gene PeWRKY83 confers salinity tolerance in transgenic Arabidopsis plants. *Scientific Reports*. 2017;7(1):1–16.
60. Lee H, Cha J, Choi C, Choi N, Ji H-S, Park SR, et al. Rice WRKY11 plays a role in pathogen defense and drought tolerance. *Rice*. 2018;11(1):1–12.
61. Yu Y, He L, Wu Y. Wheat WRKY transcription factor TaWRKY24 confers drought and salt tolerance in transgenic plants. *Plant Physiology and Biochemistry*. 2023;205:108137.
62. Sedeek KE, Mahas A, Mahfouz M. Plant genome engineering for targeted improvement of crop traits. *Frontiers in Plant Science*. 2019;10:114.
63. Salih II, Seddiq SH, Hashim SS, Dheeb BI, editors. Application of omics and proteomics in fungi. *AIP Conference Proceedings*: AIP Publishing LLC; 2022.
64. Campos NA, Panis B, Carpentier SC. Somatic embryogenesis in coffee: the evolution of biotechnology and the integration of omics technologies offer great opportunities. *Frontiers in Plant Science*. 2017;8:1460.
65. Keurentjes JJ, Koornneef M, Vreugdenhil D. Quantitative genetics in the age of omics. *Current Opinion in Plant Biology*. 2008;11(2):123–8.
66. Scott MF, Ladejobi O, Amer S, Bentley AR, Biernaskie J, Boden SA, et al. Multi-parent populations in crops: a toolbox integrating genomics and genetic mapping with breeding. *Heredity*. 2020;125(6):396–416.
67. Stapley J, Reger J, Feulner PG, Smadja C, Galindo J, Ekblom R, et al. Adaptation genomics: the next generation. *Trends in Ecology & Evolution*. 2010;25(12):705–12.
68. Rasheed A, Hao Y, Xia X, Khan A, Xu Y, Varshney RK, et al. Crop breeding chips and genotyping platforms: progress, challenges, and perspectives. *Molecular Plant*. 2017;10(8):1047–64.
69. Arif M, Jan T, Munir H, Rasul F, Riaz M, Fahad S, et al. Climate-smart agriculture: assessment and adaptation strategies in changing climate. In: Venkatramanan, V., Shah, S., Prasad, R. (eds) *Global Climate Change and Environmental Policy: Agriculture Perspectives*. 2020:351–77. Springer, Singapore. https://doi.org/10.1007/978-981-13-9570-3_12
70. Ortiz R. Leveraging agricultural biodiversity for crop improvement and food security. *Routledge handbook of agricultural biodiversity*: Routledge; 2017. p. 285–97.
71. Uffelmann E, Huang QQ, Munung NS, de Vries J, Okada Y, Martin AR, et al. Genome-wide association studies. *Nature Reviews Methods Primers*. 2021;1(1):59.
72. Sahu PK, Sao R, Mondal S, Vishwakarma G, Gupta SK, Kumar V, et al. Next generation sequencing based forward genetic approaches for identification and mapping of causal mutations in crop plants: A comprehensive review. *Plants*. 2020;9(10):1355.

73. Lee S, Abecasis GR, Boehnke M, Lin X. Rare-variant association analysis: study designs and statistical tests. *The American Journal of Human Genetics*. 2014;95(1):5–23.
74. Liu HJ, Yan J. Crop genome-wide association study: a harvest of biological relevance. *The Plant Journal*. 2019;97(1):8–18.
75. Challa S, Neelapu NR. Genome-wide association studies (GWAS) for abiotic stress tolerance in plants. Biochemical, physiological and molecular avenues for combating abiotic stress tolerance in plants: Elsevier; 2018. p. 135–50.
76. Fatemi F, Kianersi F, Pour-Aboughadareh A, Poczai P, Jadidi O. Overview of identified genomic regions associated with various agronomic and physiological traits in barley under abiotic stresses. *Applied Sciences*. 2022;12(10):5189.
77. Liu Y, Wang L, Mao S, Liu K, Lu Y, Wang J, et al. Genome-wide association study of 29 morphological traits in Aegilops tauschii. *Scientific Reports*. 2015;5(1):1–12.
78. Krishnamurthy S, Pundir P, Warraich AS, Rathor S, Lokeshkumar B, Singh NK, et al. Introgressed saltol QTL lines improves the salinity tolerance in rice at seedling stage. *Frontiers in Plant Science*. 2020;11:833.
79. Elango D, Xue W, Chopra S. Genome wide association mapping of epi-cuticular wax genes in Sorghum bicolor. *Physiology and Molecular Biology of Plants*. 2020;26(8):1727–37.
80. Mohammadi M, Xavier A, Beckett T, Beyer S, Chen L, Chikssa H, et al. Identification, deployment, and transferability of quantitative trait loci from genome-wide association studies in plants. *Current Plant Biology*. 2020;24:100145.
81. Mackay TF, Stone EA, Ayroles JF. The genetics of quantitative traits: challenges and prospects. *Nature Reviews Genetics*. 2009;10(8):565–77.
82. Powder KE. Quantitative Trait Loci (QTL) Mapping. *Methods Molecular Biology*. 2020;2082:211–29.
83. Dormatey R, Sun C, Ali K, Coulter JA, Bi Z, Bai J. Gene pyramiding for sustainable crop improvement against biotic and abiotic stresses. *Agronomy*. 2020;10(9):1255.
84. Hasan N, Choudhary S, Naaz N, Sharma N, Laskar RA. Recent advancements in molecular marker-assisted selection and applications in plant breeding programmes. *Journal of Genetic Engineering and Biotechnology*. 2021;19(1):1–26.
85. Liu S, Qin F. Genetic dissection of maize drought tolerance for trait improvement. *Molecular Breeding*. 2021;41(2):1–13.
86. Wu F, Yang J, Yu D, Xu P. Identification and validation a major QTL from "Sea Rice 86" seedlings conferred salt tolerance. *Agronomy*. 2020;10(3):410.
87. Mwadzingeni L, Shimelis H, Dube E, Laing MD, Tsilo TJ. Breeding wheat for drought tolerance: progress and technologies. *Journal of Integrative Agriculture*. 2016;15(5):935–43.
88. Abdallah NA, Prakash CS, McHughen AG. Genome editing for crop improvement: challenges and opportunities. *GM Crops & Food*. 2015;6(4):183–205.
89. Wijerathna-Yapa A, Pathirana R. Sustainable agro-food systems for addressing climate change and food security. *Agriculture*. 2022;12(10):1554.
90. Pandey P, Irulappan V, Bagavathiannan MV, Senthil-Kumar M. Impact of combined abiotic and biotic stresses on plant growth and avenues for crop improvement by exploiting physio-morphological traits. *Frontiers in Plant Science*. 2017;8:537.
91. Manghwar H, Li B, Ding X, Hussain A, Lindsey K, Zhang X, et al. CRISPR/Cas systems in genome editing: methodologies and tools for sgRNA design, off-target evaluation, and strategies to mitigate off-target effects. *Advanced Science*. 2020;7(6):1902312.
92. Bhatia S, Yadav SK. CRISPR-Cas for genome editing: Classification, mechanism, designing and applications. *International Journal of Biological Macromolecules*. 2023:124054.

93. Zhang Y, Showalter AM. CRISPR/Cas9 genome editing technology: a valuable tool for understanding plant cell wall biosynthesis and function. *Frontiers in Plant Science*. 2020;11:589517.
94. Zhang H, Li Y, Zhu J-K. Developing naturally stress-resistant crops for a sustainable agriculture. *Nature Plants*. 2018;4(12):989–96.
95. Shahriar SA, Islam MN, Chun CNW, Rahim MA, Paul NC, Uddain J, et al. Control of plant viral diseases by CRISPR/Cas9: resistance mechanisms, strategies and challenges in food crops. *Plants*. 2021;10(7):1264.
96. El-Mounadi K, Morales-Floriano ML, Garcia-Ruiz H. Principles, applications, and biosafety of plant genome editing using CRISPR-Cas9. *Frontiers in Plant Science*. 2020;11:56.
97. Anwar A, Kim J-K. Transgenic breeding approaches for improving abiotic stress tolerance: recent progress and future perspectives. *International Journal of Molecular Sciences*. 2020;21(8):2695.
98. Zhang J-B, Wang X-P, Wang Y-C , Chen Y-H, Luo J-W, Li D-D, et al. Genome-wide identification and functional characterization of cotton (Gossypium hirsutum) MAPKKK gene family in response to drought stress. *BMC Plant Biology*. 2020;20(1):1–14.
99. Klap C, Yeshayahou E, Bolger AM, Arazi T, Gupta SK, Shabtai S, et al. Tomato facultative parthenocarpy results from Sl AGAMOUS-LIKE 6 loss of function. *Plant Biotechnology Journal*. 2017;15(5):634–47.
100. Zeng Y, Wen J, Zhao W, Wang Q, Huang W. Rational improvement of rice yield and cold tolerance by editing the three genes OsPIN5b, GS3, and OsMYB30 with the CRISPR–Cas9 system. *Frontiers in Plant Science*. 2020;10:1663.

8 Sweetpotato and Climate Resilience

Strategies and Prospects

Wei Tang, Dandan Wang,
Muhammad Qadir Ahmad, and Qiang Li

8.1 INTRODUCTION

Sweetpotatoes have a long history, originating in the tropical regions of Peru and evolving from wild types. They are primarily found in Asia, Africa, and South America [1,2]. Sweetpotato is a vital food and cash crop, accounting for approximately one-third of the total global food crop production, and ranking as the fourth largest food crop after wheat, corn, and rice in China [3]. The storage roots of sweetpotato can be processed into starch, sucrose, and alcohol, serving as essential raw materials in the food industry. They are rich in nutrients, including starch, pectin, cellulose, calcium, iron, potassium, phosphorus, polyphenols, flavonoids, carotenoids, and B vitamins [4,5]. Sweetpotatoes also exhibit strong resistance to drought and waterlogging, making them well-suited for cultivation in dry and arid areas. Their cultivation has low soil requirements and minimal dependence on fertilizers, contributing to improved soil fertility and making them an ideal green crop [6]. Cultivating sweetpotato can raise farmers' income, enhance living standards, and foster rural economic development. Furthermore, sweetpotato cultivation benefits ecological protection by reducing the use of chemical pesticides, lowering environmental pollution, and promoting sustainable agriculture. Sweetpotato's versatility and resilience make it an essential crop for future food security in the face of climate change. Ongoing research and development will further enhance its potential, paving the way for innovative agricultural practices and improved global food systems.

8.2 CLIMATE CHANGE AND CHALLENGES TO SWEETPOTATO PRODUCTION

Over the past few decades, rising global temperatures have had different impacts on different regions of the world [7]. Studies have shown that since the 1960s, the global average temperature has risen more than in any other period in the past 100–200 years. The rise in global average temperature is related to the El Niño phenomenon [8–10]. The increase in global average temperature has led to more severe droughts. This warming has also caused an increase in the frequency and intensity of extreme weather events [11–13].

DOI: 10.1201/9781032720142-10

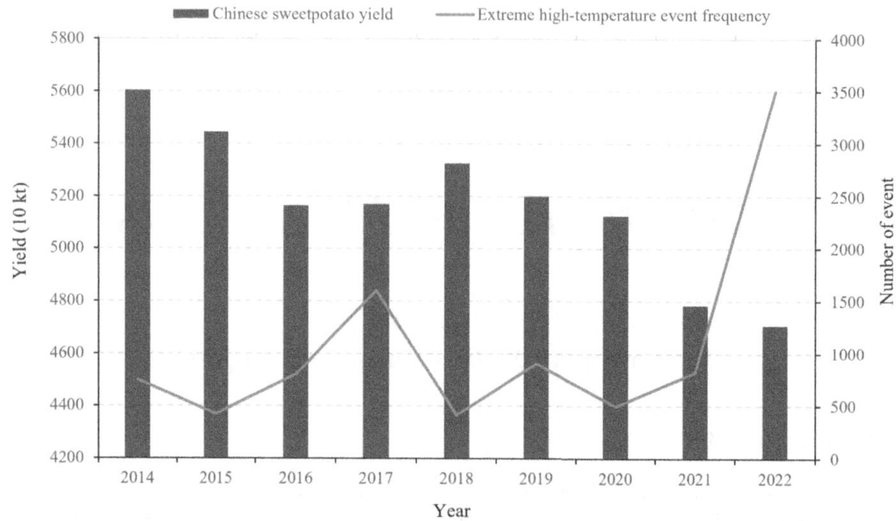

FIGURE 8.1 Extreme high temperature and sweetpotato production from 2014 to 2022 in China.

For sweetpotato production, the main impact of global climate change on its production is to reduce sweetpotato yield. With the frequent occurrence of extreme high-temperature events, the average annual sweetpotato yield declined by 8.94% from 2014 to 2022 in China (Figure 8.1). Similarly, in Uganda, the average annual yield decrease was 4.35% from 1960 to 2015 and 10.28% from 2010 to 2015 [14]. At present, the impact of global climate change on sweetpotato production is mainly assessed through observation data and model simulation. In the 1950s, there was a clear correlation between temperature and precipitation changes over the previous 60 years in the United States and Canada; after the 1960s, however, temperature and precipitation changes showed the opposite trend [15,16]. In addition, temperature, precipitation, and other climatic factors in China from 1982 to 2009 increased and decreased significantly, but the trend of temperature increase and precipitation decrease was not obvious in the past 20 years [17]. Although the current global climate change has caused a certain degree of adverse impacts on sweetpotato production, compared with other food crops, sweetpotato is less affected by climate change. The adverse effects of current climate change on sweetpotato production will be even greater if measures are not taken to address them. Therefore, it is critical to study the effects of global climate change on the growth, development, and agricultural production of sweetpotato.

8.3 IMPACT OF ABIOTIC STRESSES ON SWEETPOTATO

Abiotic stresses (drought, heat, salinity, and flooding) are significant threats to agricultural production worldwide (Figures 8.2 and 8.3). With the dramatically increasing

FIGURE 8.2 Extreme drought resulted in wilted sweetpotato plants.

global population and worsening environmental conditions [18], the development of highly nutritious, stress-tolerant crops is direly needed to ensure a sustainable future. Sweetpotato is highly valued for its adaptability, high yield potential, and rich nutrient contents, making it a vital resource for securing future food supplies.

Abiotic stresses can have a significant impact on sweetpotato production. Among these stresses, temperature variations have more adverse effects on plant development and crop production [19,20]. Sweetpotato is highly sensitive to temperature fluctuations. Both high and low temperatures significantly affect sweetpotato yield and biomass. Extreme temperatures regulate reactive oxygen metabolism leading to many cellular and physical changes such as oxidative stress, cell membrane lipid peroxidation, denaturation of proteins, and nucleotide damage [21]. When plants are subjected to heat stress, excess excitation energy is produced, which leads to the production of reactive oxygen species (ROS) and potential damage to the photosystem [22]. *Ipomea batatas* orange protein (IbOrP) from sweetpotato is predominantly localized in the nucleus. However, it may be found in chloroplast under high temperature, and overexpressing *IbOr* causes carotenoids accumulation and maintenance of higher photosystem II (PSII) efficiency [23]. IbOr also interacts with IbPsbP to regulate the stability of IbPsbP and finally stabilization of PSII [24]. Transcriptomic and proteomic analysis showed that 47 DEGs/DEPs were consistently expressed under heat stress, which involved the processing of proteins in the endoplasmic reticulum, secondary metabolites, starch, and cellulose metabolism [25]. Sweetpotato has tropical and subtropical origins, therefore, it is vulnerable to low temperatures. Cold stress significantly hinders sweetpotato production and cultivation [26]. Many low-temperature-responsive genes will change at low temperatures, enabling the sweetpotato to undergo a variety of physiological and biochemical reactions [21, 27].

Among the mechanisms involved in response to cold stress, APETALA2/Ethylene response factor (AP2/ERF) has been extensively studied. It is known as the C-repeat-binding factor/dehydration-responsive element-binding factor (CBF/DREB). It regulates cold-regulated (*COR*) genes transcription [28]. The ectopic expression of *IbCBF3* confers chilling stress by affecting several *COR* genes, such as cold-regulated 15A (*COR15A*), responsive to dehydration 29A (*Rd29A*), and COLD-RESPONSIVE 6.6 (*KIN2*) [29]. Mechanisms involved in the response to cold storage have been studied by transcription profiling and revealed that long-term exposure to low temperature results in a series of physiological and biochemical changes [30–32]. To ensure that sweetpotato can maintain a certain yield and quality under extreme temperatures, molecular mechanisms involved in the response to low temperature should be further explored.

Drought stress results in serious physiological changes such as affecting leaf water content, osmotic protectants, chlorophyll content, antioxidants (enzymatic and non-enzymatic), membrane peroxidation, and the production of some other primary and secondary metabolites. These changes can lead to wilting of plants, growth retardation, discoloration, abnormal ripening, and other issues (Figure 8.3) [33–35]. Sweetpotato stands as a drought-tolerant crop, even though extreme drought can cause significant production losses. About 30% yield losses have been observed under drought stress in sweetpotato in various sweetpotato-growing regions [36, 37]. Drought tolerance can be improved by genetic engineering methods. The expression of certain enzyme

FIGURE 8.3 Shriveled sweetpotato due to severe drought stress.

genes at the molecular level, such as *XySap*, *SoBADH*, and *AtHDG1*, improves drought tolerance in sweetpotato [38–40]. Similarly, overexpression of structural and endogenous enzyme genes, such as *IbC4H*, *IbNHX2*, *IbBT4*, and *IbARF5* [41–43], or transcription factors, such as *IbbZIP1*, *IbABF4*, *IbMYB116*, and *IbNAC3* [44–47], or microRNAs (miRNAs), such as *IbmiR319a* [48], results in enhanced drought tolerance in transgenic sweetpotato plants. Therefore, the identification of drought-tolerant genes and further utilization of these genes in breeding programs is of great importance to improve sweetpotato quality and yield.

Soil salinization induced by NaCl is a serious threat worldwide which restricts crop reproduction and production. High salt levels affect various plant mechanisms such as photosynthesis, protein synthesis, signal transduction, and lipid metabolism through ionic stress and osmotic stress. Sweetpotato is moderately tolerant to salinity, but high salt concentrations in the soil can impact its growth, root and shoot weight, antioxidant activity, and health-promoting compounds. Therefore, understanding the salt tolerance mechanism in sweetpotato is very important for breeding salt-tolerant varieties to utilize marginal and saline lands. Overexpressing *IbCAR1* in sweetpotato upregulates the genes involved in the biosynthesis of ABA, stress response, and the ROS-scavenging system, enhancing the plants' ability to tolerate salt stress, whereas RNAi of *IbCAR1* significantly increased the plants' sensitivity to salt stress [49].

Overexpression of *IbBBX24* improved the ability of transgenic sweetpotato to resist salt stress. This mechanism involves IbBBX24 directly binding to the promoter of the *IbPRX17* gene, regulating its expression and improving peroxidase activity. Additionally, IbTOE3 can interact with IbBBX24 to enhance its ability to activate *IbPRX17* transcription, thereby increasing salt tolerance in sweetpotato [50]. High salt induces the expression of the *IbWRKY2* gene, and heterologous expression of *IbWRKY2* in *Arabidopsis thaliana* increased ABA, proline content, and SOD activity in transgenic plants, thereby improving their resistance to salt stress [51]. Overexpression of *IbC3H18* increased the expression of ROS clearance genes, ABA signaling pathway-related genes, and ion transporter genes in transgenic sweetpotato plants to maintain their normal growth under high salt stress [52]. Although there have been many studies on the mining of salt-tolerance genes, little is known about the molecular mechanism of the salt response in sweetpotato.

In addition to drought, extreme temperature, and high salt, sweetpotato also faces the threat of flooding. Flooding stress ceases growth and can result in early death of plants by reducing the exchange of gases between the atmosphere and plant cells [53]. Under water-logged environments, the oxygen diffusion rate reduces by 10,000-folds compared with normal conditions [54]. Therefore, to cope with flooding stress, metabolic shifts by sensing oxygen depletion and scavenging reactive oxygen species (ROS) are very important. Despite being tolerant to other abiotic stresses, sweetpotato is vulnerable to flooding stress [55]. A reduction by 57% in storage yield of sweetpotato was observed under mid-season flooding [56]. Additionally, under hypoxia conditions, morphogenesis of sweetpotato storage roots was also changed [57]. By comparing the flooding tolerance in different sweetpotato cultivars, Park et al. [58] found that Yeonjami (YJM) cultivar was the most flooding tolerant, and it showed high ascorbate peroxidase activity in leaves, high stability in membranes, and strong antioxidant production. Furthermore, several genes involved in flooding tolerance were related to sucrose and starch metabolism, fermentation, and loosening of the cell wall, depicting a high transcript level and early induction. Presently, only a few studies on the mechanism of the sweetpotato response to flooding stress are available.

The impact of these abiotic stresses on sweetpotato can vary depending on the duration and intensity of the stress, as well as the specific variety and its inherent tolerance. By identifying and selecting for traits associated with heat, drought, salinity, and flooding tolerance, breeders aim to develop climate-resilient sweetpotato varieties that can withstand and adapt to challenging environmental conditions. Additionally, agronomic practices such as proper irrigation management, soil improvement, and crop rotation can also help mitigate the impacts of these stresses on sweetpotato production.

8.4 CLIMATE CHANGE AND DISEASE OCCURRENCE

Sweetpotato growth is inhibited not only by abiotic stresses, such as high- and low-temperature stress, salt stress, flooding, and drought stress, but also by biotic stresses such as viruses, fungi, bacteria, and stem nematodes. Climate change can significantly impact the occurrence and spread of opportunistic pathogens and insects, seriously affecting the yield and quality of the crop. Rising temperatures, changes in precipitation patterns, and increased humidity can create a more conducive environment for the development and proliferation of diseases. This can lead to increased disease pressure on sweetpotato. To mitigate the impacts of climate change on disease occurrence and pest outbreaks, integrated pest management (IPM) strategies can be employed [59]. IPM involves combining various approaches, including biological control, cultural practices, and the use of pesticides, to manage diseases and pests effectively. Additionally, breeding programs can focus on developing sweetpotato varieties with enhanced resistance or tolerance to diseases and pests, as well as improved resilience to changing environmental conditions. Preliminary progress has been made in exploring disease resistance genes and regulatory mechanisms of sweetpotato under various biotic stresses.

Sweetpotato black rot severely impairs sweetpotato production and post-harvest storage. Overexpression of the α-hordothionin (*α-HT*) gene of barley in sweetpotato showed a decrease in the degree of yellowing of transgenic sweetpotato after infection, as well as a reduction in the lesion area around sites inoculated with *Ceratocystis fimbriata* spores in the storage roots [60]. The application of endophytic *Bacillus amyloliquefaciens* YTB1407 accelerated resistance to root rot disease (*Fusarium solani* Mart. Sacc.f. sp. *batatas* McClure) and black rot disease (*C. fimbriata* Ell. & Halst) by activating the expression of the salicylic acid (SA)-responsive *PR-1* gene, increasing the SA content and decreasing the peroxide (H_2O_2) content in sweetpotato [61]. *Fusarium* wilt causes the leaves to turn yellow and fall off, the stem vascular bundles to turn brown, and eventually die, resulting in a 10–50% reduction in sweetpotato yield [62]. Comparative transcriptome analysis showed that a large number of differentially expressed genes (DEGs) were concentrated in the SA and jasmonic acid (JA) pathways. *IbMAPKK9* influences plant resistance to *Fusarium oxysporum* by mediating the SA signaling pathway [63,64]. Similarly, overexpression of *IbBBX24* increased the JA content in transgenic sweetpotato and significantly improved resistance to *Fusarium* wilt. The mechanism is as follows: *IbBBX24* regulates the JA response by antagonizing the JA signaling repressor IbJAZ10, thereby alleviating the inhibition of the JA signaling activator IbMYC2 by IbJAZ10 [65]. Sugar plays an important role in plant–pathogen interactions. Overexpression of *IbSWEET10* significantly reduced the sugar content and increased the resistance of transgenic sweetpotato to *Fusarium* wilt [66]. Additionally, *IbWRKY7*, *IbERF1*, and *IbPIF3.1* have also been reported to be involved in the response to sweetpotato wilt [67–69].

Sweetpotato stem nematode disease significantly affects the production and quality of sweetpotato, often resulting in 20–50% yield losses or complete crop failure, posing a major threat to sweetpotato production [70]. Overexpression of Oryza cystatin-I (*OCI*) enhances the resistance of transgenic sweetpotatoes to stem nematode [71,72]. Similarly, the overexpression of the *IbMIPS1* gene enhanced stem nematode resistance significantly in transgenic sweetpotato by upregulating genes involved in phosphatidylinositol (PI), inositol biosynthesis, and ABA signaling pathways, photosynthesis, stress response, and ROS clearance systems [73]. Despite the significant impact of sweetpotato stem nematode disease, there are few studies on the mechanism of resistance to sweetpotato stem nematode disease. In the future, it will be crucial to focus on the identification of insect resistance genes.

Sweetpotato pests represent a significant challenge to its normal growth, and affect the yield and quality. Among these pests, the sweetpotato weevil (SPW) stands out as one of the most formidable adversaries in sweetpotato production, particularly in tropical and subtropical regions [74]. Adult weevils feed on both the leaves and roots of sweetpotato, while larvae primarily target the stem base and root storage, resulting in a substantial decrease in the quality and edibility of sweetpotato. Liu et al. [67] identified two sweetpotato germplasm lines with high levels of resistance to sweetpotato weevils (SPWR) and uncovered two major SPW resistance genes, *SPWR1* and SPWR2, through map-based cloning. These genes play crucial roles as natural defense regulators against SPWs. The SPW-induced WRKY transcriptional factor SPWR1 directly activates the expression of *SPWR2*. Subsequently, *SPWR2* promotes the accumulation of quinate derivative metabolites, which confer SPW resistance in sweetpotato.

Sweetpotato viruses represent the most devastating disease in sweetpotato cultivation. Most sweetpotato viruses manifest in complex forms, often involving two or more viral strains, with single-form infections being rare. Influenced by changes in the ecological environment, these severe viral diseases have led to significant yield losses worldwide [75,76]. However, our current understanding of sweetpotato's molecular response to viruses remains limited. Li et al. [77] employed short tandem target mimic (STTM) technology to enhance the expression of laccase (*IbLACs*) by suppressing miR397, thereby enhancing sweetpotato resistance to sweetpotato viral disease. With the escalation of global climate change, the prevalences of sweetpotato fungi, nematodes, insects, viruses, and bacteria are on the rise, severely constraining yield and variety improvement in sweetpotato. Breeders must accelerate research into sweetpotato's response mechanism to biotic stress, and expedite the breeding of stress-resistant and superior varieties to adapt to changing climatic conditions.

8.5 GENETIC RESOURCES AND PROSPECTS OF CONVENTIONAL/PRE-BREEDING APPROACHES

Sweetpotato genetic resources are unique assets selected and improved over time from natural resources. In sweetpotato breeding, it is important to collect, preserve, and utilize these germplasms. Sweetpotato germplasm resources, encompassing both wild and cultivated species, hold significant value due to their genetic variation. At present, the collection and preservation of sweetpotato germplasm resources primarily involve artificial hybridization, molecular markers, and systematic breeding. Artificial hybridization involves selecting excellent individual plants from wild species with a certain genetic basis and potential for variation. These selected plants, suitable as breeding parents, are hybridized, and the hybrid seeds are then cultivated and propagated in the field. Molecular markers, on the other hand, involve the transfer of specific genes to other plants using biotechnological methods for assisted breeding. This approach allows for precise genetic improvements and the development of new sweetpotato varieties with desirable traits. Collecting and preserving sweetpotato germplasm resources is a challenging task, requiring significant attention and effort. Therefore, it is essential to prioritize the collection and preservation of these valuable genetic resources. Ensuring the conservation of sweetpotato germplasm is crucial for maintaining genetic diversity, which is vital for the continued improvement and resilience of sweetpotato varieties in response to environmental and biotic stresses.

Conventional breeding approaches have been used for centuries to improve crop plants, including sweetpotato. These approaches involve selecting and crossing plants with desirable traits to create new varieties. Parent selection is a conventional breeding approach that relies on visually assessing the physical characteristics, or phenotypes, of plants. Breeders select plants with desired traits, such as high yield, disease resistance, or improved quality, and use them as parents for the next generation. This approach has been successful in improving sweetpotato varieties with traits like high yield, pest resistance, and desirable root characteristics. Li et al. [78] analyzed the genetic diversity of 62 main sweetpotato parents in China using ISSR markers, revealing a rich genetic diversity among these parents, and emphasized that in future sweetpotato breeding in China, combinations of domestic and externally introduced parents, as well as Asian and non-Asian externally introduced parents, can be utilized to broaden the genetic background of Chinese sweetpotato varieties. Rahajeng et al. [79] evaluated 15 sweetpotato clones for higher root yield and dry matter content by phenotypic selection. In order to explore the breeding methods for edible sweetpotato varieties and speed up the breeding process of excellent edible sweetpotato varieties, Xie et al. [80] analyzed the main characteristics and parental lineage of 22 edible sweetpotato varieties bred from 2001 to 2020 and emphasized that introduced resources and innovative germplasm are indispensable parental sources for sweetpotato breeding, and directed hybridization is the primary method for breeding excellent edible sweetpotato varieties.

Hybridization involves crossing two genetically distinct parents to create offspring with a combination of their traits. In sweetpotato breeding, hybridization can be used to introduce novel traits or combine desirable traits from different varieties. By carefully selecting the parents, breeders can create hybrids with improved yield, disease resistance, and other important traits. Nowadays, according to the domestic demand of China, we at Xuzhou Institute of Agricultural Sciences have developed different types of sweetpotato varieties including high starch cultivars Shangshu19 and Xushu37, table used cultivars Yanshu25 and Pushu 32, and purple flesh cultivars Xuzishu8 and Jihei1. In Japan, researchers have also bred different cultivars suitable for table use, food processing, and ethanol and starch production such as Quick Sweet and Beniharuka, food-processing cultivars Murasakimasari and Ayakomachi, and high-starch cultivars Konahomare and Daichinoyume [81].

In conclusion, conventional breeding approaches offer a promising prospectus for sweetpotato improvement. By utilizing the genetic diversity within sweetpotato and employing traditional breeding techniques, breeders can develop new varieties with improved traits, resilience to climate change, and enhanced productivity. Conventional breeding also allows for farmer involvement and is a relatively efficient and cost-effective approach to sweetpotato improvement.

8.6 GENETIC RESOURCES AND PROSPECTS OF MODERN BREEDING APPROACHES

Genetic resources play an important role in addressing the threats posed by climate change and improving sweetpotato production. Sweetpotato possesses a rich genetic diversity that can be harnessed through modern breeding approaches.

Using tissue culture technology, genetic information in plant tissues can be preserved under clonal conditions. At present, tissue culture technology has been widely used in plant somatic hybridization, embryogenesis, and protoplast culture. For example, enhancing and accelerating somatic embryogenesis from stem segments of the Egyptian sweetpotato cultivar Abees was investigated using three different phytohormones: 6-benzyleaminopurine (6-BA), 2,4-dichlorophenoxyacetic acid (2,4-D), and indole acetic acid (IAA). The phytohormone 6-BA appeared as the best for the induction of embryogenic calli. Additionally, Beyene et al. [82] optimized a protocol for in vitro propagation of sweetpotato variety Kulfo using 6-BA and indole-3-butyric acid (IBA) phytohormones for shoot multiplication and rooting, respectively. The main process of tissue culturing involves isolating protoplasts from tender parts of plants, and then callus is obtained by callus induction technology and transformed into plants. However, this process encounters several challenges such as low callus regeneration rates and a tendency for regenerated seedlings to undergo cell death. In order to improve the regeneration rate of sweetpotato, researchers have started to use cell engineering technology to solve these problems [83,84].

Molecular markers are a powerful tool in molecular biotechnology, used for genetic diversity analysis, variety identification, and purity detection of sweetpotato germplasm resources. In sweetpotato breeding, molecular markers are primarily applied through DNA fingerprint and marker-assisted selection (MAS). DNA fingerprinting is used to analyze the genetic background of sweetpotato to determine the genetic relationship. For instance, a study screened 1294 SSR primer pairs, including 1215 genomic–SSR and 79 expressed sequence tag (EST)–SSR primer pairs, with sweetpotato cultivars Zhengshu 20 and Luoxushu 8 and their F1 progeny. From this screening, 273 and 38 primer pairs generated polymorphic bands, respectively [85]. Similarly, Gao et al. [86] employed 124 SSR primers to determine the genetic diversity among 76 varieties of purple-fleshed sweetpotatoes.

Quantitative trait locus (QTL) refers to the location of genes that control quantitative traits on the chromosome. QTL mapping is the process of detecting the linkage between molecular markers and QTL, estimating the genetic effects of QTL, and then locating individual QTL on specific chromosomes to assist in offspring selection. Due to its polyploid nature, many important agronomic traits of sweetpotato, such as fresh yield and starch content, are controlled by multiple genes. Notably, a significant negative correlation was observed between fresh yield and starch content [87]. Therefore, it is necessary to use molecular marker-assisted selection in sweetpotato breeding, and the important prerequisite for marker-assisted breeding is the localization of important trait QTL.

Currently, a large number of QTL sites related to sweetpotato quality traits, yield, and resistance have been identified. Cervantes-Flores et al. [88] located 12 QTLs for starch content, explaining 17–30% of the phenotypic variation, 13 QTLs for dry matter content, explaining 15–24% of the phenotypic variation, and eight QTLs for β-carotene content, explaining 17–35% of the phenotypic variation. Chang et al. [89] located 14 and nine QTLs related to sweetpotato yield in the F1 (Nanshuai Tiao ♀×Tainong 27 ♂) and F2 (Tainong 27 ♀×Nanshuai Tiao ♂) populations, explaining 14.1–29.7% and 16.0–29.9% of the phenotypic variation, respectively. Sasai et al. [90] discovered a candidate QTL region for high resistance to root-knot nematodes in the Ib01_6 linkage group of J-Red, contributing 29.6–36.7%. Meng et al. [91] conducted QTL mapping for dry matter content in sweetpotato. Two QTLs were identified in the cultivar Luoxu Shu 8, explaining 18.7% and 45.8% of the phenotypic variation, both showing negative effects. In the cultivar Zheng Shu 20, four QTLs were identified, explaining 24.5–29.7% of the phenotypic variation, with two QTLs showing positive effects.

Ma et al. [92] detected a total of seven QTLs related to root rot resistance, with five QTLs located on the fifth linkage group of the Jizishu 1 map, explaining 52.6–57.0% of the variation, and two QTLs located on two linkage groups of Longshu 9, explaining 57.6% and 53.6% of the variation, respectively. Using the composite interval mapping method, Ma et al. [93] located one QTL for branch number, explaining 53.2% of the total phenotypic variation. Similarly, one QTL for stem vine diameter, explaining 16.7% of the phenotypic variation, two QTLs for longest vine length, explaining 9.5% and 13.7% of the phenotypic variation, two QTLs for petiole length, explaining 8.8% and 11.3% of the phenotypic variation, and five QTLs related to internode length, explaining 9.6–28.1% of the phenotypic variation were identified. Compared with traditional breeding methods, molecular marker-assisted selection has many advantages such as high efficiency, high precision, convenience for large-scale planting, and rapid propagation. Additionally, molecular markers play an important role in the screening and identification of genes present in different germplasm lines. Kim et al. [94] developed molecular markers to distinguish four varieties of sweetpotato ('Beniharuka,' 'Annobeni,' 'Pungwonmi,' and 'Hogammi') using SNPs and genotyping-by-sequencing (GBS) analysis via a tetra-primer amplification refractory mutation system (ARMS)-PCR. Furthermore, molecular markers can also be used to identify purity and the genetic relationship in sweetpotato germplasm resources to provide a scientific basis for efficient utilization of sweetpotato germplasm resources. Wadl et al. [95] determined the population structure and genetic diversity of 417 USDA sweetpotato accessions originating from eight broad geographical regions using single nucleotide polymorphisms and GBS method. At present, molecular marker-assisted selection has been widely used in sweetpotato breeding in China.

Genetic engineering involves the introduction of specific genes into sweetpotato to confer desired traits. This approach can be used to enhance traits such as pest and disease resistance, drought tolerance, and nutritional quality. Genetic engineering offers the potential to rapidly introduce novel traits into sweetpotato by transgene or gene-editing technology. Kang et al. [96] transformed *IbOr* and *IbCHY-β* into sweetpotato, which not only significantly increased the carotenoid and beta-carotenoid contents, but also enhanced the heat and salt tolerance of sweetpotato. Kim et al. [97,98] transformed *IbTC*, *IbHPPD*, and *IbPD* genes into sweetpotato, respectively. The results showed that overexpression of *IbTC* and *IbHPD* transgenic lines increased tolerance to various environmental stresses (salt, drought, and oxidative stress) in addition to increasing α-tocopherol, while overexpression of *IbOr-R96H* increased carotenoid content and heat tolerance.

The CRISPR/Cas9 system is currently the most widely used gene-editing technology, consisting of a series of sgRNAs and the Cas9 protein. This system has the advantages of being easy to operate, precise in targeting, capable of editing multiple species, and free from the introduction of exogenous genes. However, there has been relatively little research on using the CRISPR/Cas system to edit genes in sweetpotato. Wang et al. [99] designed two sgRNA targets for the starch biosynthesis pathway genes *IbGBSSI* and *IbSBEII*, and constructed single and double sgRNA target CRISPR/Cas9 vectors for sweetpotato transformation. Analysis of the transgenic lines showed that the CRISPR/Cas9 system achieved good gene-editing effects in the sweetpotato genome, with a mutation rate of over 63%. This represents the first successful editing of sweetpotato starch synthesis genes using CRISPR/Cas9, improving the quality of the starch. This paves the way for future molecular breeding of sweetpotato and gene editing of polyploid crops.

In conclusion, genetic resources and modern breeding approaches are essential for developing climate-resilient sweetpotato varieties. By harnessing the genetic diversity within sweetpotato germplasm collections and using modern breeding techniques, breeders can develop varieties that are better adapted to the challenges posed by climate change, ultimately improving sweetpotato production and food security.

8.7 PHENOMICS OR AI-BASED TECHNOLOGIES

Phenomics is a comprehensive and detailed quantitative analysis of plant phenotypic features using a high-throughput plant phenotyping platform, sensing technologies, and computational techniques [100]. It covers a variety of characteristics of plant growth, development, morphology, physiology, and behavior by collecting extensive phenotypic data and using computational methods and statistical analysis, phenomics support to researchers in understanding plant phenotypic changes, exploring plant biological responses under diverse environmental conditions, identifying phenotypic characteristics linked to important agronomic traits, and conducting genetic studies [101]. For example, Yang et al. [102] established the High-throughput Rice Phenotyping Facility, which integrates color imaging, X-ray CT, automated controls, and an image analysis pipeline, to monitor a minimum of 15 agronomic traits in populations of up to 1920 rice plants. Wu et al. [103] integrated a high-throughput micro-CT-RGB imaging system with genome-wide association studies (GWAS) to assess rice tiller and tiller growth traits, along with plant traits across nine growth stages. This approach led to the identification of 402 significantly associated loci.

Artificial intelligence (AI) is the process of processing, analyzing, and modeling large-scale and complex data using computer systems that simulate human intelligence processes. These systems enable tasks similar to those performed by human intelligence. Through the data analysis of AI, researchers can find patterns and laws within data, make predictions, classify information, and extract useful insights. In the field of agriculture, AI data analysis can help breeders speed up the breeding process, optimize crop planting management, predict plant diseases and insect pests, and more. Parmley et al. [104] developed predictive models for in-season seed yield prediction, facilitating the integration of phenomics data with advanced machine learning for predictive and prescriptive plant breeding. Sandhu et al. [105] confirmed that deep learning (DL) approaches achieved superior prediction accuracy for various traits and recommended their integration into a plant breeder's toolkit for application in large-scale breeding programs.

Although there has not been any report about the application of phenomics or AI-based technologies on sweetpotato, they will surely be the direction of future development to improve breeding and production in the face of climate change. With AI technology, big data analysis and machine learning algorithms can be used to speed up the process of selecting and breeding sweetpotato. First, a large amount of sweetpotato image data is collected from different growth stages and varieties. These images can include the growth status of sweetpotato, leaf conditions, pest and disease conditions, and mature yields. Then, the collected image data are labeled, marking the growth characteristics, pest and disease conditions, and yield information of sweetpotato to provide training data for subsequent machine learning algorithms. Next, machine learning and deep learning technology are used to establish an image recognition model, such as a convolutional neural network (CNN), to extract features and identify sweetpotato images. By training the model, it can accurately identify the growth status, pest and disease conditions, and yield information of sweetpotato. Finally, the trained image recognition model is applied to real-time monitoring in the field, using cameras or other image-capture devices to obtain images of sweetpotato and identify the growth status, pest and disease conditions, and yield information through the model.

On the other hand, we need to obtain large-scale sweetpotato genomic data, which can be obtained through high-throughput sequencing technologies such as whole-genome sequencing or RNA sequencing to obtain sweetpotato genome sequences and expression data. Then, bioinformatics tools and statistical methods are used to perform QTL analysis on the preprocessed genomic data. This includes finding associated trait loci on the genome and determining their correlation with phenotypic features. Functional annotation of the genes identified in the QTL, screening out candidate genes related to the target traits, can assist in breeding selection. Finally, machine learning algorithms are used, combined with sweetpotato phenotype and genotype data, to predict the genetic performance of sweetpotato materials, helping to select the most promising parents for hybridization and accelerate sweetpotato breeding. The application of these technologies can greatly shorten the breeding cycle, improve breeding efficiency, and bring new development opportunities to sweetpotato breeding.

Overall, although there is no report about phenomics and AI-based technologies used in sweetpotato research, they offer promising opportunities to improve sweetpotato breeding and production in the face of climate change. By leveraging these tools, breeders and farmers can develop and manage sweetpotato varieties that are more resilient, productive, and sustainable.

8.8 FUTURE PROSPECTS

In recent years, the increased frequency, intensity, and prolonged duration of extreme weather events globally have posed significant risks and challenges to the sweetpotato industry. Future sweetpotato breeding efforts need to focus on several key directions. First, leveraging modern biotechnology, genetic engineering, and gene-editing techniques, among other molecular biology methods, to conduct genetic improvement research targeting key traits such as high yield, quality, and resistance to biotic and abiotic stresses. The goal is to develop specialized varieties suitable for production in different regions, with traits including drought and water-logging tolerance, high quality and yield, and broad adaptability. Second, employing molecular marker-assisted selection methods such as whole-genome selection or high-throughput sequencing to enhance existing varieties and improve sweetpotato cultivation efficiency. Additionally, utilizing advanced sweetpotato cultivation techniques tailored to different types of sweetpotato varieties can improve overall production efficiency and provide economic benefits to the industry. Furthermore, there is a need to strengthen the collection, preservation, and utilization of sweetpotato germplasm resources to better explore potential genetic diversity and provide more options and possibilities for breeding. Ultimately, the sweetpotato industry should collaborate with governments, research institutions, and international organizations to collectively address the challenges posed by climate change and drive the sustainable development of sweetpotato industry.

REFERENCES

[1] Winnicki E, Kagawa-Viviani A, Perez K, et al. Characterizing the diversity of Hawai'i sweet potatoes (*Ipomoea batatas* (L.) Lam.). *Econ Bot*. 2021;75(7):48–62.
[2] Yang J, Moeinzadeh MH, Kuhl H, et al. Haplotype-resolved sweet potato genome traces back its hexaploidization history. *Nat Plants*. 2017;3:696–703.
[3] Li Q, Zhao H, Jin YL, et al. Analysis and perspectives of sweet potato industry contributing to national food security in China. *Jiangsu J Agr Sci*. 2022;38(6):1484–91.
[4] Nguyen HC, Chen CC, Lin KH, et al. Bioactive compounds, antioxidants, and health benefits of sweet potato leaves. *Molecules*. 2021;26(7):1820.
[5] Laveriano-Santos EP, López-Yerena A, Jaime-Rodríguez C, et al. Sweet potato is not simply an abundant food crop: a comprehensive review of its phytochemical constituents, biological activities, and the effects of processing. *Antioxidants*. 2022;11(9):1648.
[6] Kwak SS. Biotechnology of the sweet potato: ensuring global food and nutrition security in the face of climate change. *Plant Cell Rep*. 2019;38(11):1361–3.
[7] Zhang C, Li S, Luo F, et al. The global warming hiatus has faded away: an analysis of 2014–2016 global surface air temperatures. *Int J Biometeorol*. 2019;39(12):4853–68.
[8] Iwakiri T, Watanabe M. Mechanisms linking multi-year La Niña with preceding strong El Niño. *Springer Sci Bus Media LLC*. 2021;11(1):17465.
[9] Martinez VC, Newman M, Vimont DJ, et al. Observed El Niño-La Niña asymmetry in a linear model. *Geophys Res Lett*. 2019;46: 9909–19.
[10] Thorne PW. Discussion on "A combined estimate of global temperature". *Environmetrics*. 2022;33(3):e2722 .
[11] Niggli EA. Towards improved understanding of cascading and interconnected risks from concurrent weather extremes: analysis of historical heat and drought extreme events. *PLOS Clim*. 2022;8:e0000057.
[12] Hettiarachchi S, Wasko C, Sharma A. Do longer dry spells associated with warmer years compound the stress on global water resources? *Earth's Future*. 2022;10(2):e2021EF002392.
[13] Hao ZC, Hao FH, Singh VP, et al. Changes in the severity of compound drought and hot extremes over global land areas. *Environ Res Lett*. 2018;13(12):124022.
[14] Young HR, Cornforth RJ, Todman LC, et al. Assessing the impact of climate change on sweet potato in Uganda. *GCRF Sweet Potato Catalyst, Policy Brief No.* 1, Walker Institute. 2020;WIBN/0520/01.
[15] Yang R, Wang L, Shen GZ, et al. Long-term trend analysis of precipitation, temperature, and extreme weather in Kunming. *South-to-North Water Diversion Water Sci Technol*. 2016;8(12):142.
[16] Jiang R, Gan TY, Xie J, et al. Historical and potential changes of precipitation and temperature of Alberta subjected to climate change impact: 1900–2100. *Theor Appl Climatol*. 2017;127:725–39.
[17] Sun Y, Yin H, Tian QH, et al. Recent progress in studies of climate change detection and attribution in the globe and China in the past 50 years. *Progressus Inquisitiones DE Mutatione Climatis*. 2013;9(4):235–45.
[18] Tilman D, Fargione J, Wolff B, et al. Forecasting agriculturally driven global environmental change. *Science*. 2001;292:281–4.
[19] Iba K. Acclimative response to temperature stress in higher plants: approaches of gene engineering for temperature tolerance. *Annu Rev Plant Biol*. 2002;53(1):225–45.
[20] Ahuja I, de Vos RC, Bones AM, et al. Plant molecular stress responses face climate change. *Trends Plant Sci*. 2010;15(12):664–74.

[21] Yu J, Su D, Yang D, et al. Chilling and heat stress-induced physiological changes and microRNA-related mechanism in sweet potato (*Ipomoea batatas* L.). *Front Plant Sci.* 2020;11:687.
[22] Pitzschke A, Forzani C, Hirt H. Reactive oxygen species signaling in plants. *Antioxid Redox Signal.* 2006;8(9–10):1757–64.
[23] Park S, Kim H, Jung Y, et al. Orange protein has a role in phytoene synthase stabilization in sweet potato. *Sci Rep.* 2016;6:33563.
[24] Kang L, Kim HS, Kwon YS, et al. IbOr regulates photosynthesis under heat stress by stabilizing IbPsbP in sweet potato. *Front Plant Sci.* 2017;8:989.
[25] Tang W, Arisha MH, Zhang ZY, et al. Comparative transcriptomic and proteomic analysis reveals common molecular factors responsive to heat and drought stresses in sweet potato (*Ipomoea batatas*). *Front Plant Sci.* 2023;13:1081948.
[26] Edmunds BA, Boyette MD, Clark CA, et al. *Postharvest handling of sweet potatoes.* North Carolina State University Cooperative Extension Service. 2008.
[27] Ji CY, Bian XF, Lee CJ, et al. De novo transcriptome sequencing and gene expression profiling of sweet potato leaves during low temperature stress and recovery. *Gene.* 2019;700:23–30.
[28] Chinnusamy V, Zhu J, Zhu JK. Cold stress regulation of gene expression in plants. *Trends Plant Sci.* 2007;12(10):444–51.
[29] Jin R, Kim BH, Ji CY, et al. Overexpressing IbCBF3 increases low temperature and drought stress tolerance in transgenic sweet potato. *Plant Physiol Biochem.* 2017;118:45–54.
[30] Xie ZY, Wang AM, Li HM, et al. High throughput deep sequencing reveals the important roles of microRNAs during sweet potato storage at chilling temperature. *Sci Rep.* 2017;7(1):16578.
[31] Xie ZY, Zhou ZL, Li HM, et al. High throughput sequencing identifies chilling responsive genes in sweet potato (*Ipomoea batatas* Lam.) during storage. *Genomics.* 2019;111(5):1006–17.
[32] Ji CY, Kim HS, Lee CJ, et al. Comparative transcriptome profiling of tuberous roots of two sweet potato lines with contrasting low temperature tolerance during storage. *Gene.* 2020;727:144244.
[33] Rontein D, Basset G, Hanson AD. Metabolic engineering of osmoprotectant accumulation in plants. *Metab Eng.* 2002;4(1):49–56.
[34] Yooyongwech S, Theerawitaya C, Samphumphuang T, et al. Water-deficit tolerant identification in sweet potato genotypes (*Ipomoea batatas* (L.) Lam.) in vegetative developmental stage using multivariate physiological indices. *Sci Hortic.* 2013;162:242–51.
[35] Arunyanark A, Jogloy S, Akkasaeng C, et al. Chlorophyll stability is an indicator of drought tolerance in peanut. *J Agron Crop Sci.* 2008;194(2):113–25.
[36] Motsa NM, Modi AT, Mabhaudhi T. Sweet potato (*Ipomoea batatas* L.) as a drought tolerant and food security crop. *S Afr J Sci.* 2015;111(11–12):1–8.
[37] Ramírez DA, Grüneberg W, Andrade MI, et al. Phenotyping of productivity and resilience in sweet potato under water stress through UAV-based multispectral and thermal imagery in Mozambique. *J Agron Crop Sci.* 2023;209(1):41–55.
[38] Ruan L, Chen LJ, Chen YH, et al. Expression of Arabidopsis homeodomain glabrous 11 enhances tolerance to drought stress in transgenic sweet potato plants. *J Plant Biol.* 2012;55:151–8.
[39] Fan WJ, Zhang M, Zhang HX, et al. Improved tolerance to various abiotic stresses in transgenic sweet potato (*Ipomoea batatas*) expressing spinach betaine aldehyde dehydrogenase. *PLoS One.* 2012;7(5):e37344.

[40] Mbinda W, Dixelius C, Oduor R. Induced expression of Xerophyta viscosa XvSap1 gene enhances drought tolerance in transgenic sweet potato. *Front Plant Sci.* 2019;10:1119.
[41] Wang B, Zhai H, He SZ, et al. A vacuolar Na+/H+ antiporter gene, IbNHX2, enhances salt and drought tolerance in transgenic sweet potato. *Sci Hortic.* 2016;201:153–66.
[42] Wang AM, Zhu MK, Luo YH, et al. A sweet potato cinnamate 4-hydroxylase gene, IbC4H, increases phenolics content and enhances drought tolerance in tobacco. *Acta Physiol Plant.* 2017;39:1–12.
[43] Zhou YY, Zhai H, He SZ, et al. The sweet potato BTB-TAZ protein gene, IbBT4, enhances drought tolerance in transgenic *Arabidopsis. Front Plant Sci.* 2020;11:877.
[44] Kang C, Zhai H, He SZ, et al. A novel sweet potato bZIP transcription factor gene, IbbZIP1, is involved in salt and drought tolerance in transgenic Arabidopsis. *Plant Cell Rep.* 2019;38:1373–82.
[45] Wang WB, Qiu XP, Yang YX, et al. Sweet potato bZIP transcription factor IbABF4 confers tolerance to multiple abiotic stresses. *Front Plant Sci.* 2019;10:630.
[46] Zhou YY, Zhu H, He SZ, et al. A novel sweet potato transcriptional factor IbMYB116 enhances drought tolerance in transgenic Arabidopsis. *Front Plant Sci.* 2019;10:1025.
[47] Meng XQ, Liu SY, Zhang CB, et al. The unique sweet potato NAC transcription factor IbNAC3 modulates combined salt and drought stresses. *Plant Physiol.* 2023;191(1):747–71.
[48] Ren L, Zhang TT, Wu HX, et al. Blocking IbmiR319a impacts plant architecture and reduces drought tolerance in sweet potato. *Genes.* 2022;13(3):404.
[49] You C, Li C, Ma M, et al. A C2-domain abscisic acid-related gene, IbCAR1, positively enhances salt tolerance in sweet potato (*Ipomoea batatas* (L.) Lam.). *Int J Mol Sci.* 2022;23(17):9680.
[50] Zhang H, Wang Z, Li X, et al. The IbBBX24-IbTOE3-IbPRX17 module enhances abiotic stress tolerance by scavenging reactive oxygen species in sweet potato. *New Phytol.* 2022;233(3):1133–52.
[51] Zhu H, Zhou YY, Zhai H, et al. A novel sweet potato WRKY transcription factor, IbWRKY2, positively regulates drought and salt tolerance in transgenic Arabidopsis. *Biomolecules.* 2020;10(4):506.
[52] Zhang H, Gao XR, Zhi YH, et al. A non-tandem CCCH-type zinc-finger protein, IbC3H18, functions as a nuclear transcriptional activator and enhances abiotic stress tolerance in sweet potato. *New Phytol.* 2019;223(4):1918–36.
[53] Park SU, Lee CJ, Kim SE, et al. Selection of flooding stress tolerant sweet potato cultivars based on biochemical and phenotypic characterization. *Plant Physiol Biochem.* 2020;155:243–51.
[54] Armstrong W. Aeration in higher plants. *Adv Bot Res.* 1980;7:225–332.
[55] Lin KH, Tsou CC, Hwang SY, et al. Paclobutrazol pre-treatment enhanced flooding tolerance of sweet potato. *J Plant Physiol.* 2006;163(7):750–60.
[56] Roberts W, Russo V. Time of flooding and cultivar affect sweet potato yield. *HortScience.* 1991;26(12):1473–4.
[57] Eguchi T, Yoshida S. Effects of gas exchange inhibition and hypoxia on tuberous root morphogenesis in sweet potato (*Ipomoea batatas* (L.) Lam.). *Environ Control Biol.* 2007;45(2):103–11.
[58] Park HJ, Kim SJ, Nie HL, et al. Molecular identification of sweet potato accessions using ARMS-PCR based on SNPs. *J Plant Biochem Biotechnol.* 2020;47(2):124–30.
[59] Sorby K, Fleischer G, Pehu E. *Integrated pest management in development: review of trends and implementation strategies.* United States of America. 2003.

[60] Muramoto N, Tanaka T, Shimamura T, et al. Transgenic sweet potato expressing thionin from barley gives resistance to black rot disease caused by Ceratocystis fimbriata in leaves and storage roots. *Plant Cell Rep*. 2012;31:987–97.
[61] Wang CJ, Wang YZ, Chu ZH, et al. Endophytic Bacillus amyloliquefaciens YTB1407 elicits resistance against two fungal pathogens in sweet potato (*Ipomoea batatas* (L.) *Lam.). J Plant Physiol*. 2020;253:153260.
[62] Ogawa K, Komada H. Biological control of Fusarium wilt of sweet potato with cross-protection by prior inoculation with nonpathogenic Fusarium oxysporum. *Jpn Agric Res Q*. 1985;19(1):20–5.
[63] Lin YL, Zou WK, Lin SQ, et al. Transcriptome profiling and digital gene expression analysis of sweet potato for the identification of putative genes involved in the defense response against Fusarium oxysporum f. sp. batatas. *PLoS One*. 2017;12(11):e0187838.
[64] Jing XJ, Jin XJ, Liu Y, et al. Cloning and characterization of lbMAPKK9 gene associated with Fusarium oxysporum f. sp. batatas in sweet potato. *Acta Agron Sin*. 2023;49:3289–301.
[65] Zhang H, Zhang Q, Zhai H, et al. IbBBX24 promotes the jasmonic acid pathway and enhances fusarium wilt resistance in sweet potato. *Plant Cell*. 2020;32(4):1102–23.
[66] Li Y, Wang YN, Zhang H, et al. The plasma membrane-localized sucrose transporter IbSWEET10 contributes to the resistance of sweet potato to Fusarium oxysporum. *Front Plant Sci*. 2017;8:197.
[67] Liu X, Wang YR, Zhu HB, et al. Natural allelic variation confers high resistance to sweet potato weevils in sweet potato. *Nat Plants*. 2022;8(11):1233–44.
[68] Liu Y, Chen PR, Yang XS, et al. Cloning and expression analysis of lbWRKY7 gene in response to Fusarium oxysporum f. sp. batatas infection in sweet potato. *J China Agric Univ*. 2022;27:91–9.
[69] Nie N, Huo JX, Sun SF, et al. Genome-Wide characterization of the PIFs family in sweet potato and functional identification of IbPIF3.1 under drought and fusarium wilt stresses. *Int J Mol Sci*. 2023;24(4):4092.
[70] Xu R, Jin H, Liu Q, et al. Anesthetic activity of (6aR, 11aR)-trifolirhizin from Trifolium pratense to Ditylenchus destructor. *Nat Prod Res Dev*. 2011;23(5):820.
[71] Gao S, Yu B, Zhai H, et al. Enhanced stem nematode resistance of transgenic sweet potato plants expressing oryzacystatin-I gene. *Agric Sci China*. 2011;10(4):519–25.
[72] Gao S, Yu B, Yuan L, et al. Production of transgenic sweet potato plants resistant to stem nematodes using oryzacystatin-I gene. *Sci Hortic*. 2011;128(4):408–14.
[73] Zhai H, Wang FB, Si ZZ, et al. A myo-inositol-1-phosphate synthase gene, IbMIPS1, enhances salt and drought tolerance and stem nematode resistance in transgenic sweet potato. *Plant Biotechnol J*. 2016;14(2):592–602.
[74] Yang YH, Chen YQ, Bo YX, et al. Research progress in the mechanisms of resistance to biotic stress in sweet potato. *Genes*. 2023;14(11):2106.
[75] Yamasaki S, Sakai J, Fuji S, et al. Comparisons among isolates of Sweet potato feathery mottle virus using complete genomic RNA sequences. *Arch Virol*. 2010;155:795–800.
[76] Wanjala BW, Ateka EM, Miano DW, et al. Storage root yield of sweet potato as influenced by sweet potato leaf curl virus and its interaction with sweet potato feathery mottle virus and sweet potato chlorotic stunt virus in Kenya. *Plant Dis*. 2020;104(5):1477–86.
[77] Li C, Liu XX, Abouelnasr H, et al. Inhibition of miR397 by STTM technology to increase sweet potato resistance to SPVD. *J Integr Agric*. 2022;21(10):2865–75.
[78] Li Q, Liu QC, Zhai H, et al. Genetic diversity in main parents of sweet potato in China as revealed by ISSR marker. *Acta Agron Sin*. 2008;34(6):972–7.

[79] Rahajeng W, Restuono J, Indriani FC, et al. Evaluation of promising sweet potato clones for higher root yield and dry matter content. *Planta TROPIKA*. 2021;9(1):42–7.

[80] Xie YZ, Jia ZD, Bian XF, et al. Breeding and pedigree analysis of edible sweet potato. *Bull Chin Agron*. 2021;37(30):32–8.

[81] Katayama K, Kobayashi A, Sakai T, et al. Recent progress in sweet potato breeding and cultivars for diverse applications in Japan. *Breed Sci*. 2017;67(1):3–14.

[82] Beyene B, Menamo T, Haile G. Protocol optimization for in vitro propagation of Kulfo, orange flesh sweet potato (*Ipomoea batatas*) variety using shoot tip culture. *Afr J Plant Sci*. 2020;14(10):395–401.

[83] Masekesa RT, Gasura E, Matikiti A, et al. Effect of BAP, NAA and GA3, either alone or in combination, on meristem culture and plantlet establishment in sweet potato (cv Brondal). *Afr J Food Agric Nutr Dev*. 2016;16(1):10649–65.

[84] Jahanara P, Mahbub R, Hasan MF, et al. Standardization of plant growth regulators for successful tissue culture of sweet potato. *J Bangladesh Agric Univ*. 2018;16(2):178–81.

[85] Meng YS, Zhao N, Li H, et al. SSR fingerprinting of 203 sweet potato (*Ipomoea batatas* (L.) Lam.) varieties. *J Integr Agric*. 2018;17:86–93.

[86] Gao RF, Li Q, Su ZX, et al. Genetic diversity analysis of 76 sweet potato varieties based on SSR markers. *Mol Plant Breed*. 2020;18(11):3775–85.

[87] Maroof MA, Zhang Q, Neale DB, et al. Associations between nuclear loci and chloroplast DNA genotypes in wild barley. *Genetics*. 1992;131(1):225–31.

[88] Cervantes-Flores JC, Sosinski B, Pecota KV, et al. Identification of quantitative trait loci for dry-matter, starch, and β-carotene content in sweet potato. *Mol Breed*. 2011;28:201–16.

[89] Chang KY, Lo HF, Lai YC, et al. Identification of quantitative trait loci associated with yield-related traits in sweet potato (*Ipomoea batatas*). *Bot Stud*. 2009;50(1):43–55.

[90] Meng. *Construction of high-density SSR linkage maps and mapping of QTLs for dry matter content in sweet potato, Ipomoea batatas (L.) Lam* [Doctoral dissertation]. China Agricultural University. 2018.

[91] Sasai R, Tabuchi H, Shirasawa K, et al. Development of molecular markers associated with resistance to Meloidogyne incognita by performing quantitative trait locus analysis and genome-wide association study in sweet potato. *DNA Res*. 2019;26(5):399–409.

[92] Ma Z, Gao W, Liu L, et al. Identification of QTL for resistance to root rot in sweet potato (*Ipomoea batatas* (L.) Lam) with SSR linkage maps. *BMC Genomics*. 2020;21:1–14.

[93] Ma M, Yan H, Gao RF, et al. Construction linkage maps and identification of quantitative trait loci associated with important agronomic traits in purple-fleshed sweet potato. *Acta Agron Sin*. 2021;47(11):2147–62.

[94] Kim SE, Lee CJ, Ji CY, et al. Transgenic sweet potato plants overexpressing tocopherol cyclase display enhanced α-tocopherol content and abiotic stress tolerance. *Plant Physiol Biochem*. 2019;144:436–44.

[95] Wadl PA, Olukolu BA, Branham SE, et al. Genetic diversity and population structure of the USDA sweet potato (*Ipomoea batatas*) germplasm collections using GBSpoly. *Front Plant Sci*. 2018;9:1166.

[96] Kang L, Ji CY, Kim SH, et al. Suppression of the β-carotene hydroxylase gene increases β-carotene content and tolerance to abiotic stress in transgenic sweet potato plants. *Plant Physiol Biochem*. 2017;117:24–33.

[97] Kim SE, Bian X, Lee CJ, et al. Overexpression of 4-hydroxyphenylpyruvate dioxygenase (IbHPPD) increases abiotic stress tolerance in transgenic sweet potato plants. *Plant Physiol Biochem*. 2021;167:420–9.

[98] Kim SE, Lee CJ, Park SU, et al. Overexpression of the golden SNP-carrying Orange gene enhances carotenoid accumulation and heat stress tolerance in sweet potato plants. *Antioxidants*. 2021;10(1):51.

[99] Wang H, Wu Y, Zhang Y, et al. CRISPR/Cas9-based mutagenesis of starch biosynthetic genes in sweet potato (*Ipomoea batatas*) for the improvement of starch quality. *Int J Mol Sci*. 2019;20(19):4702.

[100] Li D, Quan C, Song Z, et al. High-throughput plant phenotyping platform (HT3P) as a novel tool for estimating agronomic traits from the lab to the field. *Front Bioeng Biotechnol*. 2021;8:623705.

[101] Yang W, Guo Z, Huang C, et al. Combining high-throughput phenotyping and genome-wide association studies to reveal natural genetic variation in rice. *Nat Commun*. 2014;5:5087.

[102] Yang W, Feng H, Zhang X, et al. Crop phenomics and high-throughput phenotyping: past decades, current challenges, and future perspectives. *Mol Plant*. 2020;13(2):187–214.

[103] Wu D, Guo Z, Ye J, et al. Combining high-throughput micro-CT-RGB phenotyping and genome-wide association study to dissect the genetic architecture of tiller growth in rice. *J Exp Bot*. 2019;70(2):545–61.

[104] Parmley K, Higgins R, Ganapathy SB, et al. Machine learning approach for prescriptive plant breeding. *Sci Rep*. 2019;9:17132.

[105] Sandhu KS, Lozada DN, Zhang Z, et al. Deep learning for predicting complex traits in spring wheat breeding program. *Front Plant Sci*. 2021;11:613325.

Section III

Agronomic/Management Practices

9 Plant Metabolites and Climate Resilience

Challenges and Solutions

Asima Rasheed, Muhammad Abdul Rehman Rashid, Asad Ali Shah, Muhammad Hussnain Siddique, Muhammad Zubair, and Farrukh Azeem

9.1 INTRODUCTION

Climate change is one of the most critical global challenges, profoundly affecting ecosystems, biodiversity, and agricultural productivity. It is defined as a long-term alteration in climate indicators such as temperature, precipitation, wind patterns, and extreme weather events (1). Human activities, particularly industrial emissions, deforestation, and the excessive use of fossil fuels, have significantly increased greenhouse gas concentrations, including carbon dioxide (CO_2), methane (CH_4), and ozone (O_3) (2). These gases contribute to global warming, with climate projections indicating a temperature rise of between 1.4°C to 5.8°C by the year 2100 (3), and some extreme shifts anticipated as early as 2035 (4). Such climatic changes are expected to bring uncertainties in weather patterns, including more intense monsoon rainfall, increased frequency of heatwaves, decreased dry-season precipitation, and intensified storms (5).

Among the many consequences of climate change, its impact on plant metabolism is particularly significant. Plants, as sessile organisms, must continuously adapt to fluctuating environmental conditions. One of their primary adaptive strategies involves the modulation of metabolite production. Metabolites are essential molecular components that play crucial roles in plant growth, development, and interactions with the environment (6). In medicinal plants, these metabolites are responsible for pharmacological properties and therapeutic potential. However, climate-induced stressors such as extreme temperatures, drought, and elevated UV radiation can alter the biosynthesis of bioactive compounds, affecting their medicinal efficacy and availability (7). Plants are particularly sensitive to climatic fluctuations, as environmental stress directly influences the production of secondary metabolites – specialized compounds involved in plant defense and adaptation (8). These metabolites, including alkaloids, flavonoids, terpenoids, and phenolics, are widely utilized in pharmaceuticals, nutraceuticals, and traditional medicine. For instance, *Catharanthus roseus* produces vinblastine and vincristine, which are widely used in cancer treatment (9). Their biosynthesis is highly sensitive to temperature

DOI: 10.1201/9781032720142-12

and drought stress (9). *Artemisia annua* synthesizes artemisinin, an essential antimalarial compound, with production levels increasing under low-temperature and salinity conditions (10). *Ginkgo biloba* accumulates flavonol glycosides, which provide neuroprotective benefits. Heat stress has been found to enhance their concentration (11,12). *Hypericum perforatum* (St. John's wort) produces hypericin, an antidepressant compound, with higher concentrations observed under increased UV radiation (13).

The response of these metabolites to environmental stress is governed by complex biochemical pathways. Climate-related stressors can either enhance or suppress metabolite synthesis, affecting the medicinal quality, efficacy, and commercial viability of plant-based products (14).

Plants have evolved sophisticated mechanisms to cope with climate-induced stress, many of which involve metabolic adjustments. Water-deficit conditions lead to the accumulation of osmo-protective metabolites such as proline, glycine betaine, and trehalose, helping plants maintain cellular homeostasis (15). For example, *Glycyrrhiza glabra* (licorice) increases glycyrrhizin production under drought stress (16). High temperatures cause oxidative stress, stimulating the biosynthesis of phenolic compounds and flavonoids with antioxidant properties, enhancing stress tolerance (17). Increased atmospheric CO_2 alters carbon allocation, often enhancing the production of essential oils and terpenoids in medicinal plants such as *Thymus vulgaris* (18). Exposure to elevated UV radiation triggers the accumulation of flavonoids, which act as natural sunscreens. *Hypericum perforatum* exhibits increased hypericin production in response to UV stress (19).

Despite the importance of plant metabolites, their response to climate change remains relatively underexplored compared to staple crops (20). Understanding the molecular mechanisms governing metabolite biosynthesis under environmental stress is crucial for optimizing the production of bioactive compounds. Advancements in plant biotechnology, including metabolic engineering, CRISPR-based gene editing, and stress-resilient crop breeding, offer promising solutions to enhance metabolite production while maintaining plant adaptability (21). Additionally, integrating sustainable agricultural practices such as precision farming, agroforestry, and microbial-assisted plant resilience can help mitigate climate-induced losses in plant resources. This chapter provides a comprehensive overview of the relationship between climate change and plant metabolites. It explores how environmental stressors influence metabolite biosynthesis, examines adaptation strategies, and discusses the implications for agriculture, pharmaceutical industries, and biodiversity conservation.

9.2 CLASSIFICATION OF PLANT METABOLITES

Plant metabolites are broadly classified into primary and secondary metabolites, distinguished by their biosynthetic origin and physiological functions (22). Primary metabolites, including sugars, amino acids, and organic acids, are indispensable for basic cellular processes and are ubiquitously present in plant tissues. In contrast, secondary metabolites are specialized compounds not directly involved in primary metabolic pathways but crucial for plant adaptation and defense mechanisms. Secondary metabolites often undergo biosynthesis in response to environmental cues like herbivore attack, pathogen invasion, or abiotic stress factors such as drought, heat, or UV radiation (23). These compounds enhance plant resilience by acting as deterrents against herbivores and pathogens, antioxidants to counteract oxidative stress, and allelochemicals to inhibit the growth of competing plants. Coined by A. Kossel in 1891, the term 'secondary metabolites' denotes organic compounds evolved incidentally, lacking primary roles in plant life such as direct involvement in developmental and reproduction systems (24). They exhibit inadequate distribution and are often employed as taxonomic markers (25). Conversely, primary metabolites such as carbohydrates, acyl lipids, phytosterols, and organic acids are universally present in all plant tissues, fulfilling essential metabolic functions vital for plant growth and development (26).

In recent years, there has been growing interests in studying plant metabolites, driven by their promising applications across agriculture, medicine, industry, pharmaceuticals, and nutraceuticals. Plant-derived metabolites have been found to provide fragrances, dyes, and biopesticides, among other uses. To fully exploit their potential, it's imperative to comprehend these metabolites' biosynthesis, regulation, and ecological functions. This chapter aims to delve into the multifaceted realm of plant metabolites, elucidating their roles in shaping plant–environment interactions and their significance for climate resilience. Through this exploration, we shall address the hurdles and prospects inherent in leveraging plant metabolites for sustainable agriculture, ecosystem management, and human well-being amidst the challenges posed by climate change.

9.3 IMPORTANT PLANT SECONDARY METABOLITES AND THEIR ACTIONS

Plant secondary metabolites play a significant role in the innate defense mechanisms of plants against pathogens, pests, and various environmental stressors (27). Being natural products, they possess biodegradability, lower toxicity to humans, and a wide range of biological activities. These secondary metabolites are broadly categorized into four main groups based on their chemical structures: terpenes, phenolics, nitrogen-containing compounds, and sulfur-containing compounds (Table 9.1) (28). Among these, terpenes represent the most diverse class of secondary metabolites, characterized by their structural development from varying numbers of isoprene units, with or without oxygen (terpenoids and terpenes) (29).

Terpenes found in plants play important ecological and physiological functions, including allelopathy, insecticidal properties, attraction of pollinators, and involvement in plant hormone regulation (e.g., abscisic acid, gibberellins) (30). In particular, terpenoids like monoterpenoids and sesquiterpenoids, prevalent components of essential oils (EOs), possess high volatility and often act as insect toxins, safeguarding against fungal or bacterial attacks. These include mint plants (*Mentha* spp.), which are prolific producers of monoterpenoids such as menthol and menthone and are known for their fungicidal effects against *Alternaria alternata* (31, 32). Moreover, menthol demonstrates antibacterial properties against strains of *Xanthomonas campestris* and

TABLE 9.1
Various important nitrogenous and non-nitrogenous secondary metabolites (37)

Nitrogen-based secondary metabolites	Non-nitrogenous secondary metabolites
Alkaloids	Monoterpenes (C10)
Amines	Diterpenes (C20)
Cyanogenic glucosides	Triterpenes, steroids, saponins (C30, C27)
Nonprotein amino acids	Sesquiterpenes (C15)
Glucosinolates	Tetraterpenes (C40)
Alkyl-amides	Flavonoids, anthocyanins, catechins, tannins
Peptides, polypeptides	Phenylpropanoids, lignin, coumarins, lignans
	Polyacetylenes, fatty acids, waxes
	Polyketides
	Carbohydrates, simple acids

Pseudomonas syringae (33). Phenolic compounds, characterized by at least one aromatic ring linked to hydroxyl groups, comprise various subgroups like tannins, lignins, flavonoids, coumarins, and stilbenes (34). These compounds directly affect insects by acting as toxins and altering their growth and development (35). Furthermore, nitrogen-containing secondary metabolites, including alkaloids and amines, are vital defense mechanisms in plants. Moreover, alkaloids, such as those exhibiting insecticidal and fungicidal properties, exhibit insecticidal and fungicidal properties (36). Lastly, sulfur-containing secondary metabolites, such as glucosinolates and alliins, found in Brassicaceae and *Allium* species, respectively, are implicated in herbivore defense mechanisms, where they are metabolized into toxic substances upon injury to plant cells (34).

9.4 EVOLUTIONARY PATHWAY OF SECONDARY METABOLITES

Primary metabolites are the major precursors for the production of secondary metabolites in plants, which exhibit a wide array of biological activities. Secondary metabolites are crucial for enabling plants to establish robust interactions with their environment, contributing to their survival and fitness. Much like primary metabolites, secondary metabolites play vital roles in plant defense, growth, reproduction, and communication. There are over 100,000 known secondary metabolites synthesized in various plant species, with each metabolite typically originating within specific plant taxonomic groups. These metabolites can be broadly categorized into three main groups based on their biosynthetic pathways: terpenes (or isoprenoids), phenolic compounds, and nitrogen-containing compounds (38) (Figure 9.1). Terpenes, one of the most diverse classes of secondary metabolites, are synthesized from isoprene units via the mevalonate and methylerythritol phosphate (MEP) pathways

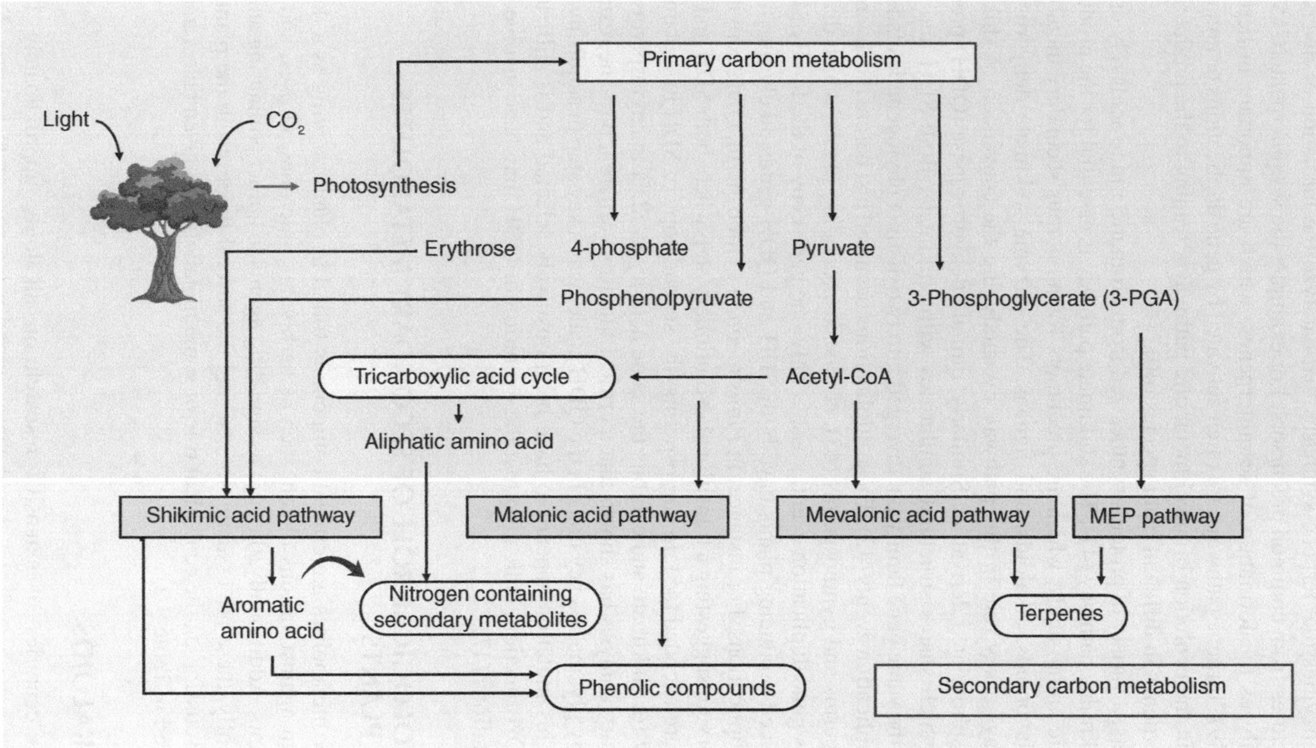

FIGURE 9.1 Schematic display of biosynthetic pathways for the production of secondary metabolites in plants.

(39). These compounds are involved in various physiological processes such as plant defense, growth regulation, and attraction of pollinators. The biosynthesis of terpenes involves enzymes like terpene synthases that catalyze the formation of diverse terpene structures from common precursors. For example, the sesquiterpene (C15) synthase catalyzes the formation of sesquiterpenes, which are important in herbivore defense (39). Terpene pathways are often activated by gene duplications or mutations, resulting in the evolution of novel metabolic pathways, leading to the production of specific terpenes in different plant species (40).

In plants, a variety of proteins synthesized from aromatic amino acids rely directly on the shikimate pathway (41). The shikimate pathway is essential for the production of aromatic amino acids, which are precursors for numerous secondary metabolites such as flavonoids, alkaloids, and lignans. The enzyme shikimate dehydrogenase (SDH) plays a key role in this pathway by catalyzing the conversion of shikimate to its phosphorylated derivative. Similarly, quinate dehydrogenase (QDH) produces quinate, which shares structural similarities with shikimate. Both SDH and QDH belong to the same gene family but diverged into two distinct phylogenetic clades due to gene duplication events (42). These duplications occurred just before the divergence of angiosperms and gymnosperms (43). Non-seed-producing plants, which diverged before this gene duplication event, possess only one gene copy of the family (44). In contrast, seed-producing plants have both SDH and QDH genes, likely due to subsequent duplications. For instance, in *Pinus taeda*, a gymnosperm, QDH activity has been observed, suggesting a transitional evolutionary step where both SDH and QDH activities are encoded (45). In angiosperms, the second copy of SDH has retained its specificity for shikimate and exhibits the same activity as SDH in angiosperms (46). Evolutionary analysis has shown that a critical mutation, where a tyrosine residue is replaced by glycine at a key node in the QDH clade, was a key step in acquiring QDH activity. This mutation appears to have been positively selected and is sufficient to confer QDH function, indicating that the evolution of QDH may have involved only a few mutations (47).

9.5 ECOLOGICAL ROLE OF SECONDARY METABOLITES IN PLANTS

Secondary metabolites are crucial compounds found in plants, serving as a defense mechanism against various threats such as herbivores, microbes, viruses, and competing plants. Additionally, they act as signaling agents to attract pollinating animals. Consequently, these are essential in ensuring the several functions in the plant (35). A brief discussion on the ecological role of some metabolites is given below and also summarized in Table 9.2.

9.6 ALKALOIDS

After two centuries of extensive research on alkaloids, approximately 10,000 compounds have been isolated and characterized using various techniques. Despite their limited distribution across kingdoms, alkaloids are widely found in bacteria,

TABLE 9.2
Ecological and medicinal significances of major groups of metabolites in various plants

Metabolite groups	Classification	Metabolite-producing plants	Secondary metabolites	Medicinal role	Ecological role	References
Alkaloids	Nitrogen-based complex group of metabolites	*Cinchona officinalis*	Quinine	Used to treat malaria	Provide protection to plants against herbivores due to toxic properties as well as lower down the decomposition rate due to the antimicrobial activity	(63–65)
		Papaver somniferum	Morphine	Used as pain reliever and cough suppressants		
Phenolic compounds	Metabolites with more than one OH aromatic ring	*Salix alba*	Salicin	Anti-inflammatory and pain-reliever properties	Lower down the decomposition rate due to the antimicrobial activity	(66, 67)
		Camellia sinensis	Catechins	Antioxidant properties		
Terpenes	Having isoprene units	*Artemisia annua*	Artemisinin	Antimalarial properties	Aromas and chemicals serve as vital signals facilitating communication between plants and animals, while also imparting resistance to both biotic (insecticidal) and abiotic stresses	(68), (69), (70)
		Taxus brevifolia	Paclitaxel	Used as a chemotherapy agent to treat cancer		

fungi, plants, and animals (48,49). They are prevalent in about 300 plant families, with specific metabolites typically confined to certain families, such as hyoscyamine in Solanaceae (50). This group of metabolites contains a single nitrogen atom and appears as clear crystalline substances, which can be found in a free state or combined with acids to form salts that are readily soluble in water (51). Bases devoid of alkaloids are barely soluble in water but highly soluble in organic solvents (51,52). Their physiological effects on animals range from analgesic properties to toxicity, as observed in substances like cocaine or morphine (49). Consequently, these have been extensively studied in pharmacology and utilized in traditional and modern medical practices for centuries. Despite this, their role in soils and as decomposers remains relatively unexplored. These compounds act as feeding deterrents and protectors against infections in plants (53). In the *Poaceae* family, alkaloid compounds are synthesized with the contribution of fungal endophytes rather than being directly produced by the plant itself. These alkaloids, produced through endophytic involvement, have the ability to influence the activity of microorganisms, thereby affecting litter decomposition rates (54).

9.7 PHENOLIC COMPOUNDS

Phenolic compounds are extensively produced by plants, fungi, and bacteria, with approximately 10,000 phenolic groups identified, which include simple structures to complex polymers such as tannins or lignins (55). In plants, their abundance can vary significantly, comprising anywhere from 1% to over 25% of the total dry mass of green leaves. These compounds typically contain at least one hydroxylated aromatic ring. However, understanding phenolic compounds can be complex due to the various criteria involved (56). The term 'polyphenol' is often used to describe compounds with multiple hydroxyl groups on one or more aromatic rings, encompassing all types of tannin or phenolic components. Later, it may also refer to the 'total phenols' fraction. These compounds can also be categorized based on their structural arrangement, including simple phenols (C6 skeleton), phenolic acids (C6-C1 or C6-C3), flavonoids (C6-C3-C6), and tannins [(C6-C3-C6)n] (55,57). Flavonoids with a C6-C3-C6 skeleton are categorized as a major group of metabolites with diverse biochemical roles. These compounds are the reasons for the shade, scent, and taste of flowers, fruits, and seeds. Additionally, they possess defensive properties against biotic factors owing to their antioxidant properties. Furthermore, flavonoids play a crucial role in signaling symbiotic bacteria to communicate effectively (58).

9.8 TERPENES

Terpenes represent a highly diverse category of plant secondary metabolites, among the more than 30,000 identified compounds. Primarily synthesized by plants, these metabolites typically constitute around 1–2% of the dry matter weight, though elevated concentrations may be found in specific plant parts such as leaves or trichomes (59). Composed of a basic five-carbon building block known as the isoprene unit, these lipophilic compounds exhibit a wide range of chemical structures and biological activities. Initially, terpenes were discovered in their volatile forms, and served as signals for organisms to communicate with each other (60). Besides, terpenes are the key factor in the formation of the possible resistance mechanisms in organisms to the biotic and abiotic stresses. With multiple functions, these compounds are also involved in attracting pollinators, safeguarding plants from herbivores (61) and producing toxic insecticides and insect repellents (62). The impact of monoterpenes on soil microorganisms is intricate, as they can both inhibit the activity and growth of certain microorganisms while also enhancing these activities in others.

9.9 ENVIRONMENTAL FACTORS AFFECTING THE PLANT METABOLITES

Plants continuously interact with their surrounding environment, and various abiotic stress factors, such as temperature fluctuations, drought, and light intensity, play a crucial role in shaping their metabolic profiles. These environmental cues influence the biosynthesis of both primary and secondary metabolites, which are essential for plant growth, development, and stress adaptation. While primary metabolites support fundamental physiological functions, secondary metabolites act as key defense compounds, helping plants cope with environmental challenges. Understanding the intricate relationship between environmental factors and plant metabolism is vital for enhancing crop resilience and optimizing metabolite production for agricultural and pharmaceutical applications (Figure 9.2).

9.10 TEMPERATURE

Temperature is a critical factor that profoundly influences plant metabolites, altering their composition and physiological functions. High-temperature stress induces the denaturation of metabolic enzymes, leading to the disruption and breakdown of numerous metabolites in plant species (71,72). Additionally, temperature stress disrupts the expression of genes involved in metabolite synthesis, thereby altering the metabolic profile of plants (73). Similar to high-temperature stress, cold stress also affects the metabolic pathways of plants, influencing their growth, development, and overall functionality. Specifically, low temperatures lead to the accumulation of osmo-protectants such as sugars, proteins, and polyols (74). These osmo-protectant compounds aid plants in adapting to and coping with different temperature stresses.

Moreover, cold stress can disrupt the metabolites involved in the plant's defense system (75). Plants exposed to cold stress respond by increasing the production of secondary metabolites to counteract oxidative damage and bolster their resilience against environmental challenges. Additionally, cold stress can impact the expression of genes related to both primary and secondary metabolic pathways, resulting in fluctuations in metabolic fluxes and the accumulation of specific metabolites. For instance, under cold stress conditions, genes encoding enzymes responsible for the biosynthesis of antifreeze proteins or other protective compounds may be upregulated. Overall, cold

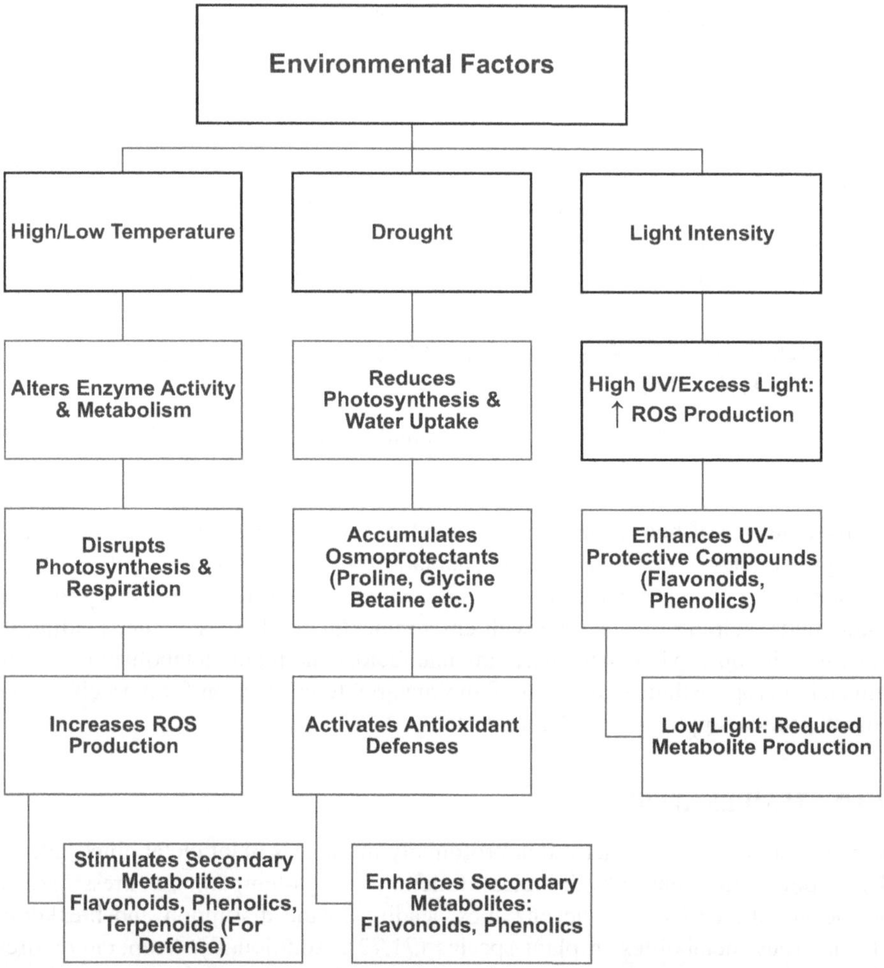

FIGURE 9.2 Illustration of the environmental factors influencing plant metabolism.

stress induces complex metabolic changes in plants, highlighting the importance of understanding and managing these responses for improving plant resilience and productivity in cold environments (76).

9.11 DROUGHT STRESS

Drought stress is one of the biggest challenges for the plant that could produce adverse effects on plant growth as well as could reduce their ability for resistance against environmental stress. The growth reduction of plants under drought conditions might be due to the lack of water that in turn can lower the photosynthesis rate (77). A decline of the water level causes the elevation of temperature, which in turn results in the high production of phenolic compounds. Furthermore, under drought conditions, plants close stomata and lower their photosynthesis activities, with both these factors negatively affecting the production of metabolites, especially secondary metabolites, e.g. phenolic compounds. Water stress also affects the relationships of tannin production and xylem pressure (78,79). During drought stress (DS), the antioxidant defense mechanism in plants becomes impaired. To counteract this, plants can enhance the production of xanthophyll pigments, which help in dissipating excess energy as heat. This increase in xanthophyll pigments serves to protect thylakoid membranes by preventing the generation of reactive oxygen species (ROS) (80). However, in cases of severe drought, the increase in zeaxanthin levels may not be adequate to maintain the photochemistry of photosystem II (PSII). Consequently, drought-induced impairments can lead to a decline in enzymatic activity and ultimately result in photoinhibition. Under drought conditions, secondary metabolites are involved in reducing the stress effects. Metabolites such as phenolics, flavonoids, and alkaloids play roles as antioxidants and prevent cell damage by scavenging ROS. They also sustain cell turgidity by maintaining osmolytes like glycine betaine. Flavonoids protect plants from UV damage and control stomatal behavior to save water. Plants use all these systems together to survive water shortages and remain productive.

9.12 LIGHT STRESS

In the plant kingdom, light stands as a foremost environmental factor dictating evolution and overall success, encompassing various aspects such as intensity, duration, and quality. Plants possess photoreceptors capable of discerning different light characteristics via photomorphogenesis (81,82). About 89% of red and blue light is absorbed by plants from incoming light, rendering them highly sensitive to alterations in their light environment (83). Both the quantity and quality of light can exert significant influences on plant morphology, development, and the synthesis of primary and secondary metabolites (84), often exhibiting species-specific responses (82). Additionally, light plays a crucial role in cellular differentiation processes, including hypocotyl expansion, chloroplast development, leaf enlargement, and flowering initiation. Blue light, in particular, can trigger phototropism, stem shortening, chloroplast movement, stomatal regulation, and genetic expression (85). It is commonly acknowledged that light intensity could be a factor that has a positive impact on the production of phytochemicals.

However, the impact of light quality yields more intricate responses, which can vary among species and light ratios (84). Light triggers the activation of numerous essential metabolites, including phenolic compounds, which serve as antimicrobial or antifungal agents in addition to functioning as scavengers of ROS (86). Carotenoids, crucial tetraterpenoid pigments, are closely associated with the thylakoid membranes of chloroplasts. Through interactions with photosystem (PS) complexes I and II, carotenoids aid in energy capture and dissipate excess energy as heat. All nitrogen-based compounds, such as six glucosinolates and three glycosides, exhibited an increase under blue light. Even a short duration of blue-light exposure, for as little as 5 days, was sufficient to enhance both aliphatic and aromatic glucosinolates in broccoli (81). Furthermore, both blue and red light were found to induce the absorption of glycoside eleutheroside B in *Siberian ginseng* (87). Additionally, red light was observed to induce the absorption of glucosinolate gluconasturtiin in *Nasturium officinale* W.T. Aiton (88).

9.13 POSSIBLE SOLUTIONS

Plant metabolites have dynamic importance in expanding climate resilience by empowering plants to tolerate and cope with the different climate changes or stresses such as drought, extreme temperatures, and salinity. Plants have a complex regulatory network for metabolite production and to perform vital functions, which is the greatest challenge to understand. To address this, comprehensive omics approaches, integrating genomics, transcriptomics, proteomics, and metabolomics, are essential. These approaches can uncover key metabolic pathways and identify critical metabolites involved in stress responses. Additionally, advanced bioinformatics tools and machine learning algorithms can help analyze large datasets, predict stress-responsive metabolites, and guide genetic engineering efforts aimed at developing climate-resilient crops. Solutions to these challenges also involve leveraging traditional breeding techniques complemented by modern biotechnological interventions. For instance, CRISPR-Cas9 technology can be employed to precisely edit genes involved in the biosynthesis of stress-responsive metabolites, enhancing plant tolerance to adverse conditions. Another solution is the development of biofortified crops through metabolic engineering, which not only improves stress resilience but also enhances nutritional value. Collaborative efforts between researchers, policymakers, and farmers are crucial for the successful implementation of these solutions. Public and private sector partnerships can facilitate the translation of scientific discoveries into practical applications, ensuring that climate-resilient crops are accessible to farmers worldwide, thus contributing to sustainable agricultural practices and food security. The process begins with various omics layers, with genomics, transcriptomics, proteomics, metabolomics, and phenomics each capturing specific biological responses to environmental stress. These datasets undergo preprocessing steps such as normalization and filtering before being integrated through multi-omics approaches to derive holistic insights. Machine learning plays a crucial role in this pipeline, applying feature selection, pattern recognition, predictive modeling, and deep learning techniques to identify key stress-responsive genes, proteins, and metabolites. The refined models enable precise predictions of plant responses under diverse climatic conditions, ultimately guiding the development of crops with enhanced drought tolerance, heat and cold resistance, improved salinity and flood resilience, and optimized yields. This integration of omics and artificial intelligence accelerates the breeding and engineering of robust crop varieties suited for future climatic challenges (Figure 9.3).

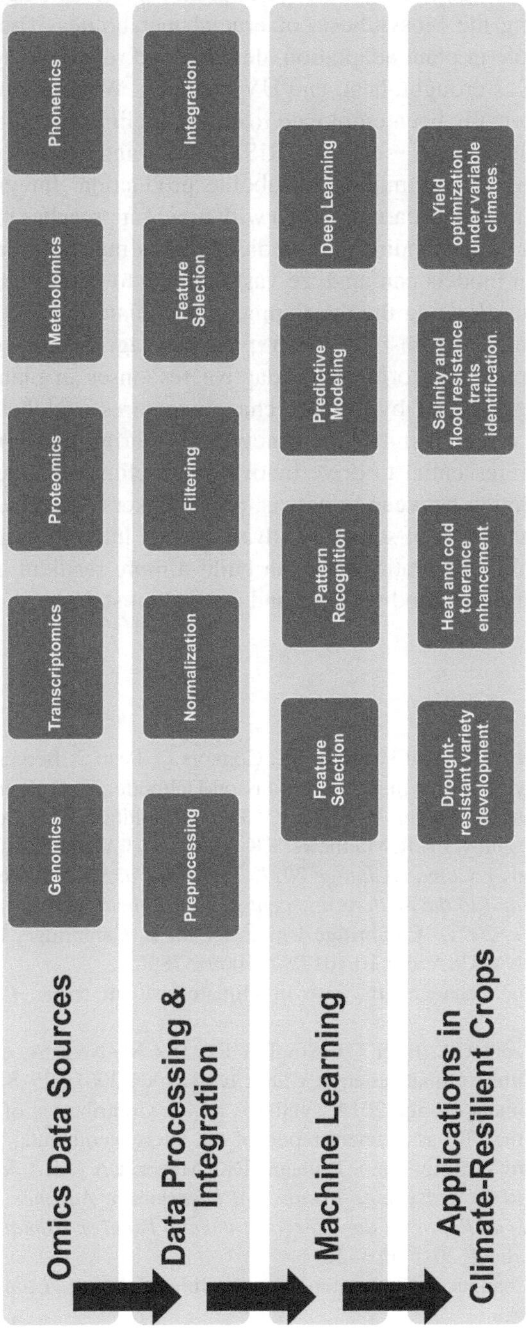

FIGURE 9.3 The integration of omics approaches with machine learning to develop climate-resilient crops by systematically linking biological data with computational analysis.

9.14 CONCLUSION

Climate change poses a significant challenge to plant survival and agricultural productivity by influencing the biosynthesis of crucial metabolites. These metabolites play a fundamental role in plant adaptation, defense, and resilience against environmental stressors such as drought, heat, and UV radiation. While traditional breeding techniques have historically been employed to enhance climate resilience, emerging biotechnological advancements, such as CRISPR-Cas9 and metabolic engineering, offer precise solutions for optimizing metabolite production. Integrating artificial intelligence (AI) and machine learning (ML) with omics approaches presents a transformative strategy for understanding and predicting plant metabolic responses to climate stress. AI-driven models can analyze vast datasets to identify key metabolites and genetic targets, accelerating the development of climate-resilient crop varieties. Furthermore, combining AI insights with precision agriculture techniques can enhance real-time stress monitoring and adaptive responses in plants. Addressing the complex challenges posed by climate change requires a globally coordinated effort. Collaborative research initiatives, knowledge-sharing platforms, and cross-sector partnerships are essential to drive innovation in climate-resilient agriculture. Strengthening cooperation between scientists, policymakers, farmers, and industries will accelerate the translation of scientific advancements into practical solutions. By fostering international collaboration, we can build a more resilient and sustainable agricultural future, ensuring food security and ecosystem stability in an era of rapid climate change.

REFERENCES

1. Masson-Delmotte VP, Zhai P, Pirani SL, Connors C, Péan S, Berger N, et al. IPCC, 2021: Summary for policymakers. In: Masson-Delmotte, V., P. Zhai, A. Pirani, S.L. Connors, C. Péan, S. Berger, N. Caud, Y. Chen, L. Goldfarb, M.I. Gomis, M. Huang, K. Leitzell, E. Lonnoy, J.B.R. Matthews, T.K. Maycock, T. Waterfield, O. Yelekçi, R. Yu, and B. Zhou (eds.). *Climate Change 2021: The Physical Science Basis. Contribution of Working Group I to the Sixth Assessment Report of the Intergovernmental Panel on Climate Change.* 2021, Cambridge University Press, Cambridge, United Kingdom and New York, NY, USA. doi:10.1017/9781009157896.
2. Kopeć P. Climate change—the rise of climate-resilient crops. *Plants.* 2024 Feb 8;13(4):490.
3. Cubasch U, Meehl GA, Boer GJ, Stouffer RJ, Dix M, Noda A, et al. Chapter 9–Projections of future climate change. *Clim Chang.* 2001;2001:525–82.
4. IPCC 2014. Climate change 2014: synthesis report. Contribution of working groups I, II and III to the fifth assessment report of the intergovernmental panel on climate change. In: Core Writing Team, Pachauri RK, Meyer, LA (eds.). Vol. 9781107025, *Managing the Risks of Extreme Events and Disasters to Advance Climate Change Adaptation: Special Report of the Intergovernmental Panel on Climate Change.* IPCC: Geneva, Switzerland; 2015. p. 3–22.
5. Trenberth K. Changes in precipitation with climate change. *Clim Res.* 2011 Mar 31;47(1):123–38.

6. Shiade SRG, Zand-Silakhoor A, Fathi A, Rahimi R, Minkina T, Rajput VD, et al. Plant metabolites and signaling pathways in response to biotic and abiotic stresses: Exploring bio stimulant applications. *Plant Stress*. 2024 Jun;12:100454.
7. Alum EU. Climate change and its impact on the bioactive compound profile of medicinal plants: implications for global health. *Plant Signal Behav*. 2024 Dec 31;19(1).
8. Chen D, Mubeen B, Hasnain A, Rizwan M, Adrees M, Naqvi SAH, et al. Role of promising secondary metabolites to confer resistance against environmental stresses in crop plants: current scenario and future perspectives. *Front Plant Sci*. 2022 May 9;13:881032.
9. Iskandar NN, Iriawati I. Vinblastine and vincristine production on madagascar periwinkle (Catharanthus roseus (L.) G. Don) callus culture treated with polethylene glycol. *Makara J Sci*. 2016 Mar 31;20(1):7-6.
10. Idrees M, Hassan IU, Naeem M, Ali A, Aftab T, Khan MMA. The accumulation and degradation of alkaloids in catharanthus roseus supported by various external agents under different environmental conditions. In: M. Naeem, Tariq Aftab, M. Masroor A. Khan (eds.). *Catharanthus roseus*. Cham: Springer International Publishing; 2017. p. 321–9.
11. Guo J, Zhou X, Wang T, Wang G, Cao F. Regulation of flavonoid metabolism in ginkgo leaves in response to different day-night temperature combinations. *Plant Physiol Biochem* . 2020 Feb;147:133–40.
12. Mao D, Zhong L, Zhao X, Wang L. Function, biosynthesis, and regulation mechanisms of flavonoids in Ginkgo biloba. *Fruit Res*. 2023;3(1):0–0.
13. Brechner ML, Albright LD, Weston LA. Effects of UV-B on secondary metabolites of St. John's Wort (*Hypericum perforatum* L.) grown in controlled environments. *Photochem Photobiol*. 2011 May 8;87(3):680–4.
14. Qaderi MM, Martel AB, Strugnell CA. Environmental factors regulate plant secondary metabolites. *Plants*. 2023 Jan 18;12(3):447.
15. Hanif S, Shakoor A, Saleem MF, Saleem I, Ali S, Ashraf MA, et al. exogenous application of proline to enhance rice tolerance against heat and drought stresses. *Pakistan J Agric Res*. 2022;35(2):324-333.
16. Nasrollahi V, Mirzaie-asl A, Piri K, Nazeri S, Mehrabi R. The effect of drought stress on the expression of key genes involved in the biosynthesis of triterpenoid saponins in liquorice (*Glycyrrhiza glabra*). *Phytochemistry*. 2014 Jul;103:32–7.
17. Zagoskina N V., Zubova MY, Nechaeva TL, Kazantseva V V., Goncharuk EA, Katanskaya VM, et al. Polyphenols in plants: structure, biosynthesis, abiotic stress regulation, and practical applications (Review). *Int J Mol Sci*. 2023 Sep 9;24(18):13874.
18. Vurro E, Bruni R, Bianchi A, Toppi L. Elevated atmospheric CO2 decreases oxidative stress and increases essential oil yield in leaves of Thymus vulgaris grown in a mini-FACE system. *Environ Exp Bot*. 2009 Jan;65(1):99–106.
19. Nocchi N, Duarte HM, Pereira RC, Konno TUP, Soares AR. Effects of UV-B radiation on secondary metabolite production, antioxidant activity, photosynthesis and herbivory interactions in *Nymphoides humboldtiana* (Menyanthaceae). *J Photochem Photobiol B Biol*. 2020 Nov;212:112021.
20. Šola I, Poljuha D, Pavičić I, Jurinjak Tušek A, Šamec D. Climate change and plant foods: the influence of environmental stressors on plant metabolites and future food sources. *Foods*. 2025 Jan 27;14(3):416.
21. Das S, Kwon M, Kim JY. Enhancement of specialized metabolites using CRISPR/Cas gene editing technology in medicinal plants. *Front Plant Sci*. 2024 Feb 21;15.
22. Hussein RA, El-Anssary AA. Plants secondary metabolites: the key drivers of the pharmacological actions of medicinal plants. *Herb Med*. 2019;1(3):11–30.

23. Erb M, Kliebenstein DJ. Plant secondary metabolites as defenses, regulators, and primary metabolites: the blurred functional trichotomy. *Plant Physiol.* 2020;184(1):39–52.
24. Thirumurugan D, Cholarajan A, Raja SS, Vijayakumar R. An introductory chapter: secondary metabolites. *Second Metab Appl.* 2018;1:13.
25. Böttger A, Vothknecht U, Bolle C, Wolf A, Böttger A, Vothknecht U, et al. Plant secondary metabolites and their general function in plants. In: *Lessons Caffeine, Cannabis Co Plant-Derived Drugs their Interact with Hum Recept.* Springer Cham; Switzerland 2018. p. 3–17.
26. Sangwan NS, Jadaun JS, Tripathi S, Mishra B, Narnoliya LK, Sangwan RS. Plant metabolic engineering. In: Debmalya Barh and Vasco Azevedo (eds.). *Omics Technologies and Bio-Engineering*. Elsevier; Netherlands, 2018. p. 143–75.
27. Zhang S, Zhu W, Wang B, Tang J, Chen X. Secondary metabolites from the invasive Solidago canadensis L. accumulation in soil and contribution to inhibition of soil pathogen *Pythium ultimum. Appl Soil Ecol.* 2011;48(3):280–6.
28. Zaynab M, Fatima M, Abbas S, Sharif Y, Umair M, Zafar MH, et al. Role of secondary metabolites in plant defense against pathogens. *Microb Pathog.* 2018;124:198–202.
29. Ntalli NG, Caboni P. Botanical nematicides: a review. *J Agric Food Chem.* 2012;60(40):9929–40.
30. Tiwari R, Rana CS. Plant secondary metabolites: a review. *Int J Eng Res Gen Sci.* 2015;3(5):661–70.
31. Doughari J. An overview of plant immunity. *J Plant Pathol Microbiol.* 2015;6(11):10–4172.
32. Perveen K, Bokahri NA. Management of Alternaria leaf blight in tomato plants by mentha essential oil. *Plant Prot Sci.* 2020;56(3):191-196.
33. Singh P, Pandey AK. Prospective of essential oils of the genus Mentha as biopesticides: a review. *Front Plant Sci.* 2018;9:409763.
34. Ahmed E, Arshad M, Khan MZ, Amjad MS, Sadaf HM, Riaz I, et al. Secondary metabolites and their multidimensional prospective in plant life. *J Pharmacogn Phytochem.* 2017;6(2):205–14.
35. Divekar PA, Narayana S, Divekar BA, Kumar R, Gadratagi BG, Ray A, et al. Plant secondary metabolites as defense tools against herbivores for sustainable crop protection. *Int J Mol Sci.* 2022;23(5):2690.
36. Wu Y, Ren D, Gao C, Li J, Du B, Wang Z, et al. Recent advances for alkaloids as botanical pesticides for use in organic agriculture. *Int J Pest Manag.* 2021;69(3):288–98.
37. Wink M. *Evolution of secondary plant metabolism.* eLS; 2016. p. 1–11.
38. Khare S, Singh NB, Singh A, Hussain I, Niharika K, Yadav V, et al. Plant secondary metabolites synthesis and their regulations under biotic and abiotic constraints. *J Plant Biol* . 2020 Jun 8;63(3):203–16.
39. Li C, Zha W, Li W, Wang J, You A. Advances in the biosynthesis of terpenoids and their ecological functions in plant resistance. *Int J Mol Sci.* 2023 Jul 17;24(14):11561.
40. Karunanithi PS, Zerbe P. Terpene synthases as metabolic gatekeepers in the evolution of plant terpenoid chemical diversity. *Front Plant Sci.* 2019 Oct 1;10:1166.
41. Shende V V., Bauman KD, Moore BS. The shikimate pathway: gateway to metabolic diversity. *Nat Prod Rep.* 2024;41(4):604–48.
42. Gritsunov A, Peek J, Diaz Caballero J, Guttman D, Christendat D. Structural and biochemical approaches uncover multiple evolutionary trajectories of plant quinate dehydrogenases. *Plant J.* 2018 Sep 11;95(5):812–22.
43. Ober D. Gene duplications and the time thereafter – examples from plant secondary metabolism. *Plant Biol.* 2010 Jul 1;12(4):570-7.

44. Fernie AR, de Vries S, de Vries J. Evolution of plant metabolism: the state-of-the-art. *Philos Trans R Soc B Biol Sci.* 2024 Nov 18;379(1914).
45. Carrington Y, Guo J, Le CH, Fillo A, Kwon J, Tran LT, et al. Evolution of a secondary metabolic pathway from primary metabolism: shikimate and quinate biosynthesis in plants. *Plant J.* 2018 Sep 11;95(5):823–33.
46. Tahara K, Milkowski C, Oda-Yamamizo C. Elucidation and reconstitution of hydrolyzable tannin biosynthesis. *Plant Biotechnol.* 2024 Sep 25;41(3):24.0601a.
47. Tahara K, Nishiguchi M, Funke E, Miyazawa SI, Miyama T, Milkowski C. Dehydroquinate dehydratase/shikimate dehydrogenases involved in gallate biosynthesis of the aluminum-tolerant tree species Eucalyptus camaldulensis. *Planta.* 2021 Jan 21;253(1):3.
48. Zenk MH, Juenger M. Evolution and current status of the phytochemistry of nitrogenous compounds. *Phytochemistry.* 2007;68(22–24):2757–72.
49. Chomel M, Guittonny-Larchevêque M, Fernandez C, Gallet C, DesRochers A, Paré D, et al. Plant secondary metabolites: a key driver of litter decomposition and soil nutrient cycling. *J Ecol.* 2016;104(6):1527–41.
50. Kohnen-Johannsen KL, Kayser O. Tropane alkaloids: chemistry, pharmacology, biosynthesis and production. *Molecules.* 2019;24(4):796.
51. Kumar S. Alkaloidal drugs–a review. *Asian J Pharm Sci Technol.* 2014;4(3):107–19.
52. Aniszewski T. *Alkaloids-Secrets of Life: Aklaloid Chemistry, Biological Significance, Applications and Ecological Role.* Elsevier; Amsterdam, Netherlands, 2007.
53. Sikder MM, Vestergård M. Impacts of root metabolites on soil nematodes. *Front Plant Sci.* 2020;10:501535.
54. Siegrist JA, McCulley RL, Bush LP, Phillips TD. Alkaloids may not be responsible for endophyte-associated reductions in tall fescue decomposition rates. *Funct Ecol.* 2010;24(2):460–8.
55. Bento C, Gonçalves AC, Jesus F, Simões M, Silva LR. Phenolic compounds: sources, properties and applications. In: RN Park (eds.). *Bioact Compd Sources, Prop Appl Porter*; 2017. p. 271–99.
56. Rahman MM, Rahaman MS, Islam MR, Rahman F, Mithi FM, Alqahtani T, et al. Role of phenolic compounds in human disease: current knowledge and future prospects. *Molecules.* 2021;27(1):233.
57. Vuolo MM, Lima VS, Junior MRM. Phenolic compounds: structure, classification, and antioxidant power. In: *Bioactive Compounds.* Elsevier; 2019. p. 33–50.
58. Gautam AK, Singh PK, Aravind M. Defensive role of plant phenolics against pathogenic microbes for sustainable agriculture. *Plant Phenolics Sustain Agric.* 2020;1:579–94.
59. Mabou FD, Yossa IBN. TERPENES: structural classification and biological activities. *IOSR J Pharm Biol Sci e-ISSN.* 2021;16:2319–7676.
60. Jha A. Plant terpenes. In: *Phytoconstituents and Antifungals.* Elsevier; 2022. p. 77–87.
61. Qiao Z, Hu H, Shi S, Yuan X, Yan B, Chen L. An update on the function, biosynthesis and regulation of floral volatile terpenoids. *Horticulturae.* 2021;7(11):451.
62. Dambolena JS, Zunino MP, Herrera JM, Pizzolitto RP, Areco VA, Zygadlo JA. Terpenes: natural products for controlling insects of importance to human health—a structure-activity relationship study. In: Muhammad Zia-Ul-Haq, Arwa Abdulkreem AL-Huqail, Muhammad Riaz, Umar Farooq Gohar (eds.). *Psyche A J Entomol.* 2016;2016:1–17.
63. Aslam S, Jabeen T, Ahmad M, AL-Huqail AA. Cinchona. In: *Essentials of Medicinal and Aromatic Crops.* Springer; 2023. p. 221–48.

64. Srivasatava P. Use of alkaloids in plant protection. In: Ravindra Soni , Deep Chandra Suyal and Reeta Goel (eds.). *Plant Prot From Chem to Biol.* De Gruyter; Berlin, Germany, 2022;337. www.degruyterbrill.com/document/doi/10.1515/9783110771558/html
65. Hedayati-Moghadam M, Moezi SA, Kazemi T, Sami A, Akram M, Zainab R, et al. The effects of *Papaver somniferum* (opium poppy) on health, its controversies and consensus evidence. *Toxin Rev.* 2022;41(3):1030–43.
66. Burge S. Salix alba: the source of salicylic acid Filipendula ulmaria the source of aspirin Gaultheria procumbens the source of methyl salicylate. In: Henry Oakeley (ed.). *Modern Medicines from Plants.* CRC Press; 2023. p. 280–92.
67. Nuryana I, Ratnakomala S, Fahrurrozi ABJ, Andriani A, Putra FJN, Rezamela E, et al. Catechin contents, antioxidant and antibacterial activities of different types of Indonesian tea (*Camellia sinensis*). *Ann Bogor.* 2020;24(2):107.
68. Septembre-Malaterre A, Lalarizo Rakoto M, Marodon C, Bedoui Y, Nakab J, Simon E, et al. Artemisia annua, a traditional plant brought to light. *Int J Mol Sci.* 2020;21(14):4986.
69. Shinyuy LM, Loe GE, Jansen O, Mamede L, Ledoux A, Noukimi SF, et al. Secondary metabolites isolated from Artemisia afra and Artemisia annua and their anti-malarial, anti-inflammatory and Immunomodulating properties-pharmacokinetics and pharmacodynamics: a review. In: Deepu Pandita, Anu Pandita (eds.). *Metabolites.* CRS Press; New York 2023;13(5):613.
70. Yadav R, Pandita A, Pandita D, Murthy KMS. Taxus brevifolia (Nutt.) Pilger. In: *Potent Anticancer Med Plants Second Metab Profiling, Act Ingredients, Pharmacol Outcomes.* 2023. p. 10.
71. Źróbek-Sokolnik A. Temperature stress and responses of plants. In: Parvaiz Ahmad, M.N.V. Prasad (eds.). *Environ Adapt Stress Toler plants era Clim Change.* Springer Nature; London, UK 2012. p. 113–34.
72. Srivastava AK, Yadava A, Singh D. Evolutionary mechanism for biosynthesis of diverse molecules. In: Akhileshwar Kumar Srivastava, Vinod Kumar Kannaujiya, Rajesh Kumar Singh, Divya Singh (eds.). *Evolutionary Diversity as a Source for Anticancer Molecules.* Elsevier; Amsterdam, Netherlands 2021. p. 1–27.
73. Yang M, Yang J, Su L, Sun K, Li D, Liu Y, et al. Metabolic profile analysis and identification of key metabolites during rice seed germination under low-temperature stress. *Plant Sci.* 2019;289:110282.
74. Zulfiqar F, Akram NA, Ashraf M. Osmoprotection in plants under abiotic stresses: New insights into a classical phenomenon. *Planta.* 2020;251(1):3.
75. Thakur P, Nayyar H. Facing the cold stress by plants in the changing environment: sensing, signaling, and defending mechanisms. In: Narendra Tuteja, Sarvajeet Singh Gill (eds.). *Plant acclimation to environmental stress.* Springer; London, UK, 2012. p. 29–69.
76. Pirzadah TB, Malik B, Rehman RU, Hakeem KR, Qureshi MI. Signaling in response to cold stress. *Plant Signal Underst Mol Crosstalk.* 2014;193–226.
77. Zargar SM, Gupta N, Nazir M, Mahajan R, Malik FA, Sofi NR, et al. Impact of drought on photosynthesis: molecular perspective. *Plant Gene.* 2017;11:154–9.
78. Ashrafi M, Azimi-Moqadam MR, Moradi P, MohseniFard E, Shekari F, Kompany-Zareh M. Effect of drought stress on metabolite adjustments in drought tolerant and sensitive thyme. *Plant Physiol Biochem.* 2018;132:391–9.
79. Kumar M, Kumar Patel M, Kumar N, Bajpai AB, Siddique KHM. Metabolomics and molecular approaches reveal drought stress tolerance in plants. *Int J Mol Sci.* 2021;22(17):9108.

80. Fini A, Guidi L, Ferrini F, Brunetti C, Di Ferdinando M, Biricolti S, et al. Drought stress has contrasting effects on antioxidant enzymes activity and phenylpropanoid biosynthesis in Fraxinus ornus leaves: an excess light stress affair? *J Plant Physiol.* 2012;169(10):929–39.
81. Kopsell DA, Sams CE, Morrow RC. Blue wavelengths from LED lighting increase nutritionally important metabolites in specialty crops. *HortScience Horts.* 2015;50(9):1285–8.
82. Estell RE, Fredrickson EL, James DK. Effect of light intensity and wavelength on concentration of plant secondary metabolites in the leaves of Flourensia cernua. *Biochem Syst Ecol.* 2016;65:108–14.
83. Fazal H, Abbasi BH, Ahmad N, Ali SS, Akbar F, Kanwal F. Correlation of different spectral lights with biomass accumulation and production of antioxidant secondary metabolites in callus cultures of medicinally important Prunella vulgaris L. *J Photochem Photobiol B Biol.* 2016;159:1–7.
84. Ahmad N, Rab A, Ahmad N. Light-induced biochemical variations in secondary metabolite production and antioxidant activity in callus cultures of Stevia rebaudiana (Bert). *J Photochem Photobiol B Biol.* 2016;154:51–6.
85. Hemm MR, Rider SD, Ogas J, Murry DJ, Chapple C. Light induces phenylpropanoid metabolism in Arabidopsis roots. *Plant J.* 2004;38(5):765–78.
86. Ouzounis T, Fretté X, Rosenqvist E, Ottosen CO. Spectral effects of supplementary lighting on the secondary metabolites in roses, chrysanthemums, and campanulas. *J Plant Physiol.* 2014;171(16):1491–9.
87. Shohael AM, Ali MB, Yu KW, Hahn EJ, Islam R, Paek KY. Effect of light on oxidative stress, secondary metabolites and induction of antioxidant enzymes in Eleutherococcus senticosus somatic embryos in bioreactor. *Process Biochem.* 2006;41(5):1179–85.
88. Engelen-Eigles G, Holden G, Cohen JD, Gardner G. The effect of temperature, photoperiod, and light quality on gluconasturtiin concentration in watercress (Nasturtium officinale R. *Br.). J Agric Food Chem.* 2006 Jan;54(2):328–34.

10 Breeding for Stress Tolerance in Crops

Muhammad Shehzad, Furqan Ahmad, Akhtar Hameed, Shoaib Ur Rehman, Saba Yaseen, and Zulqurnain Khan

10.1 INTRODUCTION

Climate change poses a significant threat to global agriculture, necessitating the development of climate-ready crops to ensure food security in the face of increasing environmental stresses. Temperature, wind, humidity, and disasters such as floods, landslides, and drought also affect agricultural production and food security conditions perceived as vulnerable (1). The projected changes in agriculture include a rise in temperature, the occurrence of heat waves, and large changes in rainfall which may make some areas experience severe drought (2). These changes interfere with crop production leading to undernutrition, malnutrition, and hunger. These global and regional changes have the effect of mitigating food production, leading to scarcity of world grain and subsequent food insecurity. These are complex and encompass difficult climate conditions such as droughts, floods, and extremely high and low temperatures that negatively impact crop yield (3). Farming today has become very difficult due to changes in temperature and rainfall patterns, thereby making traditional methods very ineffective, thus prompting the need for change.

Climate-smart agriculture aims at developing crops that would be suitable to grow under climate change. This approach includes advanced genetic technology in genomic-assisted breeding (GAB) and CRISPR/Cas systems that increase crop resistance against different abiotic and biotic stresses (4). Recent advancements in various technologies such as high-throughput phenotyping, big data, artificial intelligence (AI), and machine learning (ML) are already revolutionizing crop breeding programs because they accelerate the pace of identifying beneficial genetics (4). One of the most significant features of developing climate-resistant crops is that it would require the use of genotypes from less adaptable genomic resources. There are various genes used for enhancing photosynthesis, high carbon uptake, and tolerance to heat, salt, drought, and other extreme weather conditions. These crops normally have inherited factors concerning the environmental factors such that they may influence the ability of crops to withstand unfavorable environmental changes. These crops may be good sources for a gene pool for a breeding program aimed at developing crops with enhanced ability to produce yields in conditions that may be influenced by climate change (3). However, the change to climate-resilient agriculture is not an easy

 DOI: 10.1201/9781032720142-13

feat. The sound application of next-generation breeding strategies entails a considerable capital outlay on research and development, and partnership between scientific experts, policymakers, and farmers. Also, there are socio-economic factors such as infrastructure development and access to appropriate technologies; the latter being a major issue within developing countries that are most affected by climate change in agriculture (4).

10.2 EFFECT OF BIOTIC AND ABIOTIC STRESSES ON MAJOR CROPS

Plants perceive environmental stress and modify themselves according to drastic climatic conditions. Various molecular pathways are involved in shaping plant systems according to new stress conditions (5). The external environment, including biotic and abiotic factors, determines the plant physiology and key physiological pathways (6). The current reduction in food production worldwide is mainly due to water shortage (7). For sustainable agriculture, consumption of renewable sources must be encouraged (8). Breeding of climate-resilient varieties of major crops is also important against adverse climatic conditions (9). Climate shifts negatively affect crop production, and changes in pest attacks and soil conditions are also drastic for sustainable crop production. These changes in the environment force plant systems to alter according to changing conditions which affects yield and results in food scarcity (10). Drastic climate changes affect the crop production and cropping schemes of any area. This unpredictable change disturbs sustainable agriculture, and so the need to develop climate-resilient varieties is imminent. This change in the agroecosystem drives researchers to predict new soil conditions to control the disease cycle according to new changing climate conditions to obtain an unaffected yield (11). Plants have shown resistance to changing conditions by various mechanisms such as stomatal closure, and tolerant varieties have shown minor reductions in plant height biomass and yield (12).

Climate change is a major challenge for sustainable agriculture as it increases the production cost and decreases yields. Countries that are facing continuous climate changes are facing difficulties in food security and other aspects (13). There is a strong link between climate vulnerability and food security. Countries with large populations are comparatively more vulnerable to drastic climate changes leading to food shortages. Only the development of climate-ready crops can provide relief in these conditions. Breeding for resistant varieties can help face these conditions (13). Changing climatic conditions impose serious threats to crop production for optimum yield, and can lead to food scarcity (14). A changing environment with more UV radiation and CO_2 emissions, and less water availability is a serious challenge, and this can only be alleviated by developing climate-resilient crops which can tolerate these environmental changes (15). Constant greenhouse gas emissions and CO_2 accumulations increase the global temperature and change the climate in a way that affects plants' physiological mechanisms, growth, and flowering patterns. Droughts, heatwaves, and flooding, driven by ongoing climate change, severely impact global crop production (16). Drought, in combination with heat waves and salinity, changes the weather

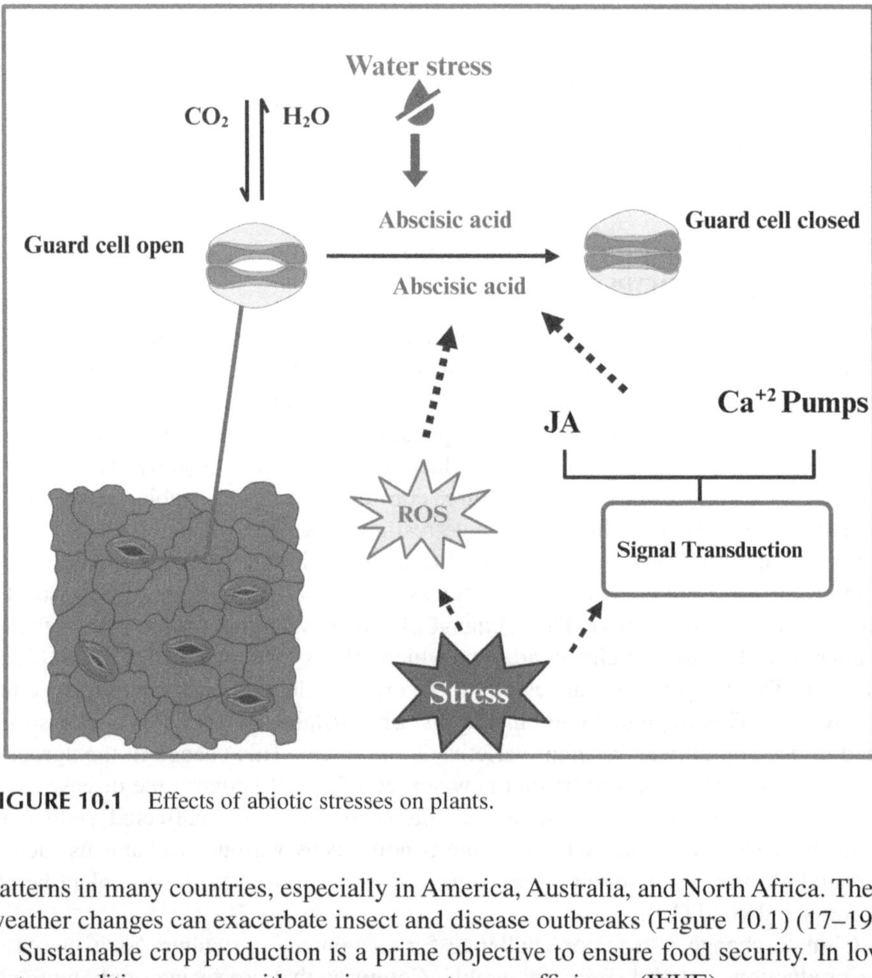

FIGURE 10.1 Effects of abiotic stresses on plants.

patterns in many countries, especially in America, Australia, and North Africa. These weather changes can exacerbate insect and disease outbreaks (Figure 10.1) (17–19).

Sustainable crop production is a prime objective to ensure food security. In low-water conditions, crops with maximum water use efficiency (WUE) can compensate for yield losses. Leaf parameters like stomatal closure are directly related to moisture conservation, and in drought conditions they become pivotal to plant survival (20). Water requirements for crop production are increasing with time to cultivate more food for the increasing population and it will be doubled by 2050, however water resources are decreasing and will decrease up to 50% due to environmental stresses (21). The need to develop crops with maximum water use efficiency and yield by consuming low moisture is increasing day by day (22).

10.3 ROLE OF ABIOTIC STRESS TOLERANCE IN SUSTAINABLE CROP PRODUCTION

Crops face various biotic and abiotic stresses during climate changes, with drought and salinity being the most significant contributors to crop damage. These abiotic stresses are major problems for cotton, wheat, rice, and maize. Frequent drought at different stages of plant growth can reduce the crop yield by more than 50%. Plant biomass and growth parameters decrease significantly during drought. Cotton is directly affected by drought stress during critical growth stages, i.e., flowering, boll formation. Drought stress is more damaging. Drought tolerance is directly related to root density and leaf thickness in cotton (23). Decreasing water availability is severely affecting cotton production and there is a strong demand for developing varieties which can withstand water shortages (24). Cotton exhibits a higher tolerance toward abiotic stresses as compared to other major crops. However, extreme environmental conditions, such as drought, affect the growth, productivity, and fiber quality of cotton (25). Similarly, salt stress is also damaging to crop normal growth and functioning. When plants encounter these stresses, various defense mechanisms are activated and concentrations of some enzymes increase to reduce ROS produced as a result of stresses. These defense mechanisms are at the cellular and subcellular levels. In salt stress, the plant faces difficulty in osmotic adjustment as the concentrations of various salts like Na^+ and Cl^- increase in root zones, and then the increase of salt concentration in leaves disturbs transpiration (26). Salt-affected areas are increasing due to changing climatic conditions. More than 800 million hectares are directly affected by salinity stress (27). It is estimated that more than 50% of currently cultivated land will be salt-affected by 2050, and so there is a strong need to develop salt-tolerant varieties (28). Drought stress affects plants and triggers various stress-related genes and pathways. Plants increase enzyme concentrations like catalase peroxidases to alleviate the effects of ROS produced as a result of stress. Various metabolites and abscisic acid (ABA) concentrations increase stress conditions. ABA mediated drought tolerance by stomatal closure and regulating water use efficiency (29).

10.4 BREEDING FOR DROUGHT TOLERANCE

Conventional breeding techniques, including selection and pedigree breeding can be used to include a plant's drought-resistance gene (1). To develop climate-ready crops, selection techniques like genetic modification (GM) and marker-assisted selection (MAS) are frequently employed. This is because these approaches enable the rapid pooling of all desired features. Several transgenic plants resistant to drought have been created, such as rice, tomato, soybean, maize, and barley, by encoding proteins with genes that regulate the networks responsible for controlling drought. Transgenic technology appears to be a more appealing method than traditional breeding as it enables the direct induction of a single or a set of desired genes (30). There are several genes involved in transgenic breeding that lead to the expression of traits that enable them to resist the effects of drought. The variety possesses the gene for the expression of drought-resistant characteristics that may be used for further introgression.

The main objective of climate-ready crop engineering is to enhance WUE, which can be achieved by two different processes (1). The first is the inclusion of genes for osmolites which are compatible, such as genes encoding sugar and amino acids. The next strategy provides a measure of resistance to dehydration by engaging in constitutive expression of embryonic proteins. The first group of genes is called the osmoprotectants mechanism, which is a cellular protectant mechanism. This occurs through the formation of osmoprotectants such as proline, mannitol, trehalose, glycerin, and betaine that protect plants from dehydration during water stress conditions. To produce the enzymes that are needed for the control and synthesis of these osmoprotectants, this process aims at fostering transcription. Transgenic wheat is stated as having an accumulation of these chemicals due to gene transfer and is postulated as being a desirable method of increasing wheat's ability to endure drought situations. Overexpression of proline in tomatoes and tobacco makes them resilient to drought stress. In the glutamate process used in plants, the proline is synthesized from glutamate.

Glutamate is converted to its semialdehyde by the enzyme P5CS (Δ^1-pyrroline-5-carboxylate synthase), which in turn is reduced to proline. The synthesis of proline is regulated by the P5CS enzyme which is feedback inhibited. Such osmoprotectants as the engineering of trehalose macromolecule, for instance, were employed in the creation of the corn variety Drought Gard TM that was released to the United States markets in 2013. Similarly, the transformation of Cold Shock Proteins (CSPs) genes from a bacterium, *E. coli*, enhanced grain yield in drought-affected maize and resistance to drought stress (31). The second method involves the use of transcription factors. The presence of environmental stress leads to the activation of several transcription factors in plants. During gene expression, the elements that bind with the enhancer or promotor region of genes include transcription factors and signaling molecules. Transgenic breeding employs a variety of transcription factors, including *WARKY2* proteins, *HDG11*, a 60-amino-acid repressor and activator of genes for drought tolerance, and dehydration-responsive element binding (*DREB*).

10.5 MOLECULAR DYNAMICS OF GREENHOUSE GAS IN THE THERMOTOLERANCE OF PLANTS

Heat-trapping greenhouse gas is causing global temperatures to rise significantly. The annual average temperature of the year 2023 was 1.47°C above the baseline temperature in the pre-industrial period of 1850–1900 (32). The impacts of high temperatures and greenhouse gases are deadly to people as well as plants and trees around the world. The global community is worried because it undermines food resources and hampers crop yields. Therefore, it may become necessary to seek means and methods for controlling high heat and greenhouse gases such as CO_2 and at the same time promote agricultural growth and production. These stressors can include environmental factors such as heat, light, and pollution, and should in the future be controlled by the use of genetic engineering and synthetic biology to reprogram the biological functioning of plants (32). Wang et al. (35) have provided clear evidence that external glucose (Gluc) treatment and high CO_2 levels can induce thermotolerance in tomato plants. Apoplastic Gluc is an energy and signaling molecule that accumulates more when the concentration of CO_2 is high (33). Some of the predictions made were that exposure to CO_2 may enhance heat tolerance when plants are grown under high temperatures. They discovered that increased levels of CO_2 give more heat tolerance than normal CO_2 levels that are in the environment. This study explored how modulation of thermotolerance at higher CO_2 and Gluc in the apoplast and all cells are connected. The results showed that apoplastic Gluc levels increased 1.93 times due to enhanced CO_2 therapy. However, it's unclear whether tomato plants treated with Gluc may develop thermotolerance. According to Wang et al. (35), tomato plants sprayed with exogenous Gluc exhibit a consistent photochemical quantum yield of PSII (Fv/Fm) and are more heat resistant than wild-type plants, even at ambient CO_2. This emphasizes that CO_2 and glucose are essential to thermotolerance (32).

The fundamental process of Gluc-mediated thermotolerance was then elucidated by the authors (35). According to reports, a Gluc sensor in *Arabidopsis* is the regulator of the G protein signaling 1 (RGS1)/heterotrimeric G protein pathway (34). It was revealed that RGS1 functions as a negative regulator in tomatoes throughout heat stress and a cell-based biolayer interferometry (cBLI) experiment was used to show a direct connection between RGS1 and Gluc (35). It was previously demonstrated that RGS1 binds to a Gα protein (AtGPA1) and Gluc causes the RGS1–GPA1 complex to dissociate (34). The authors (35) used bimolecular fluorescence complementation (BiFC) and Split Luc assays to observe the RGS1–GPA1 interaction in tomatoes and generated gpa1 gene-editing lines to reduce thermotolerance, confirming GPA1's direct role in this process.

10.6 ROLE OF MODERN TECHNIQUES IN DEVELOPING CLIMATE-RESILIENT CROPS

Conventional breeding is important for developing climate-resilient varieties. Molecular diversity can be explored to attain stable varieties. A gene pool can be utilized for backcross breeding to get the desired genome sequence through marker-assisted selection (36). Environment-resilient crop production is needed in the future. Crop biodiversity can be an important source of variation in crop breeding and can provide a link between environmental stress and resilience (36). Sustainable agriculture is the need of the hour with improved production technology and best practices focusing on sustainable solutions. Agrochemicals can be useful in producing better crop yields without affecting quality. Various omics techniques, like transcriptomics and metabolomics, can be used in sustainable agriculture. Transcriptomics has been used to identify transcription factors that are useful against environmental stresses (37). Metabolomics deals with protein functioning, metabolite profiling, and cellular metabolism (38). Transcription factors play a vital role in plant survival against drastic environmental conditions. MYC and MYB are two key transcription factors which are expressed in stress responses (Figure 10.2) (39).

Breeding methods can be used along with modern tools to get the desired genome and sequence for developing stress-tolerant and stress-resistant varieties. These varieties endure environmental challenges with a minimum decline in yield (15). Drought and salinity tolerance are complex traits, with many gene groups controlling signaling pathways that affect various physiological and molecular mechanisms. Mitogen-activated kinase protein (MAPK) networks are activated in stress responses, and several hormones like ABA biosynthesis increase, which alters stomal parameters and opening and closing of guard cells. Several other hormones, like jasmonic acid and ethylene, are also involved in stress responses (23).

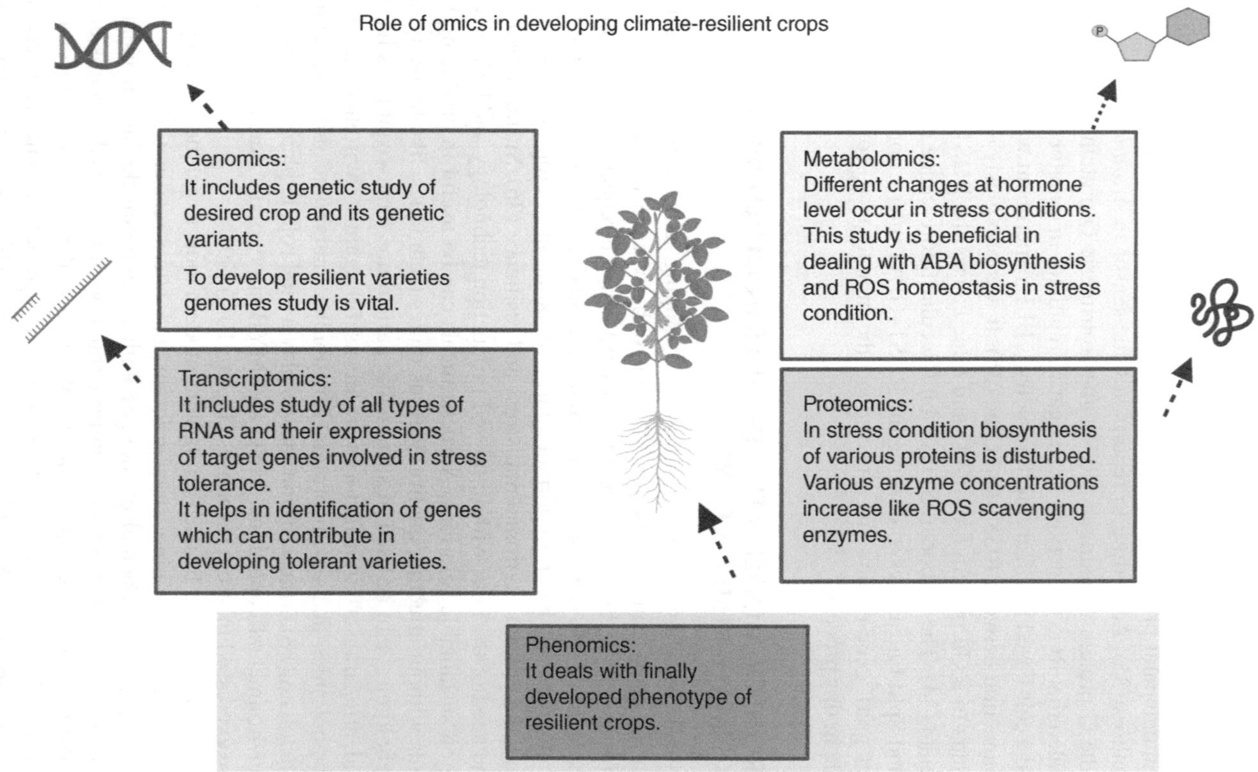

FIGURE 10.2 Role of omics in developing climate-resilient crops.

MYB, *WRKY*, *ERF*, *NAC*, and *ZIP* are important transcription factors in drought resistance in cotton. Genetic engineering of these TFs has proved that they activate stress response mechanisms in cotton in drought stress conditions. Similarly, GhABF2 is involved in drought and salt tolerance in *Arabidopsis*. This TF is involved in the regulation of ABA (40), which is directly connected with stomatal behaviors and opening and closing of guard cells. Transcriptome analysis has confirmed the increase in concentration of ROS scavenging enzymes (SOS, POD) in overexpression of GhABF2 and leads to plant survival despite environmental stresses. An *R2R3-typMYB* transcription factor gene, *GbMYB5*, is also important in abiotic stresses, like drought, as its overexpression in drought conditions increases the ABA concentration in tobacco and cotton. ROS scavenging enzymes are another stress indicator and they are also increased in transgenic plants, while lipid peroxidation was lower as a decrease in MDA concentration was reported in transgenic lines (Table 10.1) (41).

Stress-responsive genes and transcription factors affect plant defense pathways and ABA-dependent and ABA-independent pathways that are induced in stress conditions. ABRE (ABA stress response genes) is a group of transcription factors that control ABA biosynthesis in changing environment conditions and plays a role in plant defenses against stress conditions. *AREB/ABFs* and *DREB2A* transcription factors control the signaling network and regulate stress response genes (44). DNA markers aided the selection of superior climate-resistant genotypes that can be used for breeding against abiotic stress. Genomics can be helpful in crop improvement during specific stresses. The use of markers in selection can accelerate the breeding process (47). Breeding strategies can be shaped with the help of genomics for the development of environment-ready crops. This can be useful for breeding cultivars that can endure future climatic threats (48). The existing germplasm of crops can

TABLE 10.1
List of genes and transcription factors involved in abiotic stress tolerance

Sr no.	Candidate genes/TFs	Role in different areas of plant physiology	References
1	GhMKK3	Drought tolerance	(42)
2	GhMPK17	Tolerance against osmotic stress	(43)
3	AREB/ABFs and DREB2A	ABA biosynthesis and stomatal closure	(44)
4	NCED3 and GhWRKY41	Decrease lipid peroxidation	(45)
5	GbMYB5	Drought tolerance	(46)
6	MYB, WRKY, ERF, NAC, and bZIP	Drought resistance in cotton	(40)
7	P5CS, OTS, MT1D, M6PR, S6PDH, IMT1	Osmotic adjustment in salinity stress	(27)
8	R2R3-typMYB	Drought tolerance	(41)

be useful for identifying the traits that are impactful against environmental stress. The biodiversity of germplasm can be explored, and cultivars with better resistance against environmental stress can be developed. The genetics behind the traits can be examined by genetic studies. Genetic mapping and QTL mapping can be helpful in the acceleration of desired traits in cultivars (49). Genotype-by-sequencing and genome-wide association studies can be important methods in determining important traits and dissecting the genetics of traits. Additionally, next-generation sequencing can be used for the precise location of desired genes (50).

GbMYB-overexpressing tobacco showed an increase in concentration of drought-tolerant genes like *NCED3* and *GhWRKY41* and also showed higher concentrations of antioxidants and lower concentrations of lipid peroxidation. Transgenic plants showed higher drought tolerance as compared to wild-type plants (45). Antioxidant production and genes controlling antioxidants also showed an increase in their expression in drought stress. Some minor genes also have an additive effect against drought stress. The role of plant protectants and growth hormones is also very important against abiotic stress. Growth hormones like ABA and gibberellic acid jasmonic acid also play a role in stress tolerance. These hormones help in osmotic adjustment in low moisture conditions (51). Foliar application of hormones and proline can help plant survival against abiotic stresses and increase plant physiological processes like photosynthesis, stomatal conductance, etc. (41)

GbMYB5 helps in defense against drought, and virus-induced silencing of *GbMYB5* confirms that transgenic plants are more prone to drought stress as compared to wild type. Silencing of GbMYB5 increases the malonaldehyde concentration and decreases proline and soluble sugars. Other genes, like *RD22* and *NCED3*, remain unaffected by the silencing of *GbMYB5*. In another way, overexpression of the genes in tobacco increases the concentration of ABA which controls stomatal closure, hence preventing the excessive loss of moisture in drought conditions. Lipid peroxidation was also reduced and the MDA concentration was decreased along with antioxidant enzymes (46).

10.7 OPPORTUNITIES AND CHALLENGES IN CLIMATE-SMART CROP PRODUCTIONS

Climate change can be countered by developing resilient varieties of major crops. Conventional breeding methods can be useful but they are time-consuming and lengthy. If we aid them with modern tools like MAS, bioinformatics, and biotechnology then we can get the desired results within time and with more precision. Genome-wide association studies (GWAS) have been proved useful in identifying the genetic loci associated with desirable traits such as tolerance against specific stresses (Figure 10.3) (52).

The key to developing new and resilient crops is understanding the vital mechanisms and pathways involved in plant growth and development. Plant yields and stress tolerance change with the external environment, and a single stress or combined stresses can change the plant mechanism. The ultimate goal of breeding is to develop genotypes that can use nutrients efficiently and increase photosynthetic

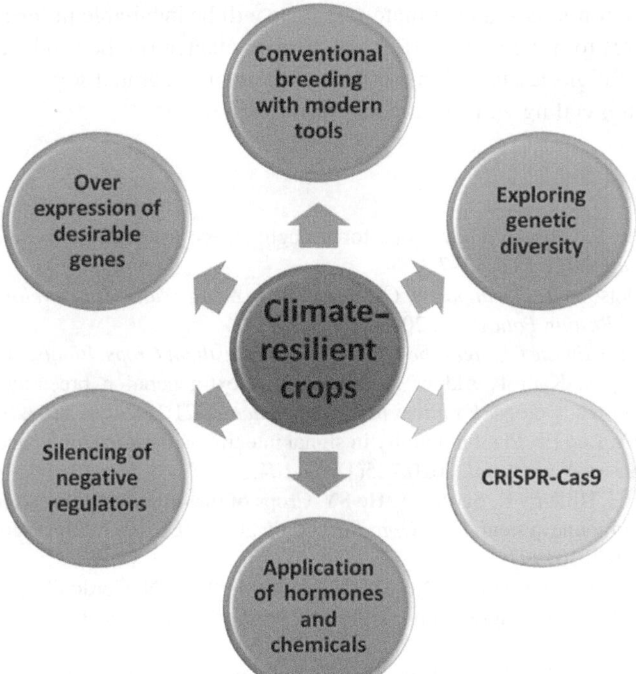

FIGURE 10.3 Techniques for developing climate resilience in crop plants.

efficiency to conserve the moisture level of plants without decreasing their yield and quality (53).

10.8 CONCLUSION

Climate resilience is one of the key approaches toward managing the complex impacts of climate change and it is for this reason that the development of climate-ready crops is such an important strategy. The rising and fluctuating temperatures of the global climate increase vulnerability, and the stability of traditional crops is in question. However, by adopting the modern tools of breeding and genetic engineering, such as genomic selection and CRISPR/Cas systems, it is feasible to grow crops that are resistant to these stresses. The use of big data analytics and AI in crop improvement programs takes crop improvement to a new level of efficiency and speeds up the process of selecting and deploying advantageous genes for the improvement of agriculture that will making the industry more robust. Focusing on the future, it is possible to note several promising lines of further development of climate-ready crops. First, further exploration of the genetic variability of underexploited genetic material can discover further characteristics that improve crop resistance. With the creation of new methods, breeders will be able to produce varieties with improved adaptability to various environmental pressures. Second, interdisciplinary collaboration among geneticists, agronomists, and climate scientists will be inevitable in formulating elaborate measures to mitigate the effects of climate change on the production of crops. Such collaborations are useful in the development of integrated solutions that incorporate modern breeding with sustainable farming methods.

REFERENCES

1. Bagale S. Climate ready crops for drought stress: a review in Nepalese. *Rev Food Agricult.* 2021; 2(2):83–7.
2. Mendelsohn R. *The Impact of Climate Change on Agriculture in Developing Countries. J. Nat. Resour. Policy Res.* 2008;1(1):5–19.
3. Kope P. *Climate Change-The Rise of Climate-Resilient Crops. Plants.* 2024;13 (4):490.
4. Razzaq A, Kaur P, Akhter N, Wani SH. Next-generation breeding strategies for climate-ready crops. *Frontiers in Plant Science.* 2021;12(July): *620420.*
5. Saijo Y, Loo EP. Plant immunity in signal integration between biotic and abiotic stress responses. *New Phytol.* 2020;225(1):87–104.
6. Kim JH, Hilleary R, Seroka A, He SY. Crops of the future: building a climate-resilient plant immune system. *Curr Opin Plant Biol.* 2021;60:101997. https://doi.org/10.1016/j.pbi.2020.101997
7. Rockström J, Williams J, Daily G, Noble A, Matthews N, Gordon L, et al. Sustainable intensification of agriculture for human prosperity and global sustainability. *Ambio.* 2017;46(1):4–17.
8. Mustafa MA, Mabhaudhi T, Massawe F. Building a resilient and sustainable food system in a changing world – a case for climate-smart and nutrient dense crops. *Glob Food Sec.* 2021;28(November 2020):100477. https://doi.org/10.1016/j.gfs.2020.100477
9. Singh JS, Koushal S, Kumar A, Vimal SR, Gupta VK. Book review: microbial inoculants in sustainable agricultural productivity- Vol. II: functional application. *Front Microbiol.* 2016;7(December).
10. Thornton PK, Jones PG, Alagarswamy G, Andresen J. Spatial variation of crop yield response to climate change in East Africa. *Global Environ Change.* 2009;19(1):54–65.
11. Sarwar N, Imran M, Shaheen MR, Ishaque W, Kamran MA, Matloob A, et al. Phytoremediation strategies for soils contaminated with heavy metals: modifications and future perspectives. *Chemosphere.* 2017;171(January):710–21.
12. Iqbal M. Vicia faba bioassay for environmental toxicity monitoring: a review. *Chemosphere.* 2016; 144:785–802. http://dx.doi.org/10.1016/j.chemosphere.2015.09.048
13. Banga SS, Kang MS. Developing climate-resilient crops. *J Crop Improv.* 2014;28(1):57–87.
14. Hatfield JL, Antle J, Garrett KA, Izaurralde RC, Mader T, Marshall E, et al. Indicators of climate change in agricultural systems. *Clim Change.* 2020;163(4):1719–32.
15. Xu J, Hou QM, Khare T, Verma SK, Kumar V. Exploring miRNAs for developing climate-resilient crops: a perspective review. *Sci Total Environ.* 2019;653:91–104.
16. Anderegg WRL, Trugman AT, Badgley G, Anderson CM. *Climate-Driven Risks to the Climate Mitigation Potential of Forests.* Science. 2020;368(6497):eaaz7005.
17. Markham KK, Greenham K. Tansley insight abiotic stress through time. *New Phytologist.* 2021;231(1):40–6.
18. Phophi MM, Mafongoya P, Lottering S. *Perceptions of Climate Change and Drivers of Insect Pest Outbreaks in Vegetable Crops in Limpopo Province of South Africa.* Climate. 2020;8(2):27.

19. Salih AAM, Baraibar M, Mwangi KK, Artan G. Climate change and locust outbreak in East Africa. *Nat Clim Chang*. 2020;10(July):2019–20. http://dx.doi.org/10.1038/s41558-020-0835-8
20. Chaves MM, Maroco JP, Pereira JS. Understanding plant responses to drought – From genes to the whole plant. *Funct Plant Biol*. 2003;30(3):239–64.
21. Gupta A, Rico-Medina A, Caño-Delgado AI. The physiology of plant responses to drought. *Science* (1979). 2020;368(6488):266–9.
22. Rui M, Chen R, Jing Y, Wu F, Jiang H, Wang Y, et al. *Guard Cell and Subsidiary Cell Sizes Are Key Determinants for Stomatal Kinetics and Drought Adaptation in Cereal Crops*. New Phytologist. 2024; 242(6):2479-94.
23. Mahmood T, Khalid S, Abdullah M, Ahmed Z, Shah MKN, Ghafoor A, et al. Insights into drought stress signaling in plants and the molecular genetic basis of cotton drought tolerance. *Cells*. 2020;9(1):105.
24. Shahzad M, Khan Z, Nazeer W, Arshad SF, Ahmad F. Effect of drought on trichome density and length in cotton (*Gossypium hirsutum*). *Journal of Bioresource Management*. 2021;8(1):154–67.
25. Parida BR, Collado WB, Borah R, Hazarika MK, Samarakoon L. Detecting drought-prone areas of rice agriculture using a MODIS-derived soil moisture index. *GIsci Remote Sens*. 2008;45(1):109–29.
26. Guo Y, Pang C, Jia X, Ma Q, Dou L, Zhao F, et al. An NAM domain gene, GhNAC79, improves resistance to drought stress in upland cotton. *Front Plant Sci*. 2017;8(September):1–15.
27. Munns R, Tester M. Mechanisms of salinity tolerance. *Annu Rev Plant Biol*. 2008;59:651–81.
28. Sikder RK, Wang X, Jin D, Zhang H, Gui H, Dong Q, et al. Screening and evaluation of reliable traits of upland cotton (*Gossypium hirsutum* L.) genotypes for salt tolerance at the seedling growth stage. *J Cotton Res*. 2020;3(1):1–13.
29. Lv X, Li Y, Chen R, Rui M, Wang Y. Stomatal responses of two drought-tolerant Barley varieties with different ROS regulation strategies under drought conditions. *Antioxidants*. 2023;12(4):790.
30. Gosal SS, Wani SH, Kang MS. *Biotechnology and Drought Tolerance*. 2009; 23(1):19-54.
31. Castiglioni P, Warner D, Bensen RJ, Anstrom DC, Harrison J, Stoecker M, et al. Bacterial RNA chaperones confer abiotic stress Tolerance in Plants and Improved Grain Yield in Maize under Water-Limited Conditions [W]. 2008;147(June):446–55.
32. Jain P. Climate-ready crops: unveiling the molecular dynamics of CO 2 and glucose in plant thermotolerance News and Views. *Plant Physiol [Internet*]. 2024;00(0):1–2. https://doi.org/10.1093/plphys/kiae131
33. Häusler RE, Heinrichs L, Schmitz J, Flügge UI. How sugars might coordinate chloroplast and nuclear gene expression during acclimation to high light intensities. 2014;(May):1121–37.
34. Urano D, Phan N, Jones JC, Yang J, Huang J, Grigston J, et al. Endocytosis of the seven-transmembrane RGS1 protein activates G-protein-coupled signaling in Arabidopsis. *Nat Cell Biol*. 2012;14(10).
35. Wang J, Luo Q, Liang X, Liu H, Wu C, Fang H, et al. Glucose-G protein signaling plays a crucial role in tomato resilience to high temperature and elevated CO 2. *Plant Physiol*. 2024;195(2):1025–37. https://doi.org/10.1093/plphys/kiae136
36. Lin BB. Resilience in agriculture through crop diversification: adaptive management for environmental change. *Bioscience*. 2011;61(3):183–93.

37. Ruotolo R, Maestri E, Pagano L, Marmiroli M, White JC, Marmiroli N. Plant response to metal-containing engineered nanomaterials: an omics-based perspective. *Environ Sci Technol.* 2018;52(5):2451–67.
38. Matich EK, Chavez Soria NG, Aga DS, Atilla-Gokcumen GE. Applications of metabolomics in assessing ecological effects of emerging contaminants and pollutants on plants. *J Hazard Mater.* 2019;373(September 2018):527–35. https://doi.org/10.1016/j.jhazmat.2019.02.084
39. Chen T, Li W, Hu X, Guo J, Liu A, Zhang B. A cotton MYB transcription factor, GbMYB5, is positively involved in plant adaptive response to drought stress. *Plant Cell Physiol.* 2014;56(5):917–29.
40. Liang C, Meng Z, Meng Z, Malik W, Yan R, Lwin KM, et al. GhABF2, a bZIP transcription factor, confers drought and salinity tolerance in cotton (*Gossypium hirsutum* L.). *Sci Rep.* 2016;6(October):1–14. http://dx.doi.org/10.1038/srep35040
41. Ullah A, Sun H, Yang X, Zhang X. Drought coping strategies in cotton: increased crop per drop. *Plant Biotechnol J.* 2017;15(3):271–84.
42. Zafar S, Afzal H, Ijaz A, Mahmood A, Ayub A, Nayab A, et al. Cotton and drought stress: an updated overview for improving stress tolerance. *South Afr J Bot.* 2023;161:258–68.
43. Sadau SB, Ahmad A, Tajo SM, Ibrahim S, Kazeem BB, Wei H, et al. Overexpression of ghmpk3 from cotton enhances cold, drought, and salt stress in arabidopsis. *Agronomy.* 2021;11(6):1–18.
44. Yoshida T, Mogami J, Yamaguchi-Shinozaki K. ABA-dependent and ABA-independent signaling in response to osmotic stress in plants. *Curr Opin Plant Biol.* 2014;21:133–9. http://dx.doi.org/10.1016/j.pbi.2014.07.009
45. Chu X, Wang C, Chen X, Lu W, Li H, Wang X, et al. The cotton WRKY gene GhWRKY41 positively regulates salt and drought stress tolerance in transgenic *Nicotiana benthamiana. PLoS One.* 2015;10(11):1–21.
46. Chen T, Zhang B. Measurements of proline and malondialdehyde content and antioxidant enzyme activities in leaves of drought stressed cotton. *Bio Protoc.* 2016;6(17):1–14.
47. Muthamilarasan M, Prasad M. Plant innate immunity: an updated insight into defense mechanism. *J Biosci.* 2013;38(2):433–49.
48. Varshney RK, Graner A, Sorrells ME. Genic microsatellite markers in plants: features and applications. *Trends Biotechnol.* 2005;23(1):48–55.
49. Kulwal PL, Thudi M, Rajeev K. *Genomics Interventions in Crop Breeding for Sustainable Agriculture.* Springer; 2012.
50. Varshney RK, Terauchi R, McCouch SR. Harvesting the promising fruits of genomics: applying genome sequencing technologies to crop breeding. *PLoS Biol.* 2014;12(6):1–8.
51. Anjum SA, Wang LC, Farooq M, Hussain M, Xue LL, Zou CM. Brassinolide application improves the drought tolerance in maize through modulation of enzymatic antioxidants and leaf gas exchange. *J Agron Crop Sci.* 2011;197(3):177–85.
52. Khan Z, Ali Z, Khan AA. Cotton breeding and biotechnology. *Cotton Breed Biotechnol.* CRC Press, 2022:47-68.
53. Rivero RM, Mittler R, Blumwald E, Zandalinas SI. Developing climate-resilient crops: improving plant tolerance to stress combination. *Plant J.* 2022;109(2):373–89.

Section IV

Techniques for Climate Mitigation

11 Mitigation of Climate Change Impacts Through the Development of Genetically Modified Crops

Muhammad Tayyab, Ali Ijaz Ahmed, Ayesha Javed, Muhammad Tehseen Azhar, Sultan Habibullah Khan, Rana Muhammad Atif, and Iqrar Ahmad Rana

11.1 INTRODUCTION

Climate change is a chronic problem caused by human activities. Pollution, deforestation, urbanization, and the emission of greenhouse gases have impacted temperatures and typical weather patterns. It is reported that greenhouse and industrial gases result in trapped heat in the atmosphere, resulting in an overall increase in the Earth's temperature. Warmer summers, prolonged winters, abrupt heavy rains, and unexpected heat spells, are due to climate change and result in crucial problems, such as rising sea levels, ocean acidification, droughts, floods, increased rainfall, storms, wildfires, landslides, etc. (Fawzy et al., 2020). Climate change affects almost every field of life including forestry, agriculture, fisheries, water resources, health, tourism, transportation, infrastructure, and construction.

In agriculture, climate change directly affects food supplies and overall food production due to irreversible fluctuations in weather. Economies dependent on agriculture are at risk of food shortages. It also alters the biodiversity, which results in disturbance in the ecosystem and variations in temperature, leading to many vector-borne and waterborne diseases (Abbass et al., 2022). About half of the total production loss is due to biotic and abiotic stresses. Apart from abiotic stress, it also promotes the growth and severity of pests and pathogens. For example, increased humidity and temperature provide the ideal conditions for insects and diseases. The average global temperature is expected to increase by 1.5°C by 2050, and every 1°C rise results in a 5–10% yield loss (Bibi and Rahman, 2023). Different strategies are used to cope with climate change in agriculture, such as genetically modified (GM) crops.

GM crops can be defined as crops with alterations in their genetic makeup done using various genetic engineering techniques. These alterations involve inserting

DOI: 10.1201/9781032720142-15

segments of foreign DNA from the same or a different species, or even from other kingdoms. These techniques have been used for more than 30 years and about 525 events in 32 different crops have been approved for yield production and to make crops tolerant to abiotic (salinity, heat, drought, and flood) and biotic stresses (insects, weeds, and diseases) (Ghimire et al., 2023; Kumar et al., 2020). Globally, about 190 million hectares of land is under cultivation of GM crops. These crops include cotton, soybean, maize, canola, and others (Turnbull et al., 2021). The global market for GM crops is $20.1 billion and projected to increase up to $30.2 billion in 2026 (Redden, 2021). Kumar et al. (2020) reported a yield increase of up to 22% and increased profits for farmers by 68% when growing GM crops.

The world's population has been increasing rapidly and will reach 9 billion in 2050. With this increase, the level of greenhouse gas emissions will reach by up to 60% (400 ppm). According to current scenarios, this will result in a temperature increase of 2°C by 2100, leading to ice melts, severe droughts, and frost. These adverse conditions will result in reductions of crop yields and quality. Traditional breeding methods are not sufficient to develop climate-resilient varieties in these changing environmental conditions. Unlike conventional breeding, GM technology offers a faster, easier, and more productive way to develop resistant cultivars (Redden, 2021).

11.2 IMPACT OF CLIMATE CHANGE ON CROP PRODUCTION

The major climatic changes include a rise in temperature, increased levels of CO_2, uneven rainfall, rise in sea level, increased frequency and intensity of extreme weather events, and negative changes in soil chemistry. Temperature is one of the most important factors for crop production, from germination to harvesting. Temperature with drought can cause more than a 50% reduction in the overall production of crops, especially at the flowering stage (Dash et al., 2022). For example, 0.5°C and 1.5°C increases in temperature could cause 0.45 t/ha and 0.75 t/ha yield losses in wheat, respectively (Bhattacharya, 2019). This affects photosynthesis, stomatal conductance, and respiration, inactivating enzymes, denaturing protein, and impairing the membrane system. For example, the yields of soybean, wheat, and rice can be decreased by 36%, 20%, and 17–35%, respectively (Batra et al., 2023; Sharma et *al.*, 2022).

Drought is another effect of climate change that could cause yield losses of up to 9% globally (Wang et al., 2022b). Crop loss also varies from area to area and crop to crop. For example, in Europe, the United States of America, and Asia, the production losses due to drought are 10%, 10%–27%, and below 10%, respectively (Naumann et al., 2021; Leng, 2021; Li et al., 2011a). Extreme water shortage could cause a yield loss of 35% in soybean and 20% in wheat and maize (Rodziewicz and Dice, 2020). Drought harms the quality of crops, for example, it reduces oil contents as well as morphological characteristics, such as the number of branches and plant height (Rad and Zandi, 2012).

No doubt, high levels of CO_2 result in an increase in photosynthesis which results in improved growth and production, but this elevated level of CO_2 leads to a reduction in protein contents and essential minerals with an increase in the carbohydrates level, as well as fluctuation in the concentrations of other nutrients, such as an 8%

average reduction in 25 nutrients including calcium, zinc, iron, and potassium. This ultimately results in lowering the quality of food and compromising human health by decreasing growth in children and immunity, malnutrition, increasing obesity, and childhood mortality (Shaheen et al., 2022; Sathee et al., 2022).

In many areas of the world, there is no source of irrigation except rain. Due to climate change, there is no predictable rainfall pattern, which causes major crop losses at any stage in the life of crop plants. Less rainfall can cause a water shortage, resulting in small leaf development which decreases photosynthesis and the ultimate yield. Climate change also affects the intensity of rainfall, which leads to soil erosion because heavy rain droplets break the soil aggregates into fine particles that can easily run off (Geneti, 2019b). On the other hand, heavy rain has also a negative impact on crop production through water logging, nutrient leaching, increased risk of flood, hindering the process of pollination, root rot, decreasing nitrogen availability and effective panicle, and disturbing crop metabolism and growth (Kannan et al., 2010; Geneti, 2019a; Ziadat et al., 2010). In addition to these factors, heavy rainfall also promotes the occurrence of diseases and insect attacks due to the increased moisture level (Kumar et al., 2018). For example, production losses in maize can reach 34% and in rice, they can reach 40% due to high rainfall (Li et al., 2019; Kumar et al., 2018). According to Kim et al. (2019), the reduction in the total dry matter yield in maize is 4745.27 kg/ha and this is expected to decrease up to 11,607 kg/ha in 2045 if rainfall patterns continue in the same way. Besides the quantity, climate change also affects the quality of crops by disturbing the concentration of nutrients such as minerals, proteins, carbohydrates, starch, and other quality parameters including taste, color, and texture. More than 2 billion people are undernourished and face many health problems, such as iron deficiency causing anemia and zinc deficiency resulting in a weak immune system, cognitive impairment, and growth retardation (Soares et al., 2019; Prasad, 2014). Apart from nutritional parameters, the quality of seeds and fruits is also affected. For instance, high temperatures increase cracking and sunburn in apples (Muchie and Assefa, 2021; Maity et al., 2023).

11.3 GENETICALLY MODIFIED ORGANISMS (GMOS) AND CLIMATE CHANGE

Climate has changed drastically since the Industrial Revolution. Atmospheric conditions have deteriorated due to pollution (air and water), waste production, increased CO_2 level, soil erosion, and overuse of fossil fuels. All these alterations have affected agricultural production negatively. According to an estimate, crop loss due to climate change is expected to rise by 2–6% every decade. No doubt, plant breeding, agronomy, and other management strategies have helped a lot in crop improvement as well as in developing climate-resilient varieties, but under this scenario, these are unable to meet the growing demand for food. For instance, production of wheat, rice, soybean, and maize has increased by 0.9%, 1.0%, 1.3%, and 1.6%, respectively, but the demand has increased by up to 2.4%. To feed the growing population, GM technology has been proven to be a valuable tool for crop improvements (Das et al., 2023).

It has positive roles in reducing the effects of climate change, such as transgenic plants which have been developed for drought tolerance, heat tolerance, herbicide tolerance, and insect and disease resistance, ultimately resulting in higher yield. GM crops do not need much maintenance in the field, for example, weeds are a major yield-limiting factor that can easily be controlled by applying herbicides instead of using machinery as these are a cheap method to control weeds without disturbing the crop. In conventional agriculture, overuse of organic fertilizers such as ammonium sulfate, calcium nitrate, sodium nitrate, and ammonium chloride contaminate the atmosphere by releasing hazardous chemicals. GM crops reduce the use of these inorganic fertilizers, for example, GM canola that uses nitrogen more effectively while requiring fewer fertilizer applications has led to the adoption of herbicide-tolerant biotech crops, like canola and soybeans, which allow farmers to replace tilling operations, which degrade the soil and reduce soil carbon sequestration, with the application of broad-spectrum herbicides like glyphosates. These resistance varieties have also decreased the application of herbicide and insecticide sprays and have decreased CO_2 emissions, resulting in a significant decrease in nitrogen fertilizer discharge into the atmosphere (Andualem and Seid, 2021).

GM technology has proved a valuable tool to deal with abiotic stresses. For example, about 60% of the African continent is desert or semi-desert. Crops face severe water shortages due to the unavailability of water and insufficient rainfall because of climate change. This resulted in the deaths of 200,000 people in Kenya in 1968–1973 and 90% of livestock died in 2008–2009 (Ngaira and Ngaira, 2017). Many studies have been conducted to make crops tolerant to drought stress by using GM technology. For instance, a study was conducted by Lv et al. (2009) that resulted in 51% higher yields than the wild type by introducing the gene *Thellungiella halophila H+-PPase (TsVP)*. Transformation of the *AtEDT1/HDG11* gene increased the production of proline and soluble sugar contents, resulting in 43% higher production of seeds in cotton (Yu et al., 2016). Apart from abiotic stress tolerance, it also significantly improved yield. For example, Bt brinjal has a 19.6% higher yield than non-Bt with minimal use of insecticides. To control the papaya ring spot virus, the replicase (rep) gene was introduced, resulting in enhanced fruit production. Transgenic crops have the potential to perform better under stressed conditions, are not labor intensive, and can help farmers earn good profits. Overall, these crops have significant economic and environmental benefits such as a 22% yield enhancement, decreased chemical use by 37%, and 68% increased profit (Sarker et al., 2023).

11.4 CLIMATE-RESILIENT GM CROPS

11.4.1 Drought-tolerant GM Crops

Drought is one of the major consequences of climate change. Several studies have been conducted to develop transgenic crops for drought tolerance. For example, wheat is badly affected due to drought. Bahieldin et al. (2005) transformed wheat with the barley *HVA1* gene that enhanced drought tolerance by elevating water use efficiency and productivity of biomass. Overexpression of *TaWRKY2* improves plant survival under water shortage conditions as well as chlorophyll contents by increasing proline and soluble sugars (Gao et al., 2018) in transgenic plants. Transgenic wheat having *DREB1* resulted in a 4.79–18.43% increase in production compared with wild-type under drought conditions (Zhou et al., 2019). Other genes, including *TaEXPA2* (Yang et al., 2020a), *AtHDG11* (Li et al., 2016), *NAC* (Saad et al., 2013), *TaNRX1* (Wang et al., 2021), *TaVQ14* (Cheng et al., 2022), *PEPC* (Qin et al., 2016), *TaWRKY46* (Liu et al., 2022), *SeCspA* (Yu et al., 2017), *TaFDL2-1A* (Wang et al., 2022a), *TaVQ4-D* (Zhang et al., 2023), *TaPRX-2A* (Su et al., 2023), *AtWRKY30* (El-Esawi et al., 2019), and *TaHDZipI-5* (Yang et al., 2018), have been successfully transformed into wheat to make it drought-tolerant.

Rice is an important staple food that provides 20% of daily calories to 3.5 billion people. Rice production decreases by up to 68% as a result of drought stress (Polthanee et al., 2014). GM technology has significantly improved the tolerance of rice to drought stress. Overexpression of the *OsAbp57* gene enhanced production under water-shortage conditions by increasing photosynthesis (Abdullah-Zawawi et al., 2022). Overexpressing *OsHsfB4b* (Zhang et al., 2022), *ZmHsf05* (Si et al., 2021), *Phehdz1* (Gao et al., 2021), *INDITTO2* (Zhao et al., 2021b), *PsGPD* (Kim et al., 2020), *SiMYB56* (Xu et al., 2020), *AtCYP78A7* (Nam et al., 2020), *OsHAK1* (Chen et al., 2019), *DRO1*, *OsNAC5*, *OsEXPA8* (Zandi et al., 2019), *AtDREB1A* (Latha et al., 2019), *OsERA1* (Tri Joko et al., 2018), and *BdRHP1* (Zeng et al., 2019) genes resulted in enhanced tolerance for drought in rice.

Maize is an important source of starch as about 50% of protein requirements are contributed by maize and wheat. Climate change, especially drought, is a major cause of yield reduction in maize and can be up to 78%. To reduce productivity losses, the transgenic approach plays a key role in developing climate-resilient varieties with higher yields. For example, overexpression of the *ZmEREBP60* gene resulted in decreased malondialdehyde levels and H_2O_2 in transgenic lines that ultimately resulted in enhanced tolerance to drought. This gene regulates the genes that are involved in H_2O_2 catabolism, the abscisic acid signaling pathway, and the water scarcity response (Zhu et al., 2022). Other studies on the development of transgenic maize are mentioned in Table 11.1.

Cotton is an important fiber crop that meets 35% of the world's fiber requirements. Due to drought, the production losses can reach 67% and it is affected at all growth stages (Zafar et al., 2023). The development of GM cotton can be very profitable under water shortage conditions. For instance, the introduction of *AmCBF1* improves drought tolerance in cotton by increasing the relative water content, soluble sugar content, and photosynthetic activity, and decreasing ion leakage. It also affects other morphological characteristics such as the root-to-shoot ratio, lower epidermal cell area, and stomatal density (Lu et al., 2022). Other examples of transgenic cotton are mentioned in Table 11.1.

Soybean is a leading oilseed crop that is mainly affected by drought. The reduction losses could be more than 30%. Many strategies are being used to make soybean drought tolerant. Transgenic crop development is most widely used in this harsh environment. Overexpression of *DREB2* increases the transcription level of *Glycine max* pyrroline-5-carboxylate synthetase (*GmP5CS*) gene and proline accumulation

TABLE 11.1
Drought tolerance responsible genes in various field crops

Sr. No.	Crop and stress	Gene	Expression	Reference
1.	Wheat	*HVA1*	Overexpression	(Bahieldin et al., 2005)
2.		*DREB1*	Overexpression	(Gao et al., 2018)
3.		*TaEXPA2*	Overexpression	(Yang et al., 2020a)
4.		*AtHDG11*	Overexpression	(Li et al., 2016)
5.		*NAC*	Overexpression	(Saad et al., 2013)
6.		*TaNRX1*	Overexpression	(Wang et al., 2021)
7.		*TaVQ14*	Overexpression	(Cheng et al., 2022)
8.		*PEPC*	Overexpression	(Qin et al., 2016)
9.		*TaWRKY46*	Overexpression	(Liu et al., 2022)
10.		*SeCspA*	Overexpression	(Yu et al., 2017)
11.		*TaFDL2-1A*	Overexpression	(Wang et al., 2022a)
12.		*TaVQ4-D*	Overexpression	(Zhang et al., 2023)
13.		*TaPRX-2A*	Overexpression	(Su et al., 2023)
14.		*AtWRKY30*	Overexpression	(El-Esawi et al., 2019)
15.		*TaHDZipI-5*	Overexpression	(Yang et al., 2018)
16.	Rice	*OsAbp57*	Overexpression	(Abdullah-Zawawi et al., 2022)
17.		*OsHsfB4b*	Overexpression	(Zhang et al., 2022)
18.		*ZmHsf05*	Overexpression	(Si et al., 2021)
19.		*Phehdz1*	Overexpression	(Gao et al., 2021)
20.		*INDITTO2*	Overexpression	(Zhao et al., 2021b)
21.		*PsGPD*	Overexpression	(Kim et al., 2020)
22.		*SiMYB56*	Overexpression	(Xu et al., 2020)
23.		*AtCYP78A7*	Overexpression	(Nam et al., 2020)
24.		*OsHAK1*	Overexpression	(Chen et al., 2019)
25.		*OsEXPA8*	Overexpression	(Zandi et al., 2019)
26.		*AtDREB1A*	Overexpression	(Latha et al., 2019)
27.		*OsERA1*	Overexpression	(Tri Joko et al., 2018)
28.		*BdRHP1*	Overexpression	(Zeng et al., 2019)
29.	Maize	*ZmEREBP60*	Overexpression	(Zhu et al., 2022)
30.		*TsCBF1*	Overexpression	(Zhang et al., 2010)
31.		*OsMYB55*	Overexpression	(Casaretto et al., 2016)
32.		*ZmPLC1*	Overexpression	(Wang et al., 2008)
33.		*TsVP* and *BetA*	Overexpression	(Wei et al., 2011)
34.		*NAC*	Overexpression	(Mao et al., 2015)
35.		*SsNHX1*	Overexpression	(Huang et al., 2018)
36.		*ARGOS*	Overexpression	(Shi et al., 2015)
37.		*TsVP*	Overexpression	(Li et al., 2008)
38.		*ZmVPP1*	Overexpression	(Jia et al., 2020)
39.		*LEA*	Overexpression	(Bui et al., 2019)
40.		*Rab28 LEA*	Overexpression	(Amara et al., 2013)

TABLE 11.1 (Continued)
Drought tolerance responsible genes in various field crops

Sr. No.	Crop and stress	Gene	Expression	Reference
41.	Cotton	*AmCBF1*	Overexpression	(Lu et al., 2022)
42.		*AtLOS5*	Overexpression	(Yue et al., 2012)
43.		*TaMnSOD*	Overexpression	(Zhang et al., 2014)
44.		*AtDREB2A-CA*	Overexpression	(Lisei-de-Sá et al., 2017)
45.		*IPT*	Overexpression	(Kuppu et al., 2013)
46.		*AVP1* and *AtNHX1*	Overexpression	(Shen et al., 2015)
47.		*CaHB12*	Overexpression	(Basso et al., 2021)
48.		*ScALDH21*	Overexpression	(Yang et al., 2016)
49.		*AREB/ABF*	Overexpression	(Kerr et al., 2018)
50.		*GaZnF*	Overexpression	(Batool et al., 2023)
51.		*SNAC1*	Overexpression	(Liu et al., 2014)
52.	Soybean	*WRKY20*	Overexpression	(Ning et al., 2017)
53.		*GmNFYA13*	Overexpression	(Ma et al., 2020a)
54.		*AtDREB1D*	Overexpression	(Guttikonda et al., 2014)
55.		*GmNFYA5*	Overexpression	(Ma et al., 2020b)
56.		*AtABF3*	Overexpression	(Kim et al., 2018)
57.		*GmCAMTA12*	Overexpression	(Noman et al., 2019)
58.		*GmTGA15*	Overexpression	(Chen et al., 2021)
59.		*PgTIP1*	Overexpression	(An et al., 2018)
60.		*AtSINA2*	Overexpression	(Yang et al., 2023)
61.	Sorghum	*SbWRKY30*	Overexpression	(Yang et al., 2020b)
62.		*SbNAC9*	Overexpression	(Jin et al., 2023)
63.		*SbMYC2*	Overexpression	(Wang et al., 2024)
64.		*mtlD*	Overexpression	(Maheswari et al., 2017)

that ultimately results in improved soybean resilience under drought conditions (Pham et al., 2020). Further genes have been discussed in Table 11.1.

Sorghum is the fifth most widely cultivated cereal crop and is a staple food for more than 500 million people and is mostly grown in arid and semi-arid regions. The *SbNAC9* gene is important for the growth of sorghum under water-shortage conditions. Its overexpression resulted in higher photochemical efficiency of PSII, increased reactive oxygen species (ROS), higher chlorophyll content, and a stronger root system, leading to better survival of sorghum under drought conditions. Furthermore, its silencing resulted in the opposite outcome, i.e., enhanced drought susceptibility (Jin et al., 2023). Additional genes are listed in Table 11.1.

11.4.2 Heat-Tolerant GM Crops

Increased heat is a result of climate change and cannot be avoided, whereas other stresses, such as drought, can be compensated for by irrigation. In the tropics and subtropics, 35°C or above is considered heat stress, whereas it is set at 25°C in winter crops. GM technology has made it possible to develop heat-tolerant varieties, such as wheat, by transforming them with HVA1 to enhance heat tolerance. In transgenic plants, there is lower oxidative stress coupled with increased yield even at higher temperatures. The GM lines are ABA-sensitive, suggesting this gene is also involved in drought tolerance (Samtani et al., 2022). Additional genes are listed in Table 11.2.

Rice production is greatly affected by higher temperatures, which can cause up to 50% or even complete crop loss and deteriorate the quality. Through GM technology, the development of heat-tolerant varieties has become attainable. Rice transformed with the *AtPLC9* gene resulted in enhanced plant survival only under heat stress and does not help plants to survive under abiotic stresses such as salt and drought stress. Moreover, transgenic rice has upregulation of some transcription factors and calcium ion-related genes (Liu et al., 2020). Further genes are covered in Table 11.2.

High temperature is a major factor in the yield reduction of maize. When temperature reaches up to 35°C, the germination rate, vegetative growth, and grain filling are affected badly. In addition to this, it alters the fluidity of the membrane, photosynthesis, respiration, and production of ROS, and decreases antioxidant activity. The application of GM technology has enabled the creation of heat-tolerant plant varieties. For example, transgenic maize with *OsMYB55* improve heat as well as drought

TABLE 11.2
Heat tolerance responsible genes in various field crops

Sr. No.	Crop	Gene	Expression	Reference
1.	Wheat	*TaPEPKR2*	Overexpression	(Zang et al., 2018)
2.	Wheat	*TaHsfA2-10*	Overexpression	(Guo et al., 2020)
3.	Wheat	*TaHsp90*	Overexpression	(Vishwakarma et al., 2018)
4.	Wheat	*Zmeftu1*	Overexpression	(Fu and Ristic, 2010)
5.	Rice	*AtPLC9*	Overexpression	(Liu et al., 2020)
6.	Rice	*OsWRKY11*	Overexpression	(Wu et al., 2009)
7.	Rice	*SBP*	Overexpression	(Feng et al., 2007)
8.	Rice	*TaMBF1c*	Overexpression	(Qin et al., 2015)
9.	Rice	Rab7	Overexpression	(El-Esawi and Alayafi, 2019)
10.	Rice	*sHSP17.7*	Overexpression	(Murakami et al., 2004)
11.	Maize	*AtGRXS17*	Overexpression	(Sprague et al., 2022)
12.	Maize	*ZmCDPK7*	Overexpression	(Zhao et al., 2021a)
13.	Maize	*ZmDREB2A*	Overexpression	(Qin et al., 2007)
14.	Cotton	*AsHSP70*	Overexpression	(Batcho et al., 2021)
15.	Cotton	*AtSAP5*	Overexpression	(Hozain et al., 2012)
16.	Cotton	*GhHSP70-26*	Overexpression	(Ni et al., 2021)
17.	Cotton	*OsSIZ1*	Overexpression	(Mishra et al., 2017)
18.	Soybean	*GmHsp90A2*	Overexpression	(Huang et al., 2019)
19.	Soybean	*GmPIF4b*	Overexpression	(Arya et al., 2024)
20.	Soybean	GmHSFA1	Overexpression	(Wei et al., 2016)

tolerance in maize (Casaretto et al., 2016). Table 11.2 contains a discussion of additional genes.

Cotton growth, production, and fiber quality are negatively affected by heat stress. It decreases the germination rate, vegetative, root, and shoot growth, flowering, and boll formation. The advancement of heat-tolerant varieties has been facilitated by genetic engineering techniques. *AsHSP70* (heat shock protein gene) introduced in cotton enhances heat tolerance by increasing the membrane stability index, boll formation, soluble sugar content, proline, and photosynthetic contents, while decreasing cell electrolyte leakage. It also increases resilience to other abiotic stresses (Batcho et al., 2021). Further genes are discussed in Table 11.2.

Soybean is one of the most susceptible crops that are badly affected by high temperatures. Imbalanced water content leads to wilting, decreased chlorophyll synthesis, reduced photosynthesis, stunted vegetative growth, and impaired flowering and pod formation, which are the main consequences of heat stress. Overexpression of *GmHsp90A2* gene improves tolerance to high temperatures by increasing the chlorophyll content and reducing the malondialdehyde content. Moreover, it also decreases lipid peroxidation and chlorophyll degradation during high temperatures (Huang et al., 2019).

11.4.3 Insect Pest-resistant GM Crops

Climate change favors insect pest infestations by expanding their geographical distribution, increasing their survival and number of generations per unit time, and altering plant–pest synchrony. They can cause 25–30% production losses. For example, increasing the temperature along with high humidity increases the populations of whitefly. GM technology provides an environment-friendly method of insect pest control without using pesticides or insecticides. In wheat, cowpea trypsin inhibitor gene (*CpTI*) was introduced to develop resistance against grain moth and resulted in a 66.76% reduction in moth-eaten seeds.

About 10% of rice production is lost due to insect pest attacks. In conventional agriculture, insect pests were controlled by chemicals because it was a cheap method, however it led to environmental pollution, pest resistance, and pesticide residues in food. The development of transgenic plants against pests could provide an easy and environment-friendly method of crop protection. Insect resistance was developed in rice through the introduction of the *Cry1Ab* gene, which resulted in the prevention of leaf and tiller damage. Huahui No.1 is the first transgenic rice with the *Bt* gene that provides resistance against stem borer and increased yield by 28.9% without using insecticides and offered resistance against many other insects (Li et al., 2023).

Maize is mainly affected by insect pests and, according to an estimate, insect pests could cause 20–40% yield losses. Lepidopteran pests are major insect pests of maize that can cause more than 10% production losses. To reduce production losses, GM maize has been developed by Du et al. (2014) against these pests through the incorporation of the *cry1C* gene. This gene provides significant resistance against pests. Cry1C* protein concentration is higher in leaves than other parts and in the vegetative phase than the reproductive phase. Other insect pest-resistance genes are discussed in Table 11.3.

Cotton is loved by insects, and more than 1326 species of insects have been reported in cotton. Thirty insect pests are considered to be most destructive, such as whitefly, jassids, caterpillars, spotted bollworms, aphids, pink and American bollworms, spider mites, and mealybugs (Shah et al., 2022). The transgenic development strategy has successfully worked against cotton insect pests. For example, *Cry2Ab*, *Cry1F*, and *Cry1Ac* genes were introduced into cotton which resulted in significant resistance against *Spodoptera litura* larvae and *Helicoverpa armigera* (Katta et al., 2020). Further insect resistance genes are described in Table 11.3.

Soybean faces many insect pests that cause major yield losses. The major insects, such as soybean pod worm, velvetbean caterpillar, and soybean looper, can cause 44% production losses. Genetic engineering has been proven to be an effective control for insect pests. For example, the *cry1A* gene was successfully introduced in soybean for resistance against the above-mentioned insects. Transgenic plants also have many beneficial agronomic characteristics (Miklos et al., 2007b). Other insect resistance-related genes are described in Table 11.3.

Insect pests are a major threat to the production of sorghum all over the world. About 150 species of insects are reported in sorghum, such as Asian corn borer, *Chilo sacchariphagus*, *Helicoverpa armigera*, sorghum strip borer, Old World bollworm, yellow peach moth, and honeydew moth. These insects can cause 50–70% yield losses. GM crops develop resistance against these pests by expressing foreign proteins. For example, the introduction of the *cry1Ac* gene made sorghum resistant to spotted stem borer with a 60% reduction in leaf damage and 40% larval mortality (Okosun et al., 2021; Girijashankar et al., 2005b). Other genes are mentioned in Table 11.3.

TABLE 11.3
Insect resistance responsible genes in various field crops

Sr. No.	Crop	Insect name	Gene	Expression	References
1.	Wheat	Angoumois grain moth	*BTI-CMe*	Overexpression	(Altpeter et al., 1999)
2.	Wheat	*Sitotroga cerealella* Olivier	*CpTI* and *nptII*	Overexpression	(Bi et al., 2006)
3.	Wheat	*Sitophilus granarius*	*avidin*	Overexpression	(Abouseadaa et al., 2015)
4.	Wheat	Aphid and Oriental armyworm	*Bt* and *PTA*	Overexpression	(Yu and Wei, 2008)
5.	Rice	Rice stem borers and leaf folder	*Bt*	Overexpression	(Chen et al., 2006)
6.	Rice	Lepidopteran pests	*cry1C*	Overexpression	(Ye et al., 2009)
7.	Rice	*Scirpophaga incertulas*	*cryIA(b)*	Overexpression	(Datta et al., 1998)
8.	Rice	*Sesamia inferens*	*pin2*	Overexpression	(Duan et al., 1996)

TABLE 11.3 (Continued)
Insect resistance responsible genes in various field crops

Sr. No.	Crop	Insect name	Gene	Expression	References
9.	Maize	Onentomophagous arthropods	*Cry1Ab*	Overexpression	(Dutton et al., 2003)
10.	Maize	Lepidopteran pests	*cry1C*	Overexpression	(Du et al., 2014)
11.	Maize	*Ostrinia nubilalis* Hübner	*Cry1Ab*	Overexpression	(Sanders et al., 2007)
12.	Maize	*Diabrotica virgifera*	*Cry3Bb1*	Overexpression	(Gassmann et al., 2014)
13.	Cotton	*Helicoverpa armigera*	*CpTI*	Overexpression	(Zhang et al., 2004)
14.	Cotton	*lepidopteran pests*	*Cry1Ac* and *Cry2Ab*	Overexpression	(Hagenbucher et al., 2017)
15.	Cotton	*Helicoverpa armigera*	*Cry1Ac+ Cry2A*	Overexpression	(Awan et al., 2015)
16.	Soybean	Lepidopteran insects	cry1A	Overexpression	(Miklos et al., 2007a)
17.	Soybean	Lepidopteran insects	cryIA(c) and *pta*	Overexpression	(Dang and Wei, 2007)
18.	Soybean	Lepidopteran insects	*Cry1Ac*	Overexpression	(Martins-Salles et al., 2017)
19.	Sorghum	*Chilo partellus*	*mpiC1*	Overexpression	(Girijashankar et al., 2005a)
20.	Sorghum	*Chilo partellus*	*cry1Aa* and *cry1B*	Overexpression	(Visarada et al., 2016)

11.4.4 Disease-resistant GM Crops

Diseases are a major biotic factor that limits crop yield and quality, and increase production costs. Some pathogens such as *Phytophthora* spp. *Fusarium* spp., *Sclerotinia* spp., *Rhizoctonia* spp., *Pythium* spp., and *Verticillium* spp. can cause 50%–75% production losses in maize, wheat, vegetables, cotton, and fruits (Panth *et al.*, 2020). The use of fungicides is a common practice to control diseases, but it has many problems, such as environmental pollution, quality reduction, labor costs, and adverse effects on beneficial soil microbes. For example, 112–253 million USD are required to invest in fungicides to control soybean rust (Ishikawa-Ishiwata and Furuya, 2021). Transgenic plants have resistance against diseases without the application of fungicides or other chemicals.

Among all biotic factors, diseases are the most serious problem of wheat and can cause about 20% production losses every year. GM technology provides an easy way to enhance disease resistance in wheat, ensuring healthier plants and minimizing environmental impacts. For example, *TaWRKY45* genes have been successfully transferred to wheat to develop resistance against *Fusarium graminearum*. In addition to this, it also confers resistance to powdery mildew and leaf rust (Bahrini *et al.*, 2011). Double gene technology was used to overexpress *Trichoderma harzianum* chitinase and chitosanase in winter wheat. Constitutive promoters (Ubiquitin-1) and inducible promoters (Vst-1) regulated the expression of genes. With up to 75% reduction in susceptibility under the inducible promoter and 60% reduction under the constitutive promoter, transgenic lines demonstrated numerous gene integrations and decreased sensitivity to the pathogen (Rana et al., 2012). Further resistant genes against different diseases are described in Table 11.4.

TABLE 11.4
Disease resistance responsible genes in various field crops

Sr. No.	Crop	Disease or pathogens	Gene	Expression	Reference
1.	Wheat	Leaf rust	*Lr34*	Overexpression	(Risk et al., 2012)
2.	Wheat	Stripe rust	*RC24*	Overexpression	(Huang et al., 2013)
3.	Wheat	Rusts and powdery mildew	*chi26*	Overexpression	(Eissa et al., 2017)
4.	Wheat	*Fusarium* head blight	*RsAFP2*	Overexpression	(Li et al., 2011b)
5.	Rice	Rice blast	*OPBP1*	Overexpression	(Chen and Guo, 2008)
6.	Rice	Rice blast	*OsWRKY31*	Overexpression	(Zhang et al., 2008)
7.	Maize	corn smut	*KP4*	Overexpression	(Allen et al., 2011a)
8.	Maize	Rust and leaf blight	*Lr34*	Overexpression	(Sucher et al., 2017)
9.	Maize	large leaf spot	*NPR1*	Overexpression	(Xi et al., 2023)
10.	Cotton	*Fusarium verticillioides* and *Verticillium dahliae*	*D4E1*	Overexpression	(Rajasekaran et al., 2005)
11.	Cotton	*Fusarium oxysporum*	*NPR1*	Overexpression	(Parkhi et al., 2010)
12.	Cotton	*Verticillium dahliae*	*NaD1*	Overexpression	(Gaspar et al., 2014)
13.	Soybean	*Phakopsora pachyrhizi* and *Colletotrichum truncatum*	*NmDef02*	Overexpression	(Soto et al., 2020)
14.	Soybean	*Phytophthora sojae* and *Cercospora sojina*	*GbNPR1* and *HrpZpsg12*	Overexpression	(Song et al., 2016)
15.	Sorghum	Anthracnose and rust	*Lr34*	Overexpression	(Schnippenkoetter et al., 2017a)

Rice production losses (20–40%) are majorly caused by diseases such as rice brown spot which can cause 4–52% yield reductions. Major diseases of rice are rice blast, bacterial leaf blight, rice brown spot, leaf smut, and sheath blight. The development of GM rice can enhance disease resistance contributing to safer agricultural production and improved food security. For instance, overexpression of the *OsRLR1* gene develops a broad range of resistance against *Xanthomonas oryzae* and *Pyricularia oryzae* (Du et al., 2021). Other disease-resistant genes are mentioned in Table 11.4.

Maize yield losses are 4–14% due to diseases. Some important diseases of maize are maize streak virus, gray leaf spot, northern corn leaf blight, southern corn leaf blight, common rust, common smut, Stewart's wilt, *Fusarium* ear rot, *Diplodia* ear rot, and maize dwarf mosaic virus. Disease resistance is crucial for agriculture and food security. Transgenic maize has valuable importance in controlling diseases, such as GM maize expressing Totivirus antifungal protein KP4 which increase resistance against *Ustilago maydis* causing corn smut fungus (Allen et al., 2011b). Additional disease-resistance genes are identified in Table 11.4.

Cotton is susceptible to a lot of diseases such as *Fusarium* wilt, cotton leaf curl disease, *Verticillium* wilt, anthracnose, boll rot, *Alternaria* leaf spot, cotton root rot, seedling disease complex, bacterial blight, and cotton rust. These diseases can cause significant yield losses, such as *Fusarium* and *Verticellum* wilt which can result in 70% crop damage. Transgenic cotton expressing the *βC1* gene result in significant resistance against cotton leaf curl disease and did not show symptoms throughout the life cycle of plant. This serious disease can cause crop losses of up to 80% (Sohrab et al., 2014). Resistant genes against different diseases are mentioned in Table 11.4.

Diseases are the major biotic factor for crop failure in soybean, and include bacterial wilt, common blight, halo blight, *Alternaria* leaf spot, powdery mildew, angular leaf, *Fusarium* root rot, and anthracnose. These diseases can cause yield reductions of up to 85%. Disease resistance against *Sclerotinia sclerotiorum* was developed in soybean by the introduction of the *chi* gene through *Agrobacterium*-mediated genetic transformation. Its overexpression elevated the levels of ROS, catalase (SOD), and peroxidase (POD) and decreased malondialdehyde content as well as ion leakage. With the increase in ROS accumulation, activities of ROS scavenging enzymes and cell death were reduced, resulting in improved resistance to *S. sclerotiorum*.

Sorghum is tolerant to many abiotic stresses but is susceptible to many biotic stresses, like diseases, which can result in about 30% production losses. Important diseases of sorghum are anthracnose, grain mold, ergot, downy mildew, head smut, leaf blight, rust, charcoal rot, bacterial stripe, and sorghum mosaic virus. These can be controlled in many ways such as breeding, agronomics, and chemicals. Transgenic plants provide a sustainable way to control diseases. For example, the *Lr34res* gene was transformed to sorghum to enhance resistance against sorghum rust and anthracnose diseases. Its expression resulted in pathogen-induced pigmentation, overexpression of flavonoid biosynthesis genes, raising the accumulation of 3-deoxy anthocyanidin metabolites, and decreasing anthracnose symptoms (Schnippenkoetter et al., 2017b).

11.4.5 Herbicide-tolerant GM Crops

In conventional agriculture, herbicides are widely used to control unwanted plants such as weeds because these plants compete for water, nutrients, space, and as an alternative host for many insect pests and diseases. For example, about 76% yield losses were recorded in wheat by six weeds: *Phalaris minor* Retz., *Coronopus didymus* (L.) Sm., *Rumex dentatus* L., *Medicago denticulata* Willd., *Poa annua* L., and *Chenopodium album* L. (Siddiqui et al., 2010). To reduce crop losses, herbicides are applied, which are cheaper, reduce labor costs, and enable tillage-less crop production. Herbicide-tolerant crops have been developed through the introduction of foreign genes. Transgenic wheat was developed by introducing the *GST-27* gene, conferring resistance not only to alachlor but also to the chloroacetanilide herbicide dimethenamid and the thiocarbamate herbicide EPTC (Milligan et al., 2001).

Weeds are indeed important yield-limiting factors in rice production, with species such as barnyard grass, duck salad, hemp sesbania, and northern joint vetch being identified as significant contributors to yield reduction. It is very difficult to remove weeds manually, but GM rice provides an easy way to control weeds because they can withstand the application of herbicides. Transgenic rice for herbicide resistance has been successfully transformed by the *AHAS* gene along with the *bar* gene conferring dual tolerance against bensulfuron methyl and glufosinate herbicides (Fartyal et al., 2018). The *EPSPS* gene has been transformed into rice by Chhapekar et al. (2015) to develop tolerance against the broad-spectrum herbicide glyphosate.

Weeds are a significant threat to maize production, decreasing yield by up to 27–60%. Important weeds of maize are common ragweed, pigweed, foxtail, Johnsongrass, barnyard grass, velvetleaf, and morning glory. The development of herbicide-resistant maize has shown significant results. For example, three genes *cry2Ab*, *vip3A*, and *cp4epsps*, were transformed to develop resistance against insect pests (*Mythimna separata*, *Spodoptera frugiperda*, and *Helicoverpa armigera*) and herbicide (glyphosate) tolerance (Liu et al., 2023). Other genes, such as *bar*, *GUS*, *ALS*, and *GFP* genes (Hong et al., 2019; Sato et al., 2011), have been successfully transformed to develop herbicide tolerance.

Weeds are a major biotic factor affecting cotton yield by up to 30%. However, severe competition between cotton and weeds could cause production losses of 40–85%. Palmer amaranth, common cocklebur, morning glory, Johnsongrass, pigweed, sicklepod, and nutgrass are major weeds of cotton. The *AHAS* gene was introduced into cotton to confer herbicide tolerance against imidazolinone and sulfonylurea herbicides (Rajasekaran et al., 1996). The *CP4-EPSPS* gene was introduced by Karthik et al. (2020), and confers tolerance to glyphosate. The *Cry1Ac+Cry2A* and *GT* genes were transformed to cotton that developed tolerance against *Helicoverpa armigera* larvae and glyphosate (Awan et al., 2019).

Production and harvesting efficiency are badly affected by weeds in soybean. The most common soybean weeds are common water hemp, giant ragweed, Canadian horseweed, common cocklebur, and ivy leaf morning glory, and these can result in yield losses of 52.1%. Many transgenic studies have been conducted for herbicide tolerance, such as *bar* and pesticidal crystal protein (*cry*)1Ac mutant 2 (M#2) transformed in soybean to confer resistance against lepidopteran insects and herbicide resistance (Nguyen et al., 2021). Transgenic soybean was developed by using the rice-derived acetolactate synthase (Os-*mALS*) gene for herbicide resistance (Tougou et al., 2009).

Weeds cause yield reductions of up to 30–61% in sorghum in different areas of the world. Under high moisture conditions, weeds compete with sorghum which is thickly planted in narrow rows. Some important weeds are Johnsongrass, pigweed species (including Palmer amaranth and redroot pigweed), crabgrass species, fall panicum, and nutsedge species. Transgenic sorghum was developed by Zhao et al. (2000) by introducing the *bar* gene in sorghum to develop herbicide tolerance. Another study for *bar* gene transformation was done using microprojectile bombardment of immature zygotic embryos to develop tolerance against herbicides (Casas et al., 1993).

11.5 SOLVING QUALITY ISSUES CREATED BY CHANGING CLIMATES

Climate change has significant impacts on the quality of crops by altering the concentration of nutrients such as sugars, flavonoid, organic acids, vitamins, carbohydrates, proteins, carotenoids, fatty acids, and antinutritional compounds. Due to increasing temperatures and CO_2 levels, the quality of crops, fruits, and vegetables is affected. This happens by altering the rate of photosynthesis, and secondary metabolite synthesis. A higher CO_2 level reduces the sugar content and causes tuber malformation in potatoes and also decreases growth, photosynthesis, and nutrient contents in strawberries and tomatoes (Mattos et al., 2014).

A study was conducted by Kawasaki and Uchida (2016) on rice production under cold and extremely high temperatures to investigate the influence of both temperatures. Their findings suggested that quantity is significantly reduced by lower temperatures and quality was compromised at higher temperatures. Climate change, such as light, temperature, and water, affects soybean quality by altering the amino acid content, crude proteins, and crude fat. Elevating the temperature by 1°C results in a 1.6% decrease in the starch content and 0.8% in protein content in wheat (He et al., 2022). Drought, CO_2, and heat stress in wheat deteriorate nutrients, protein compositions, starch granules, and fiber. These changes decrease functional properties and viscoelasticity in wheat dough. In addition, heat and drought enhanced the phytate contents that reduce the bioavailability of essential mineral elements. The impact of climate change on quality depends on the variety, type of stress, intensity, and exposure time. Increasing the CO_2 concentration could increase the global yield by 7% with 2% protein, but the concentration of grain protein reduced by 1.1%, which exhibits a relative change of −8.6% (Zahra et al., 2022).

Oilseed crops, such as soybean, when faced with water stress, resulted in oil + protein weight decreased by 57–68% of total seed production. It deteriorates fatty acid composition, such as decreased concentrations of linoleic and linolenic acids, omega-6 and omega-3 polyunsaturated fats, and 3.4–10.4% decreased oil concentrations. Other crops, such as sunflower, where climate change can reduce yields by more than 35%, and *Brassica napus* have 50% less oil content. In *B. napus*, high temperature could cause reductions of ~6% in oleic acid, 13–15% in linoleic acid, and 16–41% in linolenic acid, but increase palmitic acid by 7% and stearic acid by 21% (Attia et al., 2021). The quality of fruits and vegetables is mainly affected by climate change. For example, high temperatures can cause skin disorders in avocado, broccoli, cabbage, cranberry, grapes, lettuce, and lime. Other quality parameters, such as size, texture, malformation, taste, browning, flesh disorder, and color, are deteriorated by heat stress. The concentration of antioxidants is increased, but macronutrients and micronutrients are decreased. The elevated concentration of CO_2 also deteriorates the quality. For example, in carrots, it decreases protein, vitamin C, macronutrients, micronutrients, fatty acids, and amino acids contents. In radishes, protein, vitamin C, macronutrients, micronutrients, fatty acids, and amino acids contents are diminished due to higher concentrations of CO_2.

11.6 GENE EDITING OF CROPS FOR CLIMATE CHANGE

Gene editing (GE) is one of the most advanced technologies in plant science. This includes transcription activator-like effector nucleases (TALENs), zinc-finger nucleases (ZFNs), and more recently clustered regularly interspaced short palindromic repeats (CRISPR/Cas). These techniques lead to domains like epigenome editing, and base and prime editing. These have been proven to be the best tool to deal with biotic and abiotic stresses. For example, by using CRISPR-Cas9, resistance against viruses was develop in *Cucumis sativus* by knocking out the eukaryotic translational initiation factor *eIF4E* gene. This resulted in resistance against the zucchini yellow mosaic virus, papaya ringspot mosaic virus-W, and the cucumber vein yellowing virus. Resistance against bacterial blight was developed in rice by editing of the *SWEET11*, *SWEET13*, and *SWEET14* genes using CRISPR-Cas9 (Ahmar et al., 2020).

Drought tolerance was developed by the editing of *TaDREB2* and *TaDREB3* in wheat using CRISPR-Cas9, and *ZmARGOS8* gene in maize to increase the yield under water shortage conditions. In rice, drought tolerance developed by the knockout of *OsSAPK2* gene resulted in the closure of stomata under heat and drought conditions. Apart from stresses, gene editing can be helpful in developing ideal grain size, weight, number, and plant architecture. Unlike transgenic approaches, they have fewer regulatory procedures. GE could be revolutionary in almost all crops and so can help in rapid crop improvement in a changing climate.

11.7 RISK ASSESSMENT AND FUTURE DIRECTIONS ON GM CROPS

Genetic transformation is an important tool for the development of climate-resilient crops for improved yield, biotic, and abiotic stress tolerance. In addition, it is also used for biofortification as well as improvement of the medicinal value of plants. For example, transgenic banana for vaccines against hepatitis B, potatoes resistant to viruses, and rice and wheat rich with iron and vitamins. Despite their advantages, GM crops have some biosafety as well as environmental concerns. Acceptance of these crops for biosafety regulation varies from country to country and is a big debate among stakeholders (Ghimire et al., 2023).

Vigorous risk assessment is necessary to improve quality, as it identifies possible drawbacks and accepts that there is always some risk. Conducting a complex risk assessment for GMOs requires certain criteria for various species and characteristics. This includes assessing unexpected effects on target populations and shifts in biodiversity in non-target populations. Based on these evaluations, strategies to reduce risks are created, considering possible toxicity, crop composition, and gene expression. Ongoing assessment is essential at every step of gene development to ensure that genes are safely and stably inherited by future generations. Some international organizations, such as the International Plant Protection Convention (IPPC), the Cartagena Protocol on Biosafety, and the Convention on Biological Diversity (CBD), estimate the environmental consequences of GM organisms.

Dr. Eleanor Reed conducted a study to investigate the safety of GM corn on mice. Over 90 days, she analyzed the growth rate of the mice and recorded a 3.3% weight loss in males and a 3.7% weight gain in females. Additionally, she observed hepatorenal symptoms of toxicity, a 40% increase in triglyceride levels in females, and a 31–35% increase in urinary phosphorus and sodium excretion in males. Despite these findings, experts dismissed them as biologically insignificant. Dr. Reed uncovered potential threats posed by GM maize and suggested that these changes might be due to pesticide applications like glyphosate on GM maize. Her findings challenged existing GM protocols and highlighted the need for longer-term research on GM crops (Noushahi and Hussain, 2020). It is true that the impact of GM crops on health is yet to be determined by research. However, GM crops are considered to be safe, higher-yielding, and nutritious, as demonstrated by most studies. Many transgenic varieties of maize have proteins similar to conventional standards, but the safety of GM soybeans is still under debate. There is a need to conduct various research programs to increase public confidence in GM crops through case studies, conferences, and approvals from medical authorities and the scientific community. Almost all transgenic crops will likely prove to be free from toxic symptoms, specifically hepatic, hematological, renal, physiological, reproductive effects, and immunological consequences. Additionally, future improvements in crops according to climate change are essential. In conclusion, GM crops have the potential to meet the prerequisites of the world's increasing population under hostile climatic conditions.

REFERENCES

Abbass, K., M.Z. Qasim, H. Song, M. Murshed, H. Mahmood and I. Younis. 2022. A review of the global climate change impacts, adaptation, and sustainable mitigation measures. *Environmental Science and Pollution Research.* 29: 42539–42559.

Abdullah-Zawawi, M.-R., L.-W. Tan, Z. Ab Rahman, I. Ismail and Z. Zainal. 2022. An integration of transcriptomic data and modular gene co-expression network analysis uncovers drought stress-related hub genes in transgenic rice overexpressing OsAbp57. *Agronomy.* 12: 1959.

Abouseadaa, H.H., G.H. Osman, A.M. Ramadan, S.E. Hassanein, M.T. Abdelsattar, Y.B. Morsy, H.F. Alameldin, D.K. El-Ghareeb, H.A. Nour-Eldin, R. Salem, A.A. Gad, S.E. Elkhodary, M.M. Shehata, H.M. Mahfouz, H.F. Eissa and A. Bahieldin. 2015. Development of transgenic wheat (Triticum aestivum L.) expressing avidin gene conferring resistance to stored product insects. *BMC Plant Biology.* 15: 183.

Ahmar, S., S. Saeed, M.H.U. Khan, S. Ullah Khan, F. Mora-Poblete, M. Kamran, A. Faheem, A. Maqsood, M. Rauf, S. Saleem, W.-J. Hong and K.-H. Jung. 2020. A revolution toward gene-editing technology and its application to crop improvement. *International Journal of Molecular Sciences*. 21: 5665.

Allen, A., E. Islamovic, J. Kaur, S. Gold, D. Shah and T.J. Smith. 2011a. Transgenic maize plants expressing the Totivirus antifungal protein, KP4, are highly resistant to corn smut. *Plant Biotechnology Journal*. 9: 857–864.

Allen, A., E. Islamovic, J. Kaur, S. Gold, D. Shah and T.J. Smith. 2011b. Transgenic maize plants expressing the Totivirus antifungal protein, KP4, are highly resistant to corn smut. *Plant Biotechnology Journal*. 9: 857–64.

Altpeter, F., I. Diaz, H. Mcauslane, K. Gaddour, P. Carbonero and I.K. Vasil. 1999. Increased insect resistance in transgenic wheat stably expressing trypsin inhibitor CMe. *Molecular Breeding*. 5: 53–63.

Amara, I., M. Capellades, M.D. Ludevid, M. Pagès and A. Goday. 2013. Enhanced water stress tolerance of transgenic maize plants over-expressing LEA Rab28 gene. *Journal of Plant Physiology*. 170: 864–873.

An, J., C. Cheng, Z. Hu, H. Chen, W. Cai and B. Yu. 2018. The Panax ginseng PgTIP1 gene confers enhanced salt and drought tolerance to transgenic soybean plants by maintaining homeostasis of water, salt ions and ROS. *Environmental and Experimental Botany*. 155: 45–55.

Andualem, B. and A. Seid. 2021. The role of green biotechnology through genetic engineering for climate change mitigation and adaptation, and for food security: current challenges and future perspectives. *Journal of Advances in Biology & Biotechnology*. 24: 1–11.

Arya, H., M.B. Singh and P.L. Bhalla. 2024. Overexpression of GmPIF4b affects morpho-physiological traits to reduce heat-induced grain loss in soybean. *Plant Physiology and Biochemistry*. 206: 108233.

Attia, Z., C.S. Pogoda, S. Reinert, N.C. Kane and B.S. Hulke. 2021. Breeding for sustainable oilseed crop yield and quality in a changing climate. *Theoretical and Applied Genetics*. 134: 1817–1827.

Awan, M.F., M.A. Abass, A. Muzaffar, A. Ali, B. Tabassum, A.Q. Rao, I. Ahmad Nasir and T. Husnain. 2015. Transformation of insect and herbicide resistance genes in cotton (Gossypium hirsutum L.). *Journal of Agricultural Science and Technology*. 17: 287–298.

Awan, M.F., M.A. Abbas, A. Muzaffar, A.a.M. Ali, B. Tabassum, A.Q. Rao, I. Ahmad, Nasir and T. Husnain. Transformation of insect and herbicide resistance genes in cotton (Gossypium hirsutum L.). 2019.

Bahieldin, A., H.T. Mahfouz, H.F. Eissa, O.M. Saleh, A.M. Ramadan, I.A. Ahmed, W.E. Dyer, H.A. El-Itriby and M.A. Madkour. 2005. Field evaluation of transgenic wheat plants stably expressing the HVA1 gene for drought tolerance. *Physiologia Plantarum*. 123: 421–427.

Bahrini, I., T. Ogawa, F. Kobayashi, H. Kawahigashi and H. Handa. 2011. Overexpression of the pathogen-inducible wheat *TaWRKY45* gene confers disease resistance to multiple fungi in transgenic wheat plants. *Breeding Science*. 61: 319–326.

Basso, M.F., J.A. Costa, T.P. Ribeiro, F.B.M. Arraes, I.T. Lourenço-Tessutti, A.F. Macedo, M.R. Das Neves, S.M. Nardeli, L.W. Arge and C.E.A. Perez. 2021. Overexpression of the CaHB12 transcription factor in cotton (Gossypium hirsutum) improves drought tolerance. *Plant Physiology and Biochemistry*. 165: 80–93.

Batcho, A.A., M.B. Sarwar, B. Rashid, S. Hassan and T. Husnain. 2021. Heat shock protein gene identified from Agave sisalana (AsHSP70) confers heat stress tolerance in transgenic cotton (*Gossypium hirsutum*). *Theoretical and Experimental Plant Physiology*. 33: 141–156.

Batool, F., S. Hassan, S. Azam, Z. Sher, Q. Ali and B. Rashid. 2023. Transformation and expressional studies of GaZnF gene to improve drought tolerance in *Gossypium hirsutum*. *Scientific Reports*. 13: 5064.

Batra, D., S.B. Dhull, J. Rani, M. Meenakshi, Y. Kumar and J. Kinabo. 2023. Effect of heat stress on seed protein quality in mungbean [*Vigna radiata* (L.) Wilczek]. *Legume Science*. 5: e205.

Bhattacharya, A. 2019. *Effect of High-Temperature Stress on Crop Productivity*.

Bi, R.-M., H.-Y. Jia, D.-S. Feng and H.-G. Wang. 2006. Production and analysis of transgenic wheat (*Triticum aestivum* L.) with improved insect resistance by the introduction of cowpea trypsin inhibitor gene. *Euphytica*. 151: 351–360.

Bibi, F. and A. Rahman. 2023. An overview of climate change impacts on agriculture and their mitigation strategies. *Agriculture*. 13: 1508.

Bui, M.M., T.L. Nguyen, H.H. Hanh, X.T. Nguyen, N. Van Hai and H.T.T. Hue. 2019. A LEA gene from a vietnamese maize landrace can enhance the drought tolerance of transgenic maize and tobacco. *Agronomy*. 9: 62.

Casaretto, J.A., A. El-Kereamy, B. Zeng, S.M. Stiegelmeyer, X. Chen, Y.-M. Bi and S.J. Rothstein. 2016. Expression of OsMYB55 in maize activates stress-responsive genes and enhances heat and drought tolerance. *BMC Genomics*. 17: 312.

Casas, A.M., A.K. Kononowicz, U.B. Zehr, D.T. Tomes, J.D. Axtell, L.G. Butler, R.A. Bressan and P.M. Hasegawa. 1993. Transgenic sorghum plants via microprojectile bombardment. *Proceedings of the National Academy of Sciences of the United States of America*. 90: 11212–6.

Chen, G., J. Hu, J. Lian, Y. Zhang, L. Zhu, D. Zeng, L. Guo, L. Yu, G. Xu and Q. Qian. 2019. Functional characterization of OsHAK1 promoter in response to osmotic/drought stress by deletion analysis in transgenic rice. *Plant Growth Regulation*. 88: 241–251.

Chen, M., J.Z. Zhao, G.Y. Ye, Q. Fu and A.M. Shelton. 2006. Impact of insect-resistant transgenic rice on target insect pests and non-target arthropods in China. *Insect Science*. 13: 409–420.

Chen, X. and Z. Guo. 2008. Tobacco OPBP1 enhances salt tolerance and disease resistance of transgenic rice. *International Journal of Molecular Sciences*. 9: 2601–2613.

Chen, Z., X. Fang, X. Yuan, Y. Zhang, H. Li, Y. Zhou and X. Cui. 2021. Overexpression of transcription factor GmTGA15 enhances drought tolerance in transgenic soybean hairy roots and Arabidopsis plants. *Agronomy*. 11: 170.

Cheng, X., H. Yao, Z. Cheng, B. Tian, C. Gao, W. Gao, S. Yan, J. Cao, X. Pan and J. Lu. 2022. The wheat gene TaVQ14 confers salt and drought tolerance in transgenic Arabidopsis thaliana plants. *Frontiers in Plant Science*. 13: 870586.

Chhapekar, S., S. Raghavendrarao, G. Pavan, C. Ramakrishna, V.K. Singh, M.L.V. Phanindra, G. Dhandapani, R. Sreevathsa and P. Ananda Kumar. 2015. Transgenic rice expressing a codon-modified synthetic CP4-EPSPS confers tolerance to broad-spectrum herbicide, glyphosate. *Plant Cell Reports*. 34: 721–731.

Dang, W. and Z.-M. Wei. 2007. An optimized Agrobacterium-mediated transformation for soybean for expression of binary insect resistance genes. *Plant Science*. 173: 381–389.

Das, S., M.K. Ray, D. Panday and P.K. Mishra. 2023. Role of biotechnology in creating sustainable agriculture. *PLOS Sustainability and Transformation*. 2: e0000069.

Dash, G.K., S.K. Sahoo, J. Jena, M. Barik, S. Parida, M.J. Baig and P. Swain. 2022. Drought and high-temperature stress tolerance in field crops. *Response of Field Crops to Abiotic Stress*. CRC Press: 103–109.

Datta, K., A. Vasquez, J. Tu, L. Torrizo, M.F. Alam, N. Oliva, E. Abrigo, G.S. Khush and S.K. Datta. 1998. Constitutive and tissue-specific differential expression of the cryIA(b) gene in transgenic rice plants conferring resistance to rice insect pest. *Theoretical and Applied Genetics*. 97: 20–30.

Du, D., C. Geng, X. Zhang, Z. Zhang, Y. Zheng, F. Zhang, Y. Lin and F. Qiu. 2014. Transgenic maize lines expressing a cry1C* gene are resistant to insect pests. *Plant Molecular Biology Reporter*. 32: 549–557.

Du, D., C. Zhang, Y. Xing, X. Lu, L. Cai, H. Yun, Q. Zhang, Y. Zhang, X. Chen and M. Liu. 2021. The CC-NB-LRR OsRLR1 mediates rice disease resistance through interaction with OsWRKY19. *Plant Biotechnology Journal*. 19: 1052–1064.

Duan, X., X. Li, Q. Xue, M. Abo-Ei-Saad, D. Xu and R. Wu. 1996. Transgenic rice plants harboring an introduced potato proteinase inhibitor II gene are insect resistant. *Nature Biotechnology*. 14: 494–498.

Dutton, A., J. Romeis and F. Bigler. 2003. Assessing the risks of insect resistant transgenic plants on entomophagous arthropods Bt-maize expressing Cry1Ab as a case study. *BioControl*. 48: 611–636.

Eissa, H.F., S.E. Hassanien, A.M. Ramadan, M.M. El-Shamy, O.M. Saleh, A.M. Shokry, M. Abdelsattar, Y.B. Morsy, M.A. El-Maghraby, H.F. Alameldin, S.M. Hassan, G.H. Osman, H.T. Mahfouz, G.A. Gad El-Karim, M.A. Madkour and A. Bahieldin. 2017. Developing transgenic wheat to encounter rusts and powdery mildew by overexpressing barley chi26 gene for fungal resistance. *Plant Methods*. 13: 41.

El-Esawi, M.A., A.A. Al-Ghamdi, H.M. Ali and M. Ahmad. 2019. Overexpression of AtWRKY30 transcription factor enhances heat and drought stress tolerance in wheat (Triticum aestivum L.). *Genes*. 10: 163.

El-Esawi, M.A. and A.A. Alayafi. 2019. Overexpression of rice Rab7 gene improves drought and heat tolerance and increases grain yield in rice (*Oryza sativa* L.). *Genes*. 10: 56.

Fartyal, D., A. Agarwal, D. James, B. Borphukan, B. Ram, V. Sheri, P.K. Agrawal, V.M.M. Achary and M.K. Reddy. 2018. Developing dual herbicide tolerant transgenic rice plants for sustainable weed management. *Scientific Reports*. 8: 11598.

Fawzy, S., A.I. Osman, J. Doran and D.W. Rooney. 2020. Strategies for mitigation of climate change: a review. *Environmental Chemistry Letters*. 18: 2069–2094.

Feng, L., K. Wang, Y. Li, Y. Tan, J. Kong, H. Li, Y. Li and Y. Zhu. 2007. Overexpression of SBPase enhances photosynthesis against high temperature stress in transgenic rice plants. *Plant Cell Reports*. 26: 1635–1646.

Fu, J. and Z. Ristic. 2010. Analysis of transgenic wheat (*Triticum aestivum* L.) harboring a maize (*Zea mays* L.) gene for plastid EF-Tu: segregation pattern, expression and effects of the transgene. *Plant Molecular Biology*. 73: 339–347.

Gao, H., Y. Wang, P. Xu and Z. Zhang. 2018. Overexpression of a WRKY transcription factor TaWRKY2 enhances drought stress tolerance in transgenic wheat. *Frontiers in Plant Science*. 9: 997.

Gao, Y., H. Liu, K. Zhang, F. Li, M. Wu and Y. Xiang. 2021. A moso bamboo transcription factor, Phehdz1, positively regulates the drought stress response of transgenic rice. *Plant Cell Reports*. 40: 187–204.

Gaspar, Y.M., J.A. Mckenna, B.S. Mcginness, J. Hinch, S. Poon, A.A. Connelly, M.A. Anderson and R.L. Heath. 2014. Field resistance to Fusarium oxysporum and Verticillium dahliae in transgenic cotton expressing the plant defensin NaD1. *Journal of Experimental Botany*. 65: 1541–1550.

Gassmann, A.J., J.L. Petzold-Maxwell, E.H. Clifton, M.W. Dunbar, A.M. Hoffmann, D.A. Ingber and R.S. Keweshan. 2014. Field-evolved resistance by western corn rootworm to multiple *Bacillus thuringiensis* toxins in transgenic maize. *Proceedings of the National Academy of Sciences*. 111: 5141–5146.

Geneti, T.Z. 2019a. Review on the effect of moisture or rain fall on crop production. *Civil and Environmental Research.* 11(2): 1-7.

Geneti, T.Z. 2019b. Review on the effect of moisture or rain fall on crop production. *Civil and Environmental Research.*

Ghimire, B.K., C.Y. Yu, W.-R. Kim, H.-S. Moon, J. Lee, S.H. Kim and I.M. Chung. 2023. Assessment of benefits and risk of fenetically modified plants and products: current controversies and perspective. *Sustainability.* 15: 1722.

Girijashankar, V., H.C. Sharma, K.K. Sharma, V. Swathisree, L.S. Prasad, B.V. Bhat, M. Royer, B.S. Secundo, M.L. Narasu, I. Altosaar and N. Seetharama. 2005a. Development of transgenic sorghum for insect resistance against the spotted stem borer (*Chilo partellus*). *Plant Cell Reports.* 24: 513–522.

Girijashankar, V., H.C. Sharma, K.K. Sharma, V. Swathisree, L.S. Prasad, B.V. Bhat, M. Royer, B.S. Secundo, M.L. Narasu, I. Altosaar and N. Seetharama. 2005b. Development of transgenic sorghum for insect resistance against the spotted stem borer (*Chilo partellus*). *Plant Cell Reports.* 24: 513–522.

Guo, X.L., S.N. Yuan, H.N. Zhang, Y.Y. Zhang, Y.J. Zhang, G.Y. Wang, Y.Q. Li and G.L. Li. 2020. Heat-response patterns of the heat shock transcription factor family in advanced development stages of wheat (*Triticum aestivum* L.) and thermotolerance-regulation by TaHsfA2-10. *BMC Plant Biology.* 20: 364.

Guttikonda, S.K., B. Valliyodan, A.K. Neelakandan, L.-S.P. Tran, R. Kumar, T.N. Quach, P. Voothuluru, J.J. Gutierrez-Gonzalez, D.L. Aldrich, S.G. Pallardy, R.E. Sharp, T.-H.D. Ho and H.T. Nguyen. 2014. Overexpression of AtDREB1D transcription factor improves drought tolerance in soybean. *Molecular Biology Reports.* 41: 7995–8008.

Hagenbucher, S., M. Eisenring, M. Meissle and J. Romeis. 2017. Interaction of transgenic and natural insect resistance mechanisms against Spodoptera littoralis in cotton. *Pest Management Science.* 73: 1670–1678.

He, Q., G. Zhou and J. Liu. 2022. Progress in studies of climatic suitability of crop quality and resistance mechanisms in the context of climate warming. *Agronomy.* 12:3183.

Hong, J.K., G.-S. Lee, K.J. Park, J.-K. Kim, H.J. Jang, E.J. Suh, K.-H. Kim and Y.-H. Lee. 2019. Production of transgenic maize plants with herbicide resistance through Agrobacterium-mediated transformation. *Korean Society of Breeding Science.* 51: 290–297.

Hozain, M.D., H. Abdelmageed, J. Lee, M. Kang, M. Fokar, R.D. Allen and A.S. Holaday. 2012. Expression of AtSAP5 in cotton up-regulates putative stress-responsive genes and improves the tolerance to rapidly developing water deficit and moderate heat stress. *Journal of Plant Physiology.* 169: 1261–1270.

Huang, X., J. Wang, Z. Du, C. Zhang, L. Li and Z. Xu. 2013. Enhanced resistance to stripe rust disease in transgenic wheat expressing the rice chitinase gene RC24. *Transgenic Research.* 22: 939–947.

Huang, Y., H. Xuan, C. Yang, N. Guo, H. Wang, J. Zhao and H. Xing. 2019. GmHsp90A2 is involved in soybean heat stress as a positive regulator. *Plant Science.* 285: 26–33.

Huang, Y., X.-X. Zhang, Y.-H. Li, J.-Z. Ding, H.-M. Du, Z. Zhao, L.-N. Zhou, C. Liu, S.-B. Gao, M.-J. Cao, Y.-L. Lu and S.-Z. Zhang. 2018. Overexpression of the Suaeda salsa SsNHX1 gene confers enhanced salt and drought tolerance to transgenic Zea mays. *Journal of Integrative Agriculture.* 17: 2612–2623.

Ishikawa-Ishiwata, Y. and J. Furuya. 2021. Fungicide cost reduction with soybean rust-resistant cultivars in paraguay: a supply and demand approach. *Sustainability.* 13: 887.

Jia, T.-J., L.I. Jing-Jing, L.-F. Wang, Y.-Y. Cao, J. Ma, H. Wang, D.-F. Zhang and H.-Y. Li. 2020. Evaluation of drought tolerance in ZmVPP1-overexpressing transgenic inbred maize lines and their hybrids. *Journal of Integrative Agriculture.* 19: 2177–2187.

Jin, X., Y. Zheng, J. Wang, W. Chen, Z. Yang, Y. Chen, Y. Yang, G. Lu and B. Sun. 2023. SbNAC9 improves drought tolerance by enhancing scavenging ability of reactive oxygen species and activating stress-responsive genes of sorghum. *International Journal of Molecular Sciences.* 24: 2401.

Kannan, M., S. Prabhakaran and P. Ramachandran. 2010. *Rainfall Forecasting using Data Mining Technique.*

Karthik, K., M. Nandiganti, A. Thangaraj, S. Singh, P. Mishra, M. Rathinam, M. Sharma, N.K. Singh, P.K. Dash and R. Sreevathsa. 2020. Transgenic cotton (*Gossypium hirsutum* L.) to combat weed vagaries: utility of an apical meristem-targeted in planta transformation strategy to introgress a modified CP4-EPSPS gene for glyphosate tolerance. *Frontiers in Plant Science.* 11: 768.

Katta, S., A. Talakayala, M.K. Reddy, U. Addepally and M. Garladinne. 2020. Development of transgenic cotton (Narasimha) using triple gene Cry2Ab-Cry1F-Cry1Ac construct conferring resistance to lepidopteran pest. *Journal of Biosciences.* 45: 31.

Kawasaki, K. and S. Uchida. 2016. *Quality Matters More than Quantity: Asymmetric Temperature Effects on Crop Yield and Quality Grade.* PSN: Rural & Agricultural Development (Topic).

Kerr, T.C., H. Abdel-Mageed, L. Aleman, J. Lee, P. Payton, D. Cryer and R.D. Allen. 2018. Ectopic expression of two AREB/ABF orthologs increases drought tolerance in cotton (*Gossypium hirsutum*). *Plant, Cell & Environment.* 41: 898–907.

Kim, H.J., H.S. Cho, J.H. Pak, T. Kwon, J.-H. Lee, D.-H. Kim, D.H. Lee, C.-G. Kim and Y.-S. Chung. 2018. Confirmation of drought tolerance of ectopically expressed AtABF3 gene in soybean. *Molecules and Cells.* 41: 413–422.

Kim, M., B. Chemere and K. Sung. 2019. Effect of heavy rainfall events on the dry matter yield trend of whole crop maize (*Zea mays* L.). *Agriculture.* 9: 75.

Kim, S.Y., H. Lim, M. Kang, K.M. Jun, S.U. Ji, S.-C. Park and G.-S. Lee. 2020. Functional characterization of PsGPD in drought stress response using RNA-seq analysis of transgenic rice plant. *Plant Breeding and Biotechnology.* 8: 131–140.

Kumar, K., G. Gambhir, A. Dass, A.K. Tripathi, A. Singh, A.K. Jha, P. Yadava, M. Choudhary and S. Rakshit. 2020. Genetically modified crops: current status and future prospects. *Planta.* 251: 91.

Kumar, V., S. Jana, A. Bhardwaj, R. Deepa, S.K. Sahu, P. Pradhan and S.A. Sirdas. 2018. Greenhouse gas emission, rainfall and crop production over north-western India. *The Open Ecology Journal.* 11(1).

Kuppu, S., N. Mishra, R. Hu, L. Sun, X. Zhu, G. Shen, E. Blumwald, P. Payton and H. Zhang. 2013. Water-deficit inducible expression of a cytokinin biosynthetic gene IPT improves drought tolerance in cotton. *PLoS One.* 8: e64190.

Latha, G.M., K. Raman, J.M. Lima, D. Pattanayak, A.K. Singh, V. Chinnusamy, K.C. Bansal, K. Sambasiva Rao and T. Mohapatra. 2019. Genetic engineering of indica rice with AtDREB1A gene for enhanced abiotic stress tolerance. *Plant Cell, Tissue and Organ Culture (PCTOC).* 136: 173–188.

Leng, G. 2021. Maize yield loss risk under droughts in observations and crop models in the United States. *Environmental Research Letters.* 16: 024016.

Li, B., A. Wei, C. Song, N. Li and J. Zhang. 2008. Heterologous expression of the TsVP gene improves the drought resistance of maize. *Plant biotechnology journal.* 6: 146–159.

Li, C., J. Wang, F. Ling and A. You. 2023. Application and development of Bt insect resistance genes in rice breeding. *Sustainability.* 15: 9779.

Li, L., M. Zheng, G. Deng, J. Liang, H. Zhang, Z. Pan, H. Long and M. Yu. 2016. Overexpression of AtHDG11 enhanced drought tolerance in wheat (Triticum aestivum L.). *Molecular Breeding.* 36: 23.

Li, X., S.R. Waddington, J. Dixon, A.K. Joshi and M.C. De Vicente. 2011a. The relative importance of drought and other water-related constraints for major food crops in South Asian farming systems. *Food Security*. 3: 19–33.

Li, Y., K. Guan, G.D. Schnitkey, E. Delucia and B. Peng. 2019. Excessive rainfall leads to maize yield loss of a comparable magnitude to extreme drought in the United States. *Global Change Biology*. 25: 2325–2337.

Li, Z., M. Zhou, Z. Zhang, L. Ren, L. Du, B. Zhang, H. Xu and Z. Xin. 2011b. Expression of a radish defensin in transgenic wheat confers increased resistance to Fusarium graminearum and Rhizoctonia cerealis. *Functional & Integrative Genomics*. 11: 63–70.

Lisei-De-Sá, M.E., F. Arraes, G.G. Brito, M.A. Beneventi, I. Lourenço-Tessutti, A.M. Basso, R.M. Amorim, M.C. Silva, M. Faheem and N.G. Oliveira. 2017. AtDREB2A-CA influences root architecture and increases drought tolerance in transgenic cotton. *Agricultural Sciences*. 8: 1195.

Liu, F., Y. Liu, J. Zou, L. Zhang, H. Zheng, Y. Luo, X. Wang and L. Wang. 2023. Molecular characterization and efficacy evaluation of transgenic maize harboring cry2Ab-vip3A-cp4epsps for insect resistance and herbicide tolerance. *Plants*. 12: 612.

Liu, G., X. Li, S. Jin, X. Liu, L. Zhu, Y. Nie and X. Zhang. 2014. Overexpression of rice NAC gene SNAC1 improves drought and salt tolerance by enhancing root development and reducing transpiration rate in transgenic cotton. *PLoS One*. 9: e86895.

Liu, P., X. Tong, P. Meng, J. Zhang, J. Li, J. Zhang and Y. Zhou. 2022. Biophysical controls on water use efficiency of six plantations under different sky conditions. *Agricultural and Forest Meteorology*. 320: 108938.

Liu, Y., X. Liu, X. Wang, K. Gao, W. Qi, H. Ren, H. Hu, D. Sun, J. Bai and S. Zheng. 2020. Heterologous expression of heat stress-responsive AtPLC9 confers heat tolerance in transgenic rice. *BMC Plant Biology*. 20: 514.

Lu, G., L. Wang, L. Zhou, X. Su, H. Guo and H. Cheng. 2022. Overexpression of AmCBF1 enhances drought and cold stress tolerance, and improves photosynthesis in transgenic cotton. *PeerJ*. 10: e13422.

Lv, S.-L., L.-J. Lian, P.-L. Tao, Z.-X. Li, K.-W. Zhang and J.-R. Zhang. 2009. Overexpression of Thellungiella halophila H+-PPase (TsVP) in cotton enhances drought stress resistance of plants. *Planta*. 229: 899–910.

Ma, X.-J., J.-D. Fu, Y.-M. Tang, T.-F. Yu, Z.-G. Yin, J. Chen, Y.-B. Zhou, M. Chen, Z.-S. Xu and Y.-Z. Ma. 2020a. GmNFYA13 improves salt and drought tolerance in transgenic soybean plants. *Frontiers in Plant Science*. 11: 587244.

Ma, X.-J., T.-F. Yu, X.-H. Li, X.-Y. Cao, J. Ma, J. Chen, Y.-B. Zhou, M. Chen, Y.-Z. Ma, J.-H. Zhang and Z.-S. Xu. 2020b. Overexpression of GmNFYA5 confers drought tolerance to transgenic *Arabidopsis* and soybean plants. *BMC Plant Biology*. 20: 123.

Maheswari, M., Y. Varalaxmi, S. Yadav, N. Jyothilakshmi, M. Vanaja and B. Venkateswarlu. 2017. Enhanced tolerance of transgenic sorghum expressing mtlD gene to water-deficit stress. *Indian J. Biotechnol*. 16: 63–67.

Maity, A., D. Paul, A. Lamichaney, A. Sarkar, N. Babbar, N. Mandal, S. Dutta, P.P. Maity and S.K. Chakrabarty Chakrabarty. 2023. Climate change impacts on seed production and quality: current knowledge, implications, and mitigation strategies. *Seed Science and Technology*. 51: 7–38.

Mao, H., H. Wang, S. Liu, Z. Li, X. Yang, J. Yan, J. Li, L.-S.P. Tran and F. Qin. 2015. A transposable element in a NAC gene is associated with drought tolerance in maize seedlings. *Nature Communications*. 6: 8326.

Martins-Salles, S., V. Machado, L. Massochin-Pinto and L.M. Fiuza. 2017. Genetically modified soybean expressing insecticidal protein (Cry1Ac): management risk and perspectives. *FACETS*. 2: 496–512.

Mattos, L.M., C.L. Moretti, S. Jan, S.A. Sargent, C.E.P. Lima and M.R. Fontenelle. 2014. *Climate Changes and Potential Impacts on Quality of Fruit and Vegetable Crops*. In *Emerging Technologies and Management of Crop Stress Tolerance*. Academic Press: 467–486.

Miklos, J.A., M.F. Alibhai, S.A. Bledig, D.C. Connor-Ward, A.-G. Gao, B.A. Holmes, K.H. Kolacz, V.T. Kabuye, T.C. Macrae and M.S. Paradise. 2007a. Characterization of soybean exhibiting high expression of a synthetic *Bacillus thuringiensis* cry1A transgene that confers a high degree of resistance to lepidopteran pests. *Crop Science*. 47: 148–157.

Miklos, J.A., M.F. Alibhai, S.A. Bledig, D.S.C. Connor-Ward, A.-G. Gao, B.A. Holmes, K.H. Kolacz, V.T. Kabuye, T.C. Macrae, M.S. Paradise, A.S. Toedebusch and L.A. Harrison. 2007b. Characterization of soybean exhibiting high expression of a synthetic Bacillus thuringiensis cry1A transgene that confers a high degree of resistance to lepidopteran pests. *Crop Science*. 47: 148–157.

Milligan, A.S., A. Daly, M.a.J. Parry, P.A. Lazzeri and I. Jepson. 2001. The expression of a maize glutathione S-transferase gene in transgenic wheat confers herbicide tolerance, both in planta and in vitro. *Molecular Breeding*. 7: 301–315.

Mishra, N., L. Sun, X. Zhu, J. Smith, A. Prakash Srivastava, X. Yang, N. Pehlivan, N. Esmaeili, H. Luo, G. Shen, D. Jones, D. Auld, J. Burke, P. Payton and H. Zhang. 2017. Overexpression of the rice SUMO E3 ligase gene OsSIZ1 in cotton enhances drought and heat tolerance, and substantially improves fiber yields in the field under reduced irrigation and rainfed conditions. *Plant and Cell Physiology*. 58: 735–746.

Muchie, A. and F. Assefa. 2021. Impact of climate change on horticultural crops production and quality: a review. *Amercian Journal of Bioscience and Bioengineering*. 9: 156–161.

Murakami, T., S. Matsuba, H. Funatsuki, K. Kawaguchi, H. Saruyama, M. Tanida and Y. Sato. 2004. Over-expression of a small heat shock protein, sHSP17.7, confers both heat tolerance and UV-B resistance to rice plants. *Molecular Breeding*. 13: 165–175.

Nam, K.-H., D.Y. Kim, Y.S. Moon, I.S. Pack, S.-C. Jeong, H.B. Kim and C.-G. Kim. 2020. Performance of hybrids between abiotic stress-tolerant transgenic rice and its weedy relatives under water-stressed conditions. *Scientific Reports*. 10: 9319.

Naumann, G., C. Cammalleri, L. Mentaschi and L. Feyen. 2021. Increased economic drought impacts in Europe with anthropogenic warming. *Nature Climate Change*. 11: 485–491.

Ngaira, J.K. and K.O. Ngaira. 2017. Droughts and agricultural development in Africa: the Genetically Modified Organisms (GMOs) mitigation strategy in food security. *Current Journal of Applied Science and Technology*. 236:1–17.

Nguyen, T.D., V.H. La, V.D. Nguyen, T.-T. Bui, T.T. Nguyen, Y.H. Je, Y.S. Chung and X.B. Ngo. 2021. Convergence of bar and Cry1Ac mutant genes in soybean confers synergistic resistance to herbicide and lepidopteran insects. *Frontiers in Plant Science*. 12.

Ni, Z., N. Liu, Y. Yu, C. Bi, Q. Chen and Y. Qu. 2021. The cotton 70-kDa heat shock protein GhHSP70-26 plays a positive role in the drought stress response. *Environmental and Experimental Botany*. 191: 104628.

Ning, W., H. Zhai, J. Yu, S. Liang, X. Yang, X. Xing, J. Huo, T. Pang, Y. Yang and X. Bai. 2017. Overexpression of Glycine soja WRKY20 enhances drought tolerance and improves plant yields under drought stress in transgenic soybean. *Molecular Breeding*. 37: 1–10.

Noman, M., A. Jameel, W.-D. Qiang, N. Ahmad, W.-C. Liu, F.-W. Wang and H.-Y. Li. 2019. Overexpression of GmCAMTA12 enhanced drought tolerance in arabidopsis and soybean. *International Journal of Molecular Sciences*. 20: 4849.

Noushahi, H.A. and M. Hussain. 2020. Risk assessment of genetically modified plants: a review. *Asian Journal of Research in Crop Science*. 5: 1–10.

Okosun, O.O., K.C. Allen, J.P. Glover and G.V.P. Reddy. 2021. Biology, ecology, and management of key Sorghum insect pests. *Journal of Integrated Pest Management.* 12: 4.

Panth, M., S.C. Hassler and F. Baysal-Gurel. 2020. Methods for management of soilborne diseases in crop production. *Agriculture.* 10: 16.

Parkhi, V., V. Kumar, L.M. Campbell, A.A. Bell, J. Shah and K.S. Rathore. 2010. Resistance against various fungal pathogens and reniform nematode in transgenic cotton plants expressing Arabidopsis NPR1. *Transgenic Research.* 19: 959–975.

Pham, T.T.N., H.Q. Nguyen, T.N.L. Nguyen, X.T. Dao, D.T. Sy, L. Van Son and C. Hoang Mau. 2020. Overexpression of the "GmDREB2" gene increases proline accumulation and tolerance to drought stress in soybean plants. *Australian Journal of Crop Science.* 14: 495–503.

Polthanee, A., A. Promkhumbut and J. Bamrungrai. 2014. Drought impact on rice production and farmers' adaptation strategies in Northeast Thailand. *International Journal of Environmental and Rural Development.* 5: 45–52.

Prasad, A.S. 2014. Impact of the discovery of human zinc deficiency on health. *Journal of Trace Elements in Medicine and Biology.* 28: 357–363.

Qin, D., F. Wang, X. Geng, L. Zhang, Y. Yao, Z. Ni, H. Peng and Q. Sun. 2015. Overexpression of heat stress-responsive TaMBF1c, a wheat (*Triticum aestivum* L.) Multiprotein Bridging Factor, confers heat tolerance in both yeast and rice. *Plant Molecular Biology.* 87: 31–45.

Qin, F., M. Kakimoto, Y. Sakuma, K. Maruyama, Y. Osakabe, L.S.P. Tran, K. Shinozaki and K. Yamaguchi-Shinozaki. 2007. Regulation and functional analysis of ZmDREB2A in response to drought and heat stresses in Zea mays L. *The Plant Journal.* 50: 54–69.

Qin, N., W. Xu, L. Hu, Y. Li, H. Wang, X. Qi, Y. Fang and X. Hua. 2016. Drought tolerance and proteomics studies of transgenic wheat containing the maize C 4 phosphoenolpyruvate carboxylase (PEPC) gene. *Protoplasma.* 253: 1503–1512.

Rad, A.H.S. and P. Zandi. 2012. The effect of drought stress on qualitative and quantitative traits of spring rapeseed (*Brassica napus* L.) cultivars. *Zemdirbyste-Agriculture.* 99: 47–54.

Rajasekaran, K., J.W. Cary, J.M. Jaynes and T.E. Cleveland. 2005. Disease resistance conferred by the expression of a gene encoding a synthetic peptide in transgenic cotton (*Gossypium hirsutum* L.) plants. *Plant Biotechnology Journal.* 3: 545–554.

Rajasekaran, K., J.W. Grula, R.L. Hudspeth, S. Pofelis and D.M. Anderson. 1996. Herbicide-resistant Acala and Coker cottons transformed with a native gene encoding mutant forms of acetohydroxyacid synthase. *Molecular Breeding.* 2: 307–319.

Rana, I.A., H. Loerz, W. Schaefer and D. Becker. 2012. Overexpression of chitinase and chitosanase genes from Trichoderma harzianum under constitutive and inducible promoters in order to increase disease resistance in wheat (Triticum aestivum L). *Molecular Plant Breeding.* 3(1).

Redden, R. 2021. Genetic modification for agriculture proposed revision of GMO regulation in Australia. *Plants.* 10: 747.

Risk, J.M., L.L. Selter, S.G. Krattinger, L.A. Viccars, T.M. Richardson, G. Buesing, G. Herren, E.S. Lagudah and B. Keller. 2012. Functional variability of the Lr34 durable resistance gene in transgenic wheat. *Plant Biotechnology Journal.* 10: 477–487.

Rodziewicz, D. and J. Dice. 2020. Drought risk to the agriculture sector. *Economic Review* (01612387). 105.

Saad, A.S.I., X. Li, H.-P. Li, T. Huang, C.-S. Gao, M.-W. Guo, W. Cheng, G.-Y. Zhao and Y.-C. Liao. 2013. A rice stress-responsive NAC gene enhances tolerance of transgenic wheat to drought and salt stresses. *Plant Science.* 203–204: 33–40.

Samtani, H., A. Sharma and P. Khurana. 2022. Overexpression of HVA1 enhances drought and heat stress tolerance in *Triticum aestivum* doubled haploid plants. *Cells.* 11: 912.

Sanders, C.J., J.K. Pell, G.M. Poppy, A. Raybould, M. Garcia-Alonso and T.H. Schuler. 2007. Host-plant mediated effects of transgenic maize on the insect parasitoid *Campoletis sonorensis* (Hymenoptera: Ichneumonidae). *Biological Control.* 40: 362–369.

Sarker, P.K., A.S. Paul and D. Karmoker. 2023. Mitigating climate change and pandemic impacts on global food security: dual sustainable agriculture approach (2S approach). *Planta.* 258: 104.

Sathee, L., S.B. Adavi, B.K. Padhan, A. Sinto, A. Anand, N. Sailo, V. Jain and S.K. Jha. 2022. Chapter 3 – Elevated atmospheric CO2 induced changes in nitrogen metabolism and crop quality. *In:* Shanker, A. K., Shanker, C., Aanand, A. & Maheswari, M. (eds.) *Climate Change and Crop Stress.* Academic Press.

Sato, H., T. Takamizo, J. Horita, K. Kawai, K. Kaku and T. Shimizu. 2011. Transgenic tall fescue and maize with resistance to ALS-inhibiting herbicides. *Herbicides, Theory and Applications.* 214–224.

Schnippenkoetter, W., C. Lo, G. Liu, K. Dibley, W.L. Chan, J. White, R. Milne, A. Zwart, E. Kwong and B. Keller. 2017a. The wheat Lr34 multipathogen resistance gene confers resistance to anthracnose and rust in sorghum. *Plant Biotechnology Journal.* 15: 1387–1396.

Schnippenkoetter, W., C. Lo, G. Liu, K.E. Dibley, W.L. Chan, J.A. White, R.J. Milne, A.B. Zwart, E. Kwong, B. Keller, I.D. Godwin, S.G. Krattinger and E.S. Lagudah. 2017b. The wheat Lr34 multipathogen resistance gene confers resistance to anthracnose and rust in sorghum. *Plant Biotechnology Journal.* 15: 1387–1396.

Shah, F.S., M.A. Bhatti, A.A. Khan, I.A. Baig, M.H.N. Tahir, F. Ahmad and M.A. Bakhtavar. 2022. Cotton seed system. *In: Cotton Breeding and Biotechnology.* CRC Press: 291–309.

Shaheen, H., A. Mustafa and A. Ulfat. 2022. Chapter 6 – Crop production in response to elevated CO2: grain yield and quality. *In:* Liu, F., Li, X., Hogy, P., Jiang, D., Brestic, M. & Liu, B. (eds.) *Sustainable Crop Productivity and Quality Under Climate Change.* Academic Press.

Sharma, N., M. Thakur, P. Suryakumar, P. Mukherjee, A. Raza, C.S. Prakash and A. Anand. 2022. 'Breathing out' under heat stress—respiratory control of crop yield under high temperature. *Agronomy.* 12: 806.

Shen, G., J. Wei, X. Qiu, R. Hu, S. Kuppu, D. Auld, E. Blumwald, R. Gaxiola, P. Payton and H. Zhang. 2015. Co-overexpression of AVP1 and AtNHX1 in cotton further improves drought and salt tolerance in transgenic cotton plants. *Plant Molecular Biology Reporter.* 33: 167–177.

Shi, J., J.E. Habben, R.L. Archibald, B.J. Drummond, M.A. Chamberlin, R.W. Williams, H.R. Lafitte and B.P. Weers. 2015. Overexpression of ARGOS genes modifies plant sensitivity to ethylene, leading to improved drought tolerance in both arabidopsis and maize. *Plant Physiology.* 169: 266–282.

Si, W., Q. Liang, L. Chen, F. Song, Y. Chen and H. Jiang. 2021. Ectopic overexpression of maize heat stress transcription factor ZmHsf05 confers drought tolerance in transgenic rice. *Genes.* 12: 1568.

Siddiqui, I., R. Bajwa, Zil-E-Huma and A. Javaid. 2010. Effect of six problematic weeds on growth and yield of wheat. *Pakistan Journal of Botany.* 42: 2461–2471.

Soares, J.C., C.S. Santos, S.M.P. Carvalho, M.M. Pintado and M.W. Vasconcelos. 2019. Preserving the nutritional quality of crop plants under a changing climate: importance and strategies. *Plant and Soil.* 443: 1–26.

Sohrab, S.S., M.A. Kamal, A. Ilah, A. Husen, P.S. Bhattacharya and D. Rana. 2014. Development of Cotton leaf curl virus resistant transgenic cotton using antisense ßC1 gene. *Saudi Journal of Biological Sciences.* 23: 358 – 362.

Song, Y., X. Zhang, P. Wang, Z. Liu, H. Dong, J. Ma, Y. Fu and Z. Zhang. 2016. Characterization of transformed soybean strains expressing GbNPR1 and HrpZpsg12 genes for disease resistance. *Euphytica.* 211: 369–377.

Soto, N., Y. Hernández, C. Delgado, Y. Rosabal, R. Ortiz, L. Valencia, O. Borrás-Hidalgo, M. Pujol and G.A. Enríquez. 2020. Field resistance to Phakopsora pachyrhizi and Colletotrichum truncatum of transgenic soybean expressing the NmDef02 plant defensin gene. *Frontiers in Plant Science.* 11: 562.

Sprague, S.A., T.M. Tamang, T. Steiner, Q. Wu, Y. Hu, T. Kakeshpour, J. Park, J. Yang, Z. Peng and B. Bergkamp. 2022. Redox-engineering enhances maize thermotolerance and grain yield in the field. *Plant Biotechnology Journal.* 20: 1819–1832.

Su, P., C. Sui, Y. Niu, J. Li, S. Wang, F. Sun, J. Yan and S. Guo. 2023. Comparative transcriptomic analysis and functional characterization reveals that the class III peroxidase gene TaPRX-2A regulates drought stress tolerance in transgenic wheat. *Frontiers in Plant Science.* 14: 1119162.

Sucher, J., R. Boni, P. Yang, P. Rogowsky, H. Büchner, C. Kastner, J. Kumlehn, S.G. Krattinger and B. Keller. 2017. The durable wheat disease resistance gene Lr34 confers common rust and northern corn leaf blight resistance in maize. *Plant Biotechnology Journal.* 15: 489–496.

Tougou, M., N. Yamagishi, N. Furutani, K. Kaku, T. Shimizu, Y. Takahata, J. Sakai, S. Kanematsu and S. Hidaka. 2009. The application of the mutated acetolactate synthase gene from rice as the selectable marker gene in the production of transgenic soybeans. *Plant Cell Reports.* 28: 769–776.

Tri Joko, S., A. Aniversari Apriana, S. Atmitri Sisharmini and T. Kurniawan Rudi. 2018. Construction and Transformation of OsERA1 Gene Into Expression Vector and Response of Nipponbare-OsERA1 Transgenic Rice to Drought Stress. *Journal AgroBiogen.* 14: 23–36.

Turnbull, C., M. Lillemo and T.A. Hvoslef-Eide. 2021. Global regulation of genetically modified crops amid the gene edited crop boom–a review. *Frontiers in Plant Science.* 12: 630396.

Visarada, K.B.R.S., G.S. Prasad and M. Royer. 2016. Genetic transformation and evaluation of two sweet sorghum genotypes for resistance to spotted stemborer, Chilo partellus (Swinhoe). *Plant Biotechnology Reports.* 10: 277–289.

Vishwakarma, H., A. Junaid, J. Manjhi, G.P. Singh, K. Gaikwad and J.C. Padaria. 2018. Heat stress transcripts, differential expression, and profiling of heat stress tolerant gene TaHsp90 in Indian wheat (*Triticum aestivum* L.) cv C306. *PloS One.* 13: e0198293.

Wang, B., L. Li, M. Liu, D. Peng, A. Wei, B. Hou, Y. Lei and X. Li. 2022a. TaFDL2-1A confers drought stress tolerance by promoting ABA biosynthesis, ABA responses, and ROS scavenging in transgenic wheat. *The Plant Journal.* 112: 722–737.

Wang, C.-R., A.-F. Yang, G.-D. Yue, Q. Gao, H.-Y. Yin and J.-R. Zhang. 2008. Enhanced expression of phospholipase C 1 (ZmPLC1) improves drought tolerance in transgenic maize. *Planta.* 227: 1127–1140.

Wang, G., Y. Long, X. Jin, Z. Yang, L. Dai, Y. Yang, G. Lu and B. Sun. 2024. SbMYC2 mediates jasmonic acid signaling to improve drought tolerance via directly activating SbGR1 in sorghum. *Theoretical and Applied Genetics.* 137: 72.

Wang, Y., Z. Fu, Z. Hu and S. Niu. 2022b. Tracking global patterns of drought-induced productivity loss along severity gradient. *Journal of Geophysical Research: Biogeosciences.* 127: e2021JG006753.

Wang, Y., Y. Zhang, X. Yu, G. Jia, Z. Liu, L. Sun, P. Zheng and X. Zhu. 2021. Grassland soil moisture fluctuation and its relationship with evapotranspiration. *Ecological Indicators.* 131: 108196.

Wei, A., C. He, B. Li, N. Li and J. Zhang. 2011. The pyramid of transgenes TsVP and BetA effectively enhances the drought tolerance of maize plants. *Plant Biotechnology Journal.* 9: 216–229.

Wei, L., G. Wu, X. Tang, W. Wang, X. Wang and L. Liu. 2016. Soybean responses to high temperatures under drought stress in the presence of an over-expressed GmHSFA1 gene. *Soybean Science.* 35: 257–261.

Wu, X., Y. Shiroto, S. Kishitani, Y. Ito and K. Toriyama. 2009. Enhanced heat and drought tolerance in transgenic rice seedlings overexpressing OsWRKY11 under the control of HSP101 promoter. *Plant Cell Reports.* 28: 21–30.

Xi, Y., Z. Yang, Y. Jin, J. Qu, S. Guan, S. Liu and P. Wang. 2023. Identification and Evaluation of Insect and Disease Resistance in Transgenic Cry1Ab13–1 and NPR1 Maize. *Phyton (0031-9457).* 92(4).

Xu, W., W. Tang, C. Wang, L. Ge, J. Sun, X. Qi, Z. He, Y. Zhou, J. Chen and Z. Xu. 2020. SiMYB56 confers drought stress tolerance in transgenic rice by regulating lignin biosynthesis and ABA signaling pathway. *Frontiers in Plant Science.* 11: 785.

Yang, H., D. Zhang, X. Li, H. Li, D. Zhang, H. Lan, A.J. Wood and J. Wang. 2016. Overexpression of ScALDH21 gene in cotton improves drought tolerance and growth in greenhouse and field conditions. *Molecular Breeding.* 36: 34.

Yang, J., T. Mao, Z. Geng, W. Xue, L. Ma, Y. Jin, P. Guo, Z. Qiu, L. Wang, C. Yu, Y. Sheng, J. Zhang and H. Zhang. 2023. Constitutive expression of AtSINA2 from *Arabidopsis* improves grain yield, seed oil and drought tolerance in transgenic soybean. *Plant Physiology and Biochemistry.* 196: 444–453.

Yang, J., G. Zhang, J. An, Q. Li, Y. Chen, X. Zhao, J. Wu, Y. Wang, Q. Hao, W. Wang and W. Wang. 2020a. Expansin gene TaEXPA2 positively regulates drought tolerance in transgenic wheat (*Triticum aestivum* L.). *Plant Science.* 298: 110596.

Yang, Y., S. Luang, J. Harris, M. Riboni, Y. Li, N. Bazanova, M. Hrmova, S. Haefele, N. Kovalchuk and S. Lopato. 2018. Overexpression of the class I homeodomain transcription factor Ta HDZ ipI-5 increases drought and frost tolerance in transgenic wheat. *Plant Biotechnology Journal.* 16: 1227–1240.

Yang, Z., X. Chi, F. Guo, X. Jin, H. Luo, A. Hawar, Y. Chen, K. Feng, B. Wang, J. Qi, Y. Yang and B. Sun. 2020b. SbWRKY30 enhances the drought tolerance of plants and regulates a drought stress-responsive gene, *SbRD19, in sorghum. Journal of Plant Physiology.* 246–247: 153142.

Ye, R., H. Huang, Z. Yang, T. Chen, L. Liu, X. Li, H. Chen and Y. Lin. 2009. Development of insect-resistant transgenic rice with Cry1C*-free endosperm. *Pest Management Science: formerly Pesticide Science.* 65: 1015–1020.

Yu, L.H., S.J. Wu, Y.S. Peng, R.N. Liu, X. Chen, P. Zhao, P. Xu, J.B. Zhu, G.L. Jiao and Y. Pei. 2016. Arabidopsis EDT 1/HDG 11 improves drought and salt tolerance in cotton and poplar and increases cotton yield in the field. *Plant Biotechnology Journal.* 14: 72–84.

Yu, T.-F., Z.-S. Xu, J.-K. Guo, Y.-X. Wang, B. Abernathy, J.-D. Fu, X. Chen, Y.-B. Zhou, M. Chen and X.-G. Ye. 2017. Improved drought tolerance in wheat plants overexpressing a synthetic bacterial cold shock protein gene SeCspA. *Scientific Reports.* 7: 44050.

Yu, Y. and Z. Wei. 2008. Increased oriental armyworm and aphid resistance in transgenic wheat stably expressing *Bacillus thuringiensis* (Bt) endotoxin and Pinellia ternate agglutinin (PTA). *Plant Cell, Tissue and Organ Culture.* 94: 33–44.

Yue, Y., M. Zhang, J. Zhang, X. Tian, L. Duan and Z. Li. 2012. Overexpression of the AtLOS5 gene increased abscisic acid level and drought tolerance in transgenic cotton. *Journal of Experimental Botany.* 63: 3741–3748.

Zafar, S., H. Afzal, A. Ijaz, A. Mahmood, A. Ayub, A. Nayab, S. Hussain, M. Ul-Hussan, M.A. Sabir, U. Zulfiqar, F. Zulfiqar and A. Moosa. 2023. Cotton and drought stress: an updated overview for improving stress tolerance. *South African Journal of Botany.* 161: 258–268.

Zahra, N., M.B. Hafeez, A. Wahid, M.H.K. Al Masruri, A. Ullah, K.H.M. Siddique and M. Farooq. 2022. Impact of climate change on wheat grain composition and quality. *Journal of the Science of Food and Agriculture.* 103: 2745–2751.

Zandi, M., R. Hosseini, M. Mohsenpour, G. Hosseini Salekdeh and G. Behzad. 2019. Transformation of DRO1, OsNAC5, OsEXPA8 genes in order to improve root architecture and drought tolerance in rice. *Genetic Engineering and Biosafety Journal.* 8: 77–89.

Zang, X., X. Geng, K. He, F. Wang, X. Tian, M. Xin, Y. Yao, Z. Hu, Z. Ni, Q. Sun and H. Peng. 2018. Overexpression of the wheat (*Triticum aestivum* L.) TaPEPKR2 gene enhances heat and dehydration tolerance in both wheat and *arabidopsis*. *Frontiers in Plant Science.* 9: 1710.

Zeng, D., Z. Cai and Y. Liu. 2019. Overexpression of BdRHP1 improves drought tolerance and reduces yield loss in rice. *Biologia Plant.* 63: 371–379.

Zhang, D.-Y., H.-L. Yang, X.-S. Li, H.-Y. Li and Y.-C. Wang. 2014. Overexpression of Tamarix albiflonum TaMnSOD increases drought tolerance in transgenic cotton. *Molecular Breeding.* 34: 1–11.

Zhang, J., Y. Peng and Z. Guo. 2008. Constitutive expression of pathogen-inducible OsWRKY31 enhances disease resistance and affects root growth and auxin response in transgenic rice plants. *Cell Research.* 18: 508–521.

Zhang, J.H., C.Z. Wang, J.D. Qin and S.D. Guo. 2004. Feeding behaviour of Helicoverpa armigera larvae on insect-resistant transgenic cotton and non-transgenic cotton. *Journal of Applied Entomology.* 128: 218–225.

Zhang, L., Y. Zheng, X. Xiong, H. Li, X. Zhang, Y. Song, X. Zhang and D. Min. 2023. The wheat VQ motif-containing protein TaVQ4-D positively regulates drought tolerance in transgenic plants. *Journal of Experimental Botany.* 74: 5591–5605.

Zhang, S., N. Li, F. Gao, A. Yang and J. Zhang. 2010. Over-expression of TsCBF1 gene confers improved drought tolerance in transgenic maize. *Molecular Breeding.* 26: 455–465.

Zhang, Y., C. Wang, C. Wang, L. Yun, L. Song, M. Idrees, H. Liu, Q. Zhang, J. Yang and X. Zheng. 2022. OsHsfB4b confers enhanced drought tolerance in transgenic Arabidopsis and rice. *International Journal of Molecular Sciences.* 23: 10830.

Zhao, Y., H. Du, Y. Wang, H. Wang, S. Yang, C. Li, N. Chen, H. Yang, Y. Zhang and Y. Zhu. 2021a. The calcium-dependent protein kinase ZmCDPK7 functions in heat-stress tolerance in maize. *Journal of Integrative Plant Biology.* 63: 510–527.

Zhao, Y., L. Wu, Q. Fu, D. Wang, J. Li, B. Yao, S. Yu, L. Jiang, J. Qian and X. Zhou. 2021b. INDITTO2 transposon conveys auxin-mediated DRO1 transcription for rice drought avoidance. *Plant, Cell & Environment.* 44: 1846–1857.

Zhao, Z.Y., T. Cai, L. Tagliani, M. Miller, N. Wang, H. Pang, M. Rudert, S. Schroeder, D. Hondred, J. Seltzer and D. Pierce. 2000. *Agrobacterium*-mediated sorghum transformation. *Plant Molecular Biology.* 44: 789–98.

Zhou, Y., M. Chen, J. Guo, Y. Wang, D. Min, Q. Jiang, H. Ji, C. Huang, W. Wei, H. Xu, X. Chen, L. Li, Z. Xu, X. Cheng, C. Wang, C. Wang and Y. Ma. 2019. Overexpression of soybean DREB1 enhances drought stress tolerance of transgenic wheat in the field. *Journal of Experimental Botany.* 71: 1842–1857.

Zhu, Y., Y. Liu, K. Zhou, C. Tian, M. Aslam, B. Zhang, W. Liu and H. Zou. 2022. Overexpression of ZmEREBP60 enhances drought tolerance in maize. *Journal of Plant Physiology.* 275: 153763.

Ziadat, F., A. Al-Wadaey, Z. Masri and H. Sakai. Adaptation to heavy rainfall events: watershed-community planning of soil and water conservation technologies in Syria. *EGU General Assembly Conference Abstracts*, 2010. 9237.

12 New Frontiers in Meiotic Recombination to Develop Climate-resilient Crops

Zeeshan Shamim, Allah Bakhsh, Muhammad Naveed Shahid, Muhammad Arshad Javed, Adil Jamal, and Sue J. Armstrong

12.1 INTRODUCTION

The unprecedented challenges posed by climate change have exacerbated the vulnerability of agricultural systems worldwide, hence threatening global food security [1]. Increasing temperatures, erratic precipitation patterns, and the escalating frequency of extreme weather events have profound impacts on crop productivity [2]. Developing climate-resilient crop varieties with enhanced tolerance to abiotic stresses such as drought, heat, and salinity is imperative to sustain agricultural yields and ensure food availability for a growing global population [3]. Meiotic recombination, a fundamental process in sexual reproduction, generates genetic diversity and offers promising opportunities to accelerate the development of improved cultivars adapted to changing environmental conditions [4].

During meiosis, crossover events occur through a highly regulated process involving the formation and repair of programmed double-strand breaks (DSBs) in DNA [5]. This exchange of genetic material between homologous chromosomes reshuffles existing allelic combinations, creating novel genotypes that can be exploited through breeding and selection for desired traits [6]. Crop wild relatives (such as *Triticum dicoccoides* for wheat or teosinte for maize) and locally adapted landraces represent invaluable sources of genetic diversity harboring beneficial alleles for abiotic stress tolerance, which have been shaped by natural selection over generations of exposure to harsh environments [7]. However, the introgression of these favorable alleles into elite cultivars can be hindered by linkage drag, a phenomenon where undesirable traits are co-inherited because they are physically close to the target gene on the chromosome, making it difficult to separate them through conventional breeding. To address linkage drag, approaches such as marker-assisted selection, genomic selection, and advanced genome-editing techniques like CRISPR-Cas9 are being employed alongside meiotic recombination to precisely edit or eliminate undesirable traits [8].

 DOI: 10.1201/9781032720142-16

Meiotic recombination offers a powerful tool to break these genetic linkages, facilitating the efficient transfer of stress-resilience genes while minimizing the introduction of undesirable traits [9].

12.2 CLIMATE CHANGE CHALLENGES FOR CROP PRODUCTION

Climate change poses a significant threat to global food security, exacerbating the vulnerability of agricultural systems worldwide [10]. The impacts of climate change, including rising temperatures, changes in precipitation patterns, and increased frequency and intensity of extreme weather events, have far-reaching consequences for crop production and yield stability [11].

1. Temperature Extremes: Both high- and low-temperature extremes can have detrimental effects on crop growth, development, and productivity. For example, high temperatures can lead to accelerated crop development, reduced grain filling, and lower yields, as seen in maize, where heat stress reduces kernel set. Similarly, low temperatures can cause chilling injury, frost damage, and delayed maturation, as observed in rice, which suffers from delayed flowering and reduced grain yield under chilling conditions [12, 13].

2. Water Scarcity and Drought Stress: Climate change is projected to increase the frequency and severity of droughts in many regions, leading to water scarcity and drought stress for crops [14]. Drought conditions can severely limit crop growth, reduce yield, and increase the risk of crop failure, especially in rain-fed agricultural systems [15]. For example, drought stress in wheat can significantly affect grain filling, reducing both the quantity and quality of grain yield. Similarly, maize exhibits reduced kernel development and biomass production under drought conditions.

3. Soil Degradation: Climate change can contribute to soil degradation through processes such as erosion, driven by heavy rainfall events and increased storm intensity; salinization, caused by rising sea levels affecting coastal areas; and the loss of organic matter due to higher temperatures and altered precipitation patterns [16]. Degraded soils have reduced fertility, water-holding capacity, and nutrient availability, directly impacting crop productivity and sustainability [17].

4. Pest and Disease Pressures: Changing climatic conditions can influence the distribution, abundance, and virulence of pests and pathogens, posing new challenges for crop protection and integrated pest management strategies [18]. Emerging pests and diseases can significantly reduce crop yields and quality [19].

5. Shifts in Agroecological Zones: Climate change may lead to shifts in agroecological zones, including latitude shifts (e.g., crops previously grown in lower latitudes may need to be cultivated farther north) and altitudinal shifts (e.g., crops moving to higher elevations due to temperature changes), altering the suitability of certain regions for specific crop cultivation [20]. These changes could necessitate the relocation of crop production areas or the introduction of new crop varieties better adapted to the changing conditions. Additionally, farmers may need to adjust their sowing calendars to align with altered growing seasons and climatic patterns [21].

6. Elevated CO_2 Levels: While increased atmospheric CO_2 concentrations can enhance photosynthesis and productivity in some crops, the effects are species-specific

and can be offset by other climate change impacts, such as heat stress and water scarcity. Moreover, the "carbon–water trade-off" where increased CO_2 often reduces stomatal conductance, leading to water-use efficiency gains but potentially increasing crop sensitivity to drought further complicates these interactions [22]. Addressing these climate change challenges requires a multifaceted approach, including the development of climate-resilient crop varieties, sustainable agricultural practices, and effective adaptation and mitigation strategies [23]. Harnessing the potential of meiotic recombination can play a crucial role in generating the genetic diversity necessary for breeding climate-resilient crops with improved tolerance to abiotic stresses, pests, and diseases [24].

12.3 MEIOSIS: A FUNDAMENTAL PROCESS IN CROP BREEDING

Meiosis is a specialized cell division process that produces haploid gametes (e.g., sperm and egg cells) from diploid cells. These gametes restore diploidy upon fertilization, combining to form a diploid zygote that will develop into a new organism. It is a key biological process that generates genetic diversity and enables sexual reproduction in eukaryotic organisms, including plants [25]. During meiosis, a single round of DNA replication is followed by two rounds of cell division, resulting in the production of four haploid cells from a single diploid cell [26]. The meiotic process involves several landmark events, including the condensation of chromosomes, pairing of homologous chromosomes (synapsis), formation of programmed DSBs in DNA, reciprocal exchange of genetic material between homologous chromosomes (crossover), and segregation of chromosomes. These events are tightly regulated by a complex network of proteins and molecular pathways, ensuring the proper distribution of genetic material and the generation of genetically distinct gametes [27]. Crossover events are often concentrated in specific genomic regions known as recombination hotspots, which play a crucial role in shaping patterns of genetic diversity.

12.3.1 Meiotic Recombination Mechanism

Meiotic recombination is a highly intricate and regulated process that occurs during prophase I of meiosis, involving the formation and repair of programmed DSBs in DNA [28]. The DSB formation is catalyzed by the evolutionarily conserved SPO11 protein, which generates DSBs in the DNA, leading to the generation of free DNA ends [29]. These DSBs are subsequently processed by a complex network of proteins involved in the recombination pathway, including those involved in end resection, strand invasion, and homology search [30]. The RecA-like recombinases, such as RAD51 and DMC1, play a crucial role in promoting the invasion of the processed DNA ends into the homologous chromosome, forming a displacement loop (D-loop) structure [31]. This D-loop serves as a primer for DNA synthesis, facilitating the repair of the DSB using the homologous chromosome as a template [32]. The resolution of these joint molecule intermediates can result in either a crossover (CO) event, where flanking genetic markers are exchanged between the homologous chromosomes, or a non-crossover (NCO) event, where the genetic information remains unchanged [33].

Breeders can manipulate crossover frequencies through genetic or biotechnological approaches to increase the likelihood of desirable gene combinations, while NCO events help maintain genetic stability in critical regions. While CO events are essential for genetic diversity, NCO events can also be valuable in breeding programs, as they allow for the precise transmission of specific gene combinations without disrupting favorable linkage groups.

The frequency and distribution of COs along the chromosomes are not random but are regulated by various factors, including the chromosomal context, chromatin structure, and the activity of specific proteins involved in the recombination pathway [34, 35]. For instance, the presence of structural features like centromeres, telomeres, and heterochromatic regions can influence the recombination landscape [35]. Additionally, epigenetic modifications such as DNA methylation and histone modifications can modulate chromatin accessibility and affect the recombination machinery's ability to access and process the DNA [36]. A complete molecular mechanism has been tabulated for quick overview (Table 12.1)

TABLE 12.1
Key proteins and molecular mechanisms involved in meiotic recombination

Stage	Process	Key Proteins Involved	Outcome
Prophase I initiation	Formation of programmed double-strand breaks (DSBs) in DNA	SPO11 protein complex (including TOPOVIBL and MTOPVIB)	Creates entry points for recombination
Prophase I	DNA end resection	MRN complex (MRE11–RAD50–NBS1) and other nucleases	Generates 3' single-stranded DNA (ssDNA) overhangs
Prophase I	Homologous chromosome invasion	RecA-like recombinases (RAD51, DMC1)	ssDNA overhangs invade homologous chromosome, forming a displacement loop (D-loop)
Prophase I	D-loop extension and double Holliday junction (dHJ) formation	Replication machinery	D-loop serves as primer for DNA synthesis, creating a dHJ intermediate
Prophase I resolution	Crossover (CO) pathway	MUS81–EME1 complex, MLH1–MLH3 heterodimer	dHJ is cleaved in a specific way, resulting in exchange of genetic material between chromosomes
Prophase I resolution	Non-crossover (NCO) pathway	BTR complex (BLM–TOPOIIIα–RMI1–RMI2)	dHJ is dissolved without exchange of genetic material

12.3.1.1 Key Proteins Involved in Meiotic Recombination

Meiotic recombination involves a complex interplay of numerous proteins, each playing a critical role in the various stages of the process. Some of the key proteins involved in meiotic recombination are described in Table 12.2.

12.3.1.2 Regulation of Meiotic Recombination

Meiotic recombination is a tightly regulated process, governed by various mechanisms that control the frequency, distribution, and outcome of crossover events. Several factors contribute to the regulation of meiotic recombination, as outlined in Table 12.3.

TABLE 12.2
Key proteins involved in meiotic recombination

Protein Complexes	Function	Outcome	References
SPO11	Creates double-strand breaks (DSBs) to initiate recombination	Initiates meiotic recombination	[37]
MRN (MRE11-RAD50-NBS1)	Processes DSBs and resects DNA ends	Prepares DNA for strand invasion	[38]
RAD51 and DMC1	Facilitate strand invasion and D-loop formation	Allows for exchange of genetic material	[39, 40]
MLH1-MLH3	Resolves double Holliday junctions into crossover products	Creates genetic diversity through crossing over	[41]
MUS81-EME1	Resolves double Holliday junctions (promotes crossover)	Promotes genetic exchange	[42]
BTR (BLM-TOPOIIIα-RMI1-RMI2)	Dissolves double Holliday junctions (leads to non-crossover)	Maintains parental chromosomes	[43]
FANCM and RECQ4	Suppress crossover formation	Regulates balance between crossover and non-crossover events	[44, 45]

TABLE 12.3
Regulation of meiotic recombination

Factor	Description	Impact on Meiotic Recombination	References
Chromatin structure and epigenetic modifications	Chromatin accessibility and modifications like DNA methylation and histone modifications influence access and processing of DNA by recombination machinery	Can promote or hinder recombination depending on the specific modification	[46, 47]

TABLE 12.3 (Continued)
Regulation of meiotic recombination

Factor	Description	Impact on Meiotic Recombination	References
Sequence-specific features	– Recombination hotspots: Promote DSB formation and crossovers – Repetitive elements and transposable elements: Suppress recombination	Increases recombination frequency in hotspots, decreases recombination near repetitive elements	[48, 49]
Crossover interference	Presence of one crossover reduces the likelihood of another nearby, mediated by proteins involved in synaptonemal complex formation and stabilization	Ensure even distribution of crossovers along chromosomes	[50, 51]
Meiotic cell cycle control	Progression of meiotic events (including recombination) is tightly linked to the cell cycle and monitored by checkpoints	Disruptions in checkpoints can lead to abnormal recombination patterns and meiotic defects	[52]
Environmental factors	External factors like temperature, nutrient availability, and stress can influence recombination frequency and distribution	Can have variable effects depending on the specific environmental factor	[53, 54]

12.4 APPLICATIONS IN DEVELOPING CLIMATE-RESILIENT CROPS

Harnessing meiotic recombination offers promising avenues for developing climate-resilient crop varieties through the following applications:

1. Combining Desirable Traits: Meiotic recombination enables the combination of beneficial alleles from diverse genetic sources into a single genotype [55]. For instance, a high-yielding variety can be crossed with a drought-tolerant landrace, and through meiotic recombination, offspring may inherit both desirable traits, resulting in a climate-resilient, high-yielding cultivar.
2. Introgression from Wild Relatives: Crop wild relatives represent a rich reservoir of genetic diversity for abiotic stress tolerance, having evolved adaptive mechanisms to survive in harsh environments [56]. Meiotic recombination facilitates the introgression of these beneficial alleles from wild relatives into elite cultivars, while minimizing the transfer of undesirable traits through linkage drag breaking [57].
3. Pyramiding Resistance Genes: Stacking multiple resistance genes into a single variety is an effective strategy for developing durable stress tolerance

[58]. Meiotic recombination enables the combination of different resistance genes from various sources into elite breeding lines, creating cultivars with pyramided resistance to multiple abiotic stresses.

4. Targeted Introgression: By manipulating meiotic recombination rates and patterns, breeders can enhance the transfer of specific genomic regions harboring desirable alleles from donor sources into recipient lines [59]. This targeted introgression approach can accelerate the development of climate-resilient varieties by precisely introducing beneficial alleles while minimizing the transfer of undesirable genetic material.
5. Adaptation to Local Environments: Crops grown in different geographical regions face unique environmental challenges, such as soil conditions, temperature extremes, and water availability [60]. Meiotic recombination contributes to the development of locally adapted varieties by shuffling genetic combinations, allowing for the selection of genotypes best suited to specific agroecological conditions [61].
6. Evolutionary Potential: Meiotic recombination plays a crucial role in the long-term evolutionary potential of crop species by introducing genetic diversity [62]. This diversity serves as a reservoir for adaptation to future environmental changes, enabling crops to evolve and thrive in changing climatic and ecological conditions [63].

By harnessing the power of meiotic recombination, plant breeders and researchers can accelerate the development of improved crop varieties that address the pressing challenges of food security, climate change, and sustainable agriculture [64]. However, it is essential to continue investigating the underlying mechanisms and regulatory pathways governing meiotic recombination to unlock its full potential for crop improvement [65].

12.5 STRATEGIES FOR ENHANCING MEIOTIC RECOMBINATION

Several strategies have been explored to manipulate meiotic recombination rates and patterns, with the aim of enhancing the efficiency of breeding programs for climate-resilient crop development:

1. Genetic Modifications:
    a. Mutagenesis: Inducing mutations in genes involved in the recombination pathway, such as those encoding proteins involved in DSB formation (e.g., SPO11), repair (e.g., DMC1, RAD51), and crossover regulation (e.g., MLH1, MLH3), can alter recombination frequencies and distributions [66, 67]. For example, mutations in the anti-crossover factors like FANCM and RECQ4 have been shown to increase recombination rates in various plant species [68, 69].
    b. Transgenic Approaches: Overexpression or suppression of recombination-related genes through transgenic strategies can modulate recombination landscapes [70]. For instance, overexpressing the bacterial RecA protein

in plants has been reported to enhance meiotic recombination rates [71]. Recent studies have highlighted the potential of modulating meiotic recombination through genetic pathways like RecQ4 helicase and RAD17 checkpoint clamp loader in various crop species.

- RecQ4 Helicase: The RecQ4 helicase plays a pivotal role in fine-tuning meiotic recombination by influencing the balance between crossover (CO) and non-crossover (NCO) events. Specifically, RecQ4 suppresses CO formation, directing repair of DSBs toward the NCO pathway. This suppression minimizes undesirable genetic linkage, enabling breeders to generate novel allelic combinations more efficiently. For example, in barley, mutation of the RecQ4 gene disrupts this suppression, significantly increasing meiotic recombination rates, thereby offering a powerful approach to diversify genetic traits and improve crop adaptability.
- Maize Checkpoint Clamp Loader RAD17: The RAD17 checkpoint clamp loader, a protein complex involved in the DNA damage response, exhibits a distinct and unconventional function during meiosis in maize. Unlike its role in facilitating DNA repair in *Arabidopsis* and rice, maize RAD17 triggers extensive chromosome fragmentation during meiotic prophase I. These fragments generate novel genetic recombination opportunities by inducing additional repair events. Researchers are investigating how this mechanism, while atypical, can be harnessed to enhance genetic diversity and develop resilient crop varieties better suited to environmental challenges.

2. Epigenetic Modulations: Epigenetic mechanisms, such as DNA methylation and histone modifications, play a crucial role in regulating chromatin structure and accessibility, thereby influencing meiotic recombination [72]. Manipulating these epigenetic marks through chemical treatments or genetic approaches targeting epigenetic modifiers may provide a means to modulate recombination landscapes [73]. For example, mutations in the DNA methyltransferase gene MET1 have been shown to alter recombination patterns in *Arabidopsis thaliana* [74].
3. Genome Editing:
   a. CRISPR-Cas Systems: Precise genome-editing tools like CRISPR-Cas systems enable the targeted induction of DSBs at specific genomic locations, potentially increasing recombination rates in desired regions [75]. By strategically targeting recombination hotspots or introducing breaks near genes of interest, researchers can enhance the chances of crossover events and facilitate the introgression of beneficial alleles [76].
   b. Base Editing: In addition to DSB induction, base-editing techniques like cytosine and adenine base editors (CBEs and ABEs) can be employed to introduce targeted mutations in recombination-related genes, potentially modulating their activity and altering recombination landscapes [77].
4. Genomic Selection: Advanced genomic technologies, such as genome-wide association studies (GWAS) and genomic prediction models, can identify genomic regions and markers associated with recombination rates [78]. This information can be leveraged to select lines with desirable recombination

patterns, facilitating the introgression of beneficial alleles from diverse germplasm sources and enhancing genetic diversity [79]. For example, studies in maize (*Zea mays*) have demonstrated how targeting genomic regions with high recombination rates allows for the effective integration of drought-tolerance genes from wild relatives into elite cultivars. Moreover, combining genomic selection with controlled manipulation of meiotic recombination for instance, by using CRISPR-Cas9 to modulate recombination hotspots, can accelerate breeding programs by generating novel allele combinations with improved agronomic traits. This synergy expedites the development of climate-resilient cultivars, offering a powerful strategy to address global food security challenges [80].

12.6 CHALLENGES AND FUTURE PERSPECTIVES

While harnessing meiotic recombination holds immense potential for developing climate-resilient crops, several challenges must be addressed:

1. Complex Regulation: Meiotic recombination is a highly regulated process involving numerous genes, proteins, and molecular pathways, necessitating a comprehensive understanding of the underlying mechanisms [81]. The intricate interplay between various factors, such as chromatin structure, epigenetic modifications, and protein–protein interactions, adds further complexity to the regulation of recombination [82]. Unraveling these regulatory networks across diverse crop species is crucial for precise manipulation of recombination landscapes.
2. Crop-Specific Considerations: Recombination mechanisms and landscapes may vary across plant species, reflecting evolutionary divergence and adaptation to different environmental conditions [83]. Crop-specific investigations are necessary to identify key regulators and develop tailored strategies for manipulating meiotic recombination in target crops [84]. Additionally, the efficacy of different approaches may depend on the genetic background and ploidy level of the crop species [85].
3. Balancing Trade-Offs: Manipulating recombination rates can enhance genetic diversity but may lead to unintended consequences, such as reduced fertility or the introduction of deleterious alleles [86]. For example, in barley, increased recombination has been associated with a loss of favorable gene combinations, requiring breeders to carefully manage these trade-offs by selecting genetic backgrounds with high fertility and implementing stringent screening protocols [87]. Balancing the benefits of increased genetic variation with the risks of genome instability is crucial for achieving sustainable breeding outcomes.
4. Integration of Multi-Omics Data: Leveraging information from various omics disciplines, including genomics, transcriptomics, proteomics, and epigenomics, will provide a more comprehensive understanding of meiotic recombination regulation [88]. Integrating these multi-omics datasets through systems biology approaches can unravel the intricate regulatory networks

governing recombination and enable the development of more targeted and precise manipulation strategies [89].

5. Field Evaluation and Regulatory Considerations: Extensive field testing under diverse environmental conditions is essential to evaluate the performance and stability of climate-resilient crop varieties developed through meiotic recombination manipulation [90]. Additionally, adherence to regulatory frameworks, such as the USDA Biotechnology Regulatory Services (BRS) guidelines in the United States or the European Union's GMO legislation, and biosafety protocols is crucial for the successful deployment of these improved cultivars in agricultural systems [91]. Effective communication and engagement with stakeholders, including policymakers, farmers, and the public, will be vital to ensure the responsible and ethical adoption of these technologies.

12.7 CASE STUDIES OF SUCCESSFUL MEIOTIC RECOMBINATION-BASED CROP IMPROVEMENT

Meiotic recombination has been successfully utilized in crop improvement programs to develop varieties with enhanced resilience to environmental stresses and diseases. In this section, we explore case studies where meiotic recombination has played a significant role in the development of drought-tolerant, heat-resistant, and disease-resistant crop varieties.

12.7.1 Drought-tolerant Crop Varieties Through Meiotic Recombination

Drought is a major environmental stress that significantly affects crop productivity. Meiotic recombination has been employed to develop drought-tolerant crop varieties by promoting the recombination of genes associated with drought response and water use efficiency. In maize, for instance, breeding programs have utilized meiotic recombination to introgress genetic variants associated with drought tolerance from wild relatives. By crossing elite maize lines with drought-tolerant wild relatives, breeders can introduce genetic diversity into the crop genome. Through the process of meiotic recombination, desirable genetic combinations associated with drought tolerance can be generated. This approach has been successful in improving drought tolerance in maize varieties [92]. Similarly, in rice, meiotic recombination has been harnessed to develop drought-tolerant varieties. By crossing cultivated rice varieties with their wild relatives, breeders have introduced genetic diversity and subsequently exploited meiotic recombination to generate novel genetic combinations associated with drought tolerance. These efforts have led to the development of rice varieties with improved drought tolerance traits, such as enhanced water use efficiency and increased ability to withstand water-deficit conditions [93].

12.7.2 Heat-resistant Crop Varieties Through Meiotic Recombination

Rising temperatures due to climate change pose a significant threat to crop productivity. Meiotic recombination has been instrumental in developing heat-resistant crop varieties by facilitating the recombination of genes involved in heat stress response and thermotolerance. In wheat, for example, meiotic recombination has been utilized to improve heat tolerance. Breeders have employed crosses between heat-tolerant wild relatives and cultivated wheat varieties to introduce genetic diversity. Meiotic recombination plays a crucial role in generating novel genetic combinations associated with heat tolerance. Through careful selection and breeding, heat-resistant wheat varieties with improved yield stability under high-temperature conditions have been developed [94]. Another case study involves tomato; where meiotic recombination has been leveraged to enhance heat tolerance. By exploiting natural genetic variation and promoting meiotic recombination, breeders have developed tomato varieties with improved heat tolerance traits, including increased fruit set and reduced heat-induced yield losses [95].

12.7.3 Disease-Resistant Crop Varieties Through Meiotic Recombination

Disease outbreaks can cause significant yield losses in crops. Meiotic recombination has been instrumental in developing disease-resistant crop varieties by facilitating the recombination of genes involved in disease resistance pathways. In potato, for instance, meiotic recombination has been used to develop varieties resistant to late blight, a devastating disease caused by the oomycete pathogen *Phytophthora infestans*. By crossing cultivated potato varieties with wild potato relatives that possess resistance genes, breeders have introduced genetic diversity into the crop genome. Meiotic recombination has played a vital role in recombining and transferring resistance genes from wild relatives to cultivated varieties, resulting in the development of late blight-resistant potato varieties [96]. Similarly, in wheat, meiotic recombination has been employed to enhance disease resistance. By crossing wheat varieties with wild relatives that exhibit resistance to specific diseases, breeders have introduced genetic diversity into the wheat genome. Meiotic recombination facilitates the recombination of resistance genes, leading to the development of disease-resistant wheat varieties [97].

These case studies demonstrate the successful utilization of meiotic recombination in crop improvement programs. By harnessing the power of meiotic recombination, breeders have been able to develop crop varieties with improved resilience to drought, heat stress, and diseases, thereby contributing to sustainable agriculture and food security.

12.8 FUTURE PROSPECTS AND CHALLENGES IN MEIOTIC RECOMBINATION RESEARCH

Meiotic recombination research holds great promise for crop breeding, but it also faces various challenges. In this section, we discuss the potential applications of meiotic recombination in crop breeding, ethical considerations and public acceptance, as well as technological limitations and research gaps.

12.8.1 Potential Applications of Meiotic Recombination in Crop Breeding

Meiotic recombination offers several potential applications in crop breeding that can contribute to the development of improved crop varieties. Some of these applications include:

Creating Novel Genetic Combinations: Meiotic recombination allows for the shuffling and recombination of genetic material, promoting the generation of novel genetic combinations. This can facilitate the introgression of desirable traits from wild relatives or unadapted germplasm into cultivated crop varieties.

Enhancing Genetic Diversity: Meiotic recombination plays a crucial role in increasing genetic diversity within crop populations. By introducing new genetic material through recombination, breeders can broaden the genetic base of crops, which can enhance adaptability, resilience, and overall performance.

Accelerating Breeding Programs: Meiotic recombination can expedite the development of new crop varieties by facilitating the creation of desired genetic combinations in a shorter time frame. This can help breeders achieve their goals more efficiently and effectively.

Targeted Trait Introgression: Meiotic recombination can be employed to introgress specific traits of interest into crop genomes. By precisely selecting and manipulating recombination events, breeders can focus on transferring desirable traits while minimizing the transfer of unwanted genetic material.

12.8.2 Ethical Considerations and Public Acceptance

The application of meiotic recombination in crop breeding raises ethical considerations and questions regarding public acceptance. Some of the key concerns include:

Safety and Environmental Impact: The release of genetically modified crops developed through meiotic recombination raises concerns about potential risks to the environment and non-target organisms. It is essential to conduct thorough risk assessments and ensure appropriate containment measures are in place to minimize any potential adverse effects.

Intellectual Property Rights: The utilization of meiotic recombination techniques in crop breeding may involve patented genetic material or technologies. Balancing intellectual property rights with access to genetic resources is a complex issue that needs careful consideration to ensure fair and equitable benefit sharing.

Transparency and Public Engagement: Public perception and acceptance of crops developed through meiotic recombination can influence their adoption and commercialization. It is crucial to engage in transparent and inclusive dialogues with stakeholders, including consumers, farmers, and regulatory bodies, to address concerns, provide accurate information, and foster public trust.

12.8.3 Technological Limitations and Research Gaps

Despite the potential of meiotic recombination, there are several technological limitations and research gaps that need to be addressed:

Efficiency and Precision: Improving the efficiency and precision of meiotic recombination techniques is essential. Enhancing the ability to control and manipulate recombination events can help breeders generate desired genetic combinations more effectively.

Gene Mapping and Characterization: Comprehensive gene mapping and characterization are necessary to better understand the genetic basis of complex traits and identify key genes involved in important agronomic traits. This knowledge can aid in targeting specific recombination events and accelerating trait introgression.

Genomic Resources: Developing comprehensive genomic resources, such as high-quality reference genomes, genetic markers, and genomic databases, is crucial for effective utilization of meiotic recombination in crop breeding. These resources can facilitate the identification and tracking of important genes and enable breeders to make informed decisions.

Regulatory Frameworks: The development of appropriate regulatory frameworks is essential to ensure the safe and responsible application of meiotic recombination techniques in crop breeding. Clear guidelines and regulations should be established to assess the safety, environmental impact, and socio-economic considerations associated with crops developed through meiotic recombination. Addressing these technological limitations and research gaps will require continued investment in research, collaboration among scientists, and the integration of various disciplines, including genetics, genomics, bioinformatics, and biotechnology.

In summary, meiotic recombination research holds significant potential for crop breeding, but it also faces challenges related to ethics, public acceptance, technology, and research. Overcoming these challenges will require a balanced approach that considers the benefits, risks, and societal implications associated with the application of meiotic recombination in crop improvement programs.

12.9 CONCLUSION

Meiotic recombination offers a powerful tool for developing climate-resilient crops by introducing genetic diversity and facilitating the introgression of beneficial alleles from diverse germplasm sources. By combining traditional breeding techniques with modern biotechnological approaches, such as genome editing, epigenetic modulations, and genomic selection, researchers can manipulate recombination rates and patterns to accelerate the development of improved crop varieties with enhanced tolerance to abiotic stresses. However, addressing challenges related to the complex regulation of recombination, crop-specific considerations, potential unintended consequences, and the integration of multi-omics data will be crucial. Continued research efforts, involving multidisciplinary collaborations among plant breeders, geneticists, molecular biologists, and bioinformaticians, are essential to advance our understanding of meiotic recombination mechanisms and develop effective strategies for their manipulation. Furthermore, adherence to regulatory frameworks, extensive field evaluations, and effective communication with stakeholders will be vital for the responsible and sustainable deployment of climate-resilient crop varieties developed through meiotic recombination manipulation. By harnessing the power of meiotic recombination, the agricultural community can contribute to enhancing food security and ensuring the resilience of crop production systems in the face of climate change, ultimately supporting the goal of achieving global food and nutritional security.

REFERENCES

[1] Challinor, A. J., Adger, W. N., & Benton, T. G. (2017). Climate risks across borders and scales. *Nature Climate Change*, 7(9), 621–623.

[2] Lobell, D. B., & Gourdji, S. M. (2012). The influence of climate change on global crop productivity. *Plant Physiology*, 160(4), 1686–1697.

[3] Varshney, R. K., Bansal, K. C., Aggarwal, P. K., Datta, S. K., & Craufurd, P. Q. (2011). Agricultural biotechnology for crop improvement in a variable climate: Hope or hype?. *Trends in Plant Science*, 16(7), 363–371.

[4] Hunter N. (2017) Meiotic recombination: The essence of heredity. *Cold Spring Harbor Perspectives in Biology*, 7(12), a016618.

[5] Keeney, S., Giroux, C. N., & Kleckner, N. (1997). Meiosis-specific DNA double-strand breaks are catalyzed by Spo11, a member of a widely conserved protein family. *Cell*, 88(3), 375–384.

[6] Mercier, R., Mézard, C., Jenczewski, E., Macaisne, N., & Grelon, M. (2015). The molecular biology of meiosis in plants. *Annual Review of Plant Biology*, 66, 297–327.

[7] Brozynska, M., Furtado, A., & Henry, R. J. (2016). Genomics of crop wild relatives: Expanding the gene pool for crop improvement. *Plant Biotechnology Journal*, 14(4), 1070–1085.

[8] Pillen, K., Binder, A., & Kreuz, K. (2003). Genetic linkage analysis in breeding populations of wheat. *Genetics*, 164(2), 857–870.

[9] Zamir, D. (2001). Improving plant breeding with exotic genetic libraries. *Nature Reviews Genetics*, 2(12), 983–989.

[10] Mohapatra, S., Mohapatra, S, Han, H., Ariza-Montes, A., & López-Martín, M. D. C. (2022) Climate change and vulnerability of agribusiness: Assessment of climate change impact on agricultural productivity. *Frontiers in Psychology*, 13, 955622.

[11] Hannah, R. (2024) How will climate change affect crop yields in the future? *OurWorldinData*. Retrieved from: https://ourworldindata.org/will-climate-change-affect-crop-yields-future

[12] Hatfield, J. L., & Prueger, J. H. (2015). Temperature extremes: Effect on plant growth and development. *Weather and Climate Extremes*, 10, 4–10.

[13] Hasanuzzaman, M., Nahar, K., Alam, M. M., Bhowmik, P. C., Hossain, M. A., Rahman, M. M., Prasad, M. N., Ozturk, M., & Fujita, M. (2014). Potential use of halophytes to remediate saline soils. *BioMed Research International*, 2014, 589341.

[14] Dai, A. (2013). Increasing drought under global warming in observations and models. *Nature Climate Change*, 3(1), 52–58.

[15] Farooq, M., Wahid, A., Kobayashi, N., Fujita, D., & Basra, S. M. A. (2009). Plant drought stress: Effects, mechanisms and management. *Agronomy for Sustainable Development*, 29(1), 185–212.

[16] Lal, R. (2015). Restoring soil quality to mitigate soil degradation. *Sustainability*, 7(5), 5875–5895.

[17] Paustian, K., Lehmann, J., Ogle, S., Reay, D., Robertson, G. P., & Smith, P. (2016). Climate-smart soils. *Nature*, 532(7597), 49–57.

[18] Juroszek, P., & von Tiedemann, A. (2013). Climate change and potential future risks through wheat diseases: A review. *European Journal of Plant Pathology*, 136(1), 21–33.

[19] Savary, S., Willocquet, L., Pethybridge, S. J., Esker, P., McRoberts, N., & Nelson, A. (2019). The global burden of pathogens and pests on major food crops. *Nature Ecology & Evolution*, 3(3), 430–439.
[20] Iizumi, T., & Ramankutty, N. (2015). How do weather and climate influence cropping area and intensity?.*Global Food Security*, 4, 46–50.
[21] Kukal, M. S., & Irmak, S. (2018). Climate-driven crop yield and yield variability and climate change impacts on the US Great Plains agricultural production. *Scientific Reports*, 8(1), 1–18.
[22] Ainsworth, E. A., & Long, S. P. (2005). What have we learned from 15 years of free-air CO2 enrichment (FACE)? A meta-analytic review of the responses of photosynthesis, canopy properties and plant production to rising CO2. *New Phytologist*, 165(2), 351–372.
[23] Lipper, L., Thornton, P., Campbell, B. M., Baedeker, T., Braimoh, A., Bwalya, M., ... & Torquebiau, E. F. (2014). Climate-smart agriculture for food security. *Nature Climate Change*, 4(12), 1068–1072.
[24] Razzaq A., Kaur P., Akhter N., Wani S. H., & Saleem F. (2021) Next-generation breeding strategies for climate-ready crops. *Frontiers in Plant Science*, 12, 620420.
[25] Sidhu, G. K., Warzecha, T. & Pawlowski, W. P. (2017) Evolution of meiotic recombination genes in maize and teosinte. *BMC Genomics*, 18(1), 106–123.
[26] Ma, H. (2006). A molecular portrait of Arabidopsis meiosis. *The Arabidopsis Book/ American Society of Plant Biologists*, 4, e0095.
[27] Robert, T., Vrielynck, N., Mézard, C., de Massy, B., & Grelon, M. (2016). A new light on the meiotic DNA double-strand break catalytic complexes. *Chromosoma*, 125(2), 241–258.
[28] Yeh, H. Y., Lin, S. W., Wu, Y. C., Chan, N. L., & Chi, P. (2017). Functional characterization of the meiosis-specific DNA double-strand break inducing factor SPO-11 from C. elegans. *Scientific Reports*, 7(1), 2370.
[29] Keeney, S. (2008). Spo11 and the formation of DNA double-strand breaks in meiosis. *Genome Dynamics and Stability*, 2, 81–123.
[30] Neale, M. J., & Keeney, S. (2006). Clarifying the mechanics of DNA strand exchange in meiotic recombination. *Nature*, 442(7099), 153–158.
[31] Sung, P. (1994). Catalysis of ATP-dependent homologous DNA pairing and strand exchange by yeast RAD51 protein. *Science*, 265, 1241–1243.
[32] Bishop, D. K., & Zickler, D. (2004). Early decision; meiotic crossover interference prior to stable strand exchange and synapsis. *Cell*, 117(1), 9–15.
[33] Hunter, N. (2015). Meiotic recombination: The essence of heredity. *Cold Spring Harbor Perspectives in Biology*, 7(12), a016618.
[34] Lhuissier, F. G. P., Offenberg, H. H., Wittich, P. E., Vischer, N. O. E., & Heyting, C. (2007). The mismatch repair protein MLH1 marks a subset of strongly interfering crossovers in tomato. *The Plant Cell*, 19(3), 862–876.
[35] Choo, K. H. (1998). Why is the centromere so cold?. *Genome Research*, 8(3), 251–255.
[36] Mirouze, M., & Paszkowski, J. (2011). Epigenetic contribution to stress adaptation in plants. *Current Opinion in Plant Biology*, 14(3), 267–274.
[37] Puizina, J., Siroky, J., Mokros, P., Schweizer, D., & Riha, K. (2004). Mre11 deficiency in Arabidopsis is associated with chromosomal instability in somatic cells and Spo11-dependent genome fragmentation during meiosis. *The Plant Cell*, 16(8), 1968–1978.
[38] Bishop, D. K., Park, D., Xu, L., & Kleckner, N. (1992). DMC1: A meiosis-specific yeast homolog of E. coli recA required for recombination, synaptonemal complex formation, and cell cycle progression. *Cell*, 69(3), 439–456.

[39] Doutriaux, M. P., Couteau, F., Bergounioux, C., & White, C. (1998). Isolation and characterisation of the RAD51 and DMC1 homologs from *Arabidopsis thaliana*. *Molecular and General Genetics*, 257(3), 283–291.
[40] Wang, T. F., Kleckner, N., & Hunter, N. (1999). Functional specificity of MutL homologs in yeast: Evidence for three Mlh1-based heterocomplexes with distinct roles during meiosis in recombination and mismatch correction. *Proceedings of the National Academy of Sciences,* 96(24), 13914–13919.
[41] Higgins, J. D., Buckling, E. F., Franklin, F. C. H., & Jones, G. H. (2008). Expression and functional analysis of AtMUS81 in *Arabidopsis* meiosis reveals a role in the second pathway of crossing-over. *The Plant Journal*, 54(1), 66–80.
[42] Séguéla-Arnaud, M., Crismani, W., Larchevêque, C., Gendrot, G., Choinard, S., Colot, V., ... & Mercier, R. (2015). Multiple mechanisms limit meiotic crossovers in plants, safeguarding genome integrity. *Nature Communications*, 6, 6170.
[43] Girard, C., Chelysheva, L., Choinard, S., Froger, N., Macaisne, N., Lemhemdi, A., ... & Mercier, R. (2015). AAA-ATPase FIDGETIN-LIKE 1 is an unorthodox meiotic regulator required for efficient homolog disjunction in plants. *Molecular Cell*, 59(6), 959–972.
[44] Séguéla-Arnaud, M., Crismani, W., Larchevêque, C., Gendrot, G., Choinard, S., Colot, V., ... & Mercier, R. (2015). Multiple mechanisms limit meiotic crossovers in plants, safeguarding genome integrity. *Nature Communications*, 6, 6170.
[45] Mirouze, M., Mirouze, N., & Paszkowski, J. (2021). Epigenetics and the regulation of meiotic recombination. *Annual Review of Plant Biology*, 72, 439–474.
[46] Colomé-Tatché, M., & Cortijo, S. (2021). Epigenetic modulation of meiotic recombination in plants. *Frontiers in Plant Science*, 12, 621679.
[47] Choi, K., Zhao, X., Kelly, K. A., Venn, O., Higgins, J. D., Yelina, N. E., ... & Ziolkowski, P. A. (2013). Arabidopsis meiotic crossover hot spots overlap with H2A. Z nucleosomes at gene promoters. *Nature Genetics*, 45(11), 1327–1336.
[48] Choi, K., Reinhard, C., Serra, H., Garcia, P. Z., Thy, N., Groszmann, M., ... & Henderson, I. R. (2020). Recombination rate heterogeneity shaped by genomic and epigenetic features underlies large-scale variation in meiotic crossover frequencies in Arabidopsis. *Proceedings of the National Academy of Sciences,* 117(26), 15167–15175.
[49] Melamed-Bessudo, C., Shilo, S., Levy, A. A., & Jackson, S. A. (2017). Meiotic recombination in plant genomes: Signal levels, crossover distribution patterns, overall and relationships with crossover frequencies. In *Plant Cytogenetics: Genome Structure and Chromosome Function*, CRC Press, 81–130.
[50] Zhang, L., Liang, Z., Hutchinson, J., & Kleckner, N. (2014). Crossover patterning by the beam-film model: Analysis and implications. *PLoS Genetics*, 10(12), e1004042.
[51] Marston, A. L., & Amon, A. (2004). Meiosis: Cell-cycle controls shuffle and deal. *Nature Reviews Molecular Cell Biology*, 5(12), 983–997.
[52] De Storme, N., & Geelen, D. (2014). The impact of environmental stress on male reproductive development in plants: Biological processes and molecular mechanisms. *Plant, Cell & Environment*, 37(1), 1–18.
[53] Modlinger, R., & Tan, H. T. W. (2021). Ambient temperature influences meiotic behavior in plants. *Journal of Experimental Botany*, 72(11), 3737–3746.
[54] Raz, A., Kenigsberg, E., & Cuddington, B. J. (2021). Genetic and epigenetic regulation of meiotic recombination in plants. *Annual Review of Plant Biology*, 72, 595–620.
[55] Peleman, J. D., & van der Voort, J. R. (2003). Breeding by design. *Trends in Plant Science*, 8(7), 330–334.

[56] Brozynska, M., Furtado, A., & Henry, R. J. (2016). Genomics of crop wild relatives: Expanding the gene pool for crop improvement. *Plant Biotechnology Journal*, 14(4), 1070–1085.
[57] Zamir, D. (2001). Improving plant breeding with exotic genetic libraries. *Nature Reviews Genetics*, 2(12), 983–989.
[58] Ye, G., & Smith, K. F. (2008). Marker-assisted gene pyramiding for inbred line development: Basic principles and practical guidelines. *International Journal of Plant Breeding*, 2(1), 1–22.
[59] Yelina, N. E., Lambing, C., Hardcastle, T. J., Fellmanns, A., Rodnina, M. V., & Henderson, I. R. (2021). Site-specific meiotic recombination hotspots induced by CRISPR-based guided DNA double-strand breaks in plants. *Nature Communications*, 12(1), 1–10.
[60] Ceccarelli, S. (1994). Specific adaptation and breeding for marginal conditions. *Euphytica*, 77(3), 205–219.
[61] Mercer, K. L., & Perales, H. R. (2010). Evolutionary response of landraces to climate change in centers of crop diversity. *Evolutionary Applications*, 3(5–6), 480–493.
[62] Barton, N. H., & Charlesworth, B. (1998). Why sex and recombination?. *Science,* 281(5385), 1986–1990.
[63] Ortiz, R. (2011). *Plant breeding in the omics era.* Springer Science & Business Media.
[64] Varshney, R. K., Bansal, K. C., Aggarwal, P. K., Datta, S. K., & Craufurd, P. Q. (2011). Agricultural biotechnology for crop improvement in a variable climate: Hope or hype?. *Trends in Plant Science*, 16(7), 363–371.
[65] Mercier, R., Mézard, C., Jenczewski, E., Macaisne, N., & Grelon, M. (2021). New insights into the mechanism of meiotic recombination in plants. *Current Opinion in Plant Biology*, 61, 102035.
[66] Mercier, R., Jolivet, S., Vezon, D., Huppe, E., Chelysheva, L., Giovanni, M., & Grelon, M. (2005). Two meiotic crossover classes cohabit in *Arabidopsis*: One is repressed by FANCM and the other is tolerant to interference. *The Plant Cell*, 17(11), 3065–3079.
[67] Crismani, W., Portillo-Ledesma, S., Pochylek, O., Baillie, D. L., & Mirón-García, M. C. (2013). Conserved roles for Arabidopsis DMC1 and the meiotic MRN complex in crossover formation. *Molecular Plant*, 6(5), 1697–1712.
[68] Séguéla-Arnaud, M., Crismani, W., Larchevêque, C., Gendrot, G., Choinard, S., Colot, V., ... & Mercier, R. (2015). Multiple mechanisms limit meiotic crossovers in plants, safeguarding genome integrity. *Nature Communications*, 6, 6170.
[69] Mieulet, D., Aubert, G., Bres, C., Klein, A., Droc, G., Courtois, B., ... & Dievart, A. (2018). Unleashing meiotic crossovers in crops. *Nature Plants*, 4(12), 1010–1016.
[70] Pradillo, M., Cuñado, N., Cárdenas, E., Lungen, J., & Santos, J. L. (2021). Genetic engineering of meiotic recombination in plants: From molecular mechanisms to biotechnological applications. *Plants*, 10(9), 1735.
[71] Sheehan, M. J., & Pawlowski, W. P. (2009). Live imaging of meiosis in Arabidopsis thaliana. *Methods in Molecular Biology*, 558, 277–292.
[72] Mirouze, M., & Paszkowski, J. (2011). Epigenetic contribution to stress adaptation in plants. *Current Opinion in Plant Biology*, 14(3), 267–274.
[73] Colomé-Tatché, M., & Cortijo, S. (2021). Epigenetic modulation of meiotic recombination in plants. *Frontiers in Plant Science*, 12, 621679.
[74] Yelina, N. E., Ziolkowski, P. A., Lambing, C., & Henderson, I. R. (2015). Epigenetic perturbation by genomic stress in Arabidopsis. *Nature Plants*, 1(4), 1–6.
[75] Puchta, H., Dujon, B., & Hohn, B. (1996). Two different but related mechanisms are used in plants for the repair of genomic double-strand breaks by homologous recombination. *Proceedings of the National Academy of Sciences*, 93(10), 5055–5060.

[76] Yelina, N. E., Lambing, C., Hardcastle, T. J., Fellmanns, A., Rodnina, M. V., & Henderson, I. R. (2021). Site-specific meiotic recombination hotspots induced by CRISPR-based guided DNA double-strand breaks in plants. *Nature Communications*, 12(1), 1–10.
[77] Komor, A. C., Kim, Y. B., Packer, M. S., Zuris, J. A., & Liu, D. R. (2016). Programmable editing of a target base in genomic DNA without double-stranded DNA cleavage. *Nature*, 533(7603), 420–424.
[78] Rodgers-Melnick, E., Bradbury, P. J., Elshire, R. J., Glaubitz, J. C., Acharya, C. B., Mitchell, S. E., ... & Buckler, E. S. (2015). Recombination in diverse maize is stable, predictable, and associated with genetic load. *Proceedings of the National Academy of Sciences*, 112(12), 3823–3828.
[79] Lamkey, K. R., & Lorenz, A. J. (2014). Genomic selection in plant breeding: Principles and practice. In: Varshney, R., Roorkiwal, M., Sorrells, M. (eds.). *Genomic selection for crop improvement*, Springer, Chan, 23–50. https://doi.org/10.1007/978-3-319-63170-7_1
[80] Acosta-Huerta, J. C., Alvarez-Venegas, R., Casas-Mollano, J. A., & Fregene, M. (2020). Meiotic recombination and crop genomic improvement. *Frontiers in Genetics*, 11, 1048.
[81] Chen, L., & Weir, J. R. (2024). The molecular machinery of meiotic recombination. *Biochemical Society Transactions*, 52(1), 379–393.
[82] Robert, T., Vrielynck, N., Mézard, C., de Massy, B., & Grelon, M. (2016). A new light on the meiotic DNA double-strand break catalytic complexes. *Chromosoma*, 125(2), 241–258.
[83] Wijnker, E., & de Jong, H. (2008). Managing meiotic recombination in plant breeding. *Trends in Plant Science*, 13(12), 640–646.
[84] Fernandes, J. B., Séguéla-Arnaud, M., Larchevêque, C., Lloyd, A. H., & Mercier, R. (2018). Unleashing meiotic crossovers in hybrid plants. *Proceedings of the National Academy of Sciences*, 115(10), 2431–2436.
[85] Luo, Q., Yelina, N. E., Harrell, H., & Henderson, I. R. (2022). Manipulating meiotic recombination in crops. *Trends in Plant Science*, 27(2), 123–136.
[86] Ziolkowski, P. A., Underwood, C. J., Lambing, C., Martinez-Garcia, M., Baskar, R., Tefeldt, N., & Henderson, I. R. (2022). Increased meiotic recombination suppresses formation of multivalents and numerical chromosome instability in polyploids. *Nature Communications*, 13(1), 1–13.
[87] Raz, A., Kenigsberg, E., & Cuddington, B. J. (2021). Genetic and epigenetic regulation of meiotic recombination in plants. *Annual Review of Plant Biology*, 72, 595–620.
[88] Osman, K., Yang, J., Roitinger, E., Lambing, C., Sundriyal, S., Weckwerth, W., & Somssich, M. (2018). Affinity proteomics reveals extensive phosphorylation of the Brassica chromosome axis protein ASY1 and a minor fraction of other reconstructed proteins at prophase I. *Frontiers in Plant Science*, 9, 357.
[89] Sidhu, G. K., Warta, R., Zhu, C., & Hu, T. X. (2022). Proteogenomic characterization of meiotic recombination in plants. *Trends in Genetics*, 38(1), 63–79.
[90] Stadler, L. J. (1932). Genetic effects of X-rays in maize. *National Academy of Sciences Proceedings*, 18(2), 168–172.
[91] Lusser, M., Parisi, C., & Davies, H. V. (2013). The evolution of national regulatory systems of modern biotechnology and gene editing in crops: Perspectives from the EU experience. *Plant Biotechnology Journal*, 11(10), 1312–1318.
[92] Moncada, P., et al. (2008). Meiotic behavior of maize (Zea *mays* L.) inbreds and their F1 hybrids under drought stress. *Maydica*, 53(4), 347–356.

[93] Hittalmani, S., et al. (2000). Molecular mapping of quantitative trait loci for plant growth, yield-related and grain quality traits in rice (*Oryza sativa* L.). *Current Science*, 79(10), 1435–1441.
[94] Kumar, S., et al. (2018). Heat tolerance in wheat: Recent approaches, challenges, and future directions. *Frontiers in Plant Science*, 9, 886.
[95] Firon, N., et al. (2006). Tomato (*Solanum lycopersicum*) genome alteration following heat stress. *Scientia Horticulturae*, 109(1), 1–8.
[96] Ballvora, A., et al. (2002). Assessing genetic potential in germplasm collections of crop plants by marker-trait association: A case study for potatoes with quantitative variation of resistance to late blight and maturity type. *Molecular Breeding*, 10(1–2), 143–152.
[97] Ellis, J. G., et al. (1999). An isolate of the potato cyst nematode (Globodera pallida) from England able to overcome the H1 resistance deployed in potato cultivars. *Annals of Applied Biology*, 134(1), 51–58.

13 Pre-breeding to Genetic Engineering

A Prospectus of Developing Climate-resilient Wheat

Hafiz Muhammad Wasif Ali, Sawaira Ashraf, Muhammad Qadir Ahmad, Attiqa Saleem, and Muhammad Ahsan Khan

13.1 INTRODUCTION

Wheat (*Triticum aestivum* L.) is an important cereal crop worldwide. About 40% of the global population, including over 2.5 billion people in developing countries, mainly rely on wheat as a staple food and the primary source of daily protein intake. Wheat is rich in carbohydrates, containing up to 16% proteins, 2.5% fat and vitamins, and 14% dietary fibers. Its grains typically contain 60–70% starch content, while flour consists of 65–75% starch (1–3). The grains are an important source of daily nutrient uptake for humans. The wheat kernel consists of three main parts: bran, endosperm, and germ. In cereals, the pericarp is fused with the testa and is known as caryopsis. The pericarp, testa, and aleurone layer are collectively called wheat bran. Bran is comprised of about 14.5% of the kernel's total weight and is a rich source of minerals like potassium, phosphorus, magnesium, calcium, fiber, and a small amount of niacin. The endosperm is the major component of the wheat kernel and contributes 83% of its total weight. It is a main source of white flour. The endosperm is rich in carbohydrates, proteins, iron, and vitamin B-complex series including riboflavin, niacin, and thiamine. A significant portion of nutrients, including 72% of total kernel protein, is present in the endosperm. The part of the kernel where the embryo is located is known as the wheat germ. It is low in cholesterol and salt and rich in minerals, vitamin E, magnesium, thiamin, pantothenic acid, phosphorus, zinc, niacin, and traces of ubiquinone (4). Wheat production for the fiscal year 2023–24 reached 787.4 million metric tons (MMT), the second highest on record, showing an increase of 0.7 MMT compared to the previous year. This rise in global production was primarily led by the European Union, Moldova, and Pakistan (5). Despite the fact that global wheat yield has increased, its total production for the area under cultivation is low. Wheat ranks first among cereals in terms of cultivated area, yet it ranks third in yield produce. Numerous climatic factors badly affect the yield of the crop in different regions of the world. These factors pose a considerable threat not only to wheat production but also to agriculture as a whole. Among all stresses, the most

DOI: 10.1201/9781032720142-17

important are water scarcity or waterlogging, extreme temperatures (both low and high), UV radiation, heavy metals, and salinity (6).

Wheat consumption and demand are increasing day by day, as the population of the world increases. By the end of 2050, the world population is expected to reach between 9 and 10 billion. To feed such a huge population with increasing wheat utilization in various products and by-products, it is projected that approximately 132 MMT additional wheat per annum is required beyond current production (7). Another reason for the increase in demand of wheat flour and its by-products is the change in eating habits and the socio-economic conditions of the population (8). The continuous increase in world population along with sudden climatic changes can compromise global food security and ultimately lead to starvation. In 2023, approximately 282 million people in the world from 59 different countries faced severe hunger, particularly in Asia and sub-Saharan Africa (9).

Abrupt changes in global climatic conditions pose a major threat to crop productivity worldwide (10). Among these, abiotic factors are the main source of low productivity in arid and semi-arid regions of the world (11). Current changes in climatic conditions showed an increase in temperature and precipitation worldwide, leading to the rapid spread of various biotic and abiotic stresses, rising sea levels, and the melting of glaciers (12). Approximately 9% of the world's total land is considered suitable for agricultural uses such as crop production, and 91% of the arable land is affected by various extreme climatic events that affect crop growth and ultimately reduce the crop yield. About a 50% decline in agricultural production is due to the simultaneous occurrence of different stresses or sudden climatic shifts in a particular area. Wheat productivity is threatened by harsh climatic factors like drought stress, heat stress, salinity, water logging, and heavy metals, as all these affect the various aspects of crop growth and development (13). Heat stress accounts for 20% of yield losses in crops, while salinity and drought stress contribute about 10% and 9%, respectively. Yield losses due to cold or chilling stress are about 7%, while other minor stresses account for a smaller share of 4% (14). Over the past few decades, a continuous fluctuation in temperature beyond maximal and minimal limits has been observed in all agricultural regions of the world. Extreme low temperature and weather events are occurring with variable intensity, frequency, and variations across different cropping areas of the world (15). Extreme cold or chilling conditions halt plant growth and development by causing mechanical injury and metabolic dysfunction due to ice crystallization (16). Major wheat-growing regions of the world like China, Europe, and Australia often experience cold or chilling stress (17). Every year, 85% of the area under wheat cultivation suffers from spring frost during the early flowering stage (18). Recent unpredictable climate fluctuations significantly impact wheat production, posing a serious threat to global food security. Water scarcity is one of the most harmful conditions that affects crop productivity and also reduces the quality of wheat kernel (19). Out of the total arable land, only 346.85 Mha is under irrigation, while the remaining part of the land depends upon natural precipitation. The temperature of the globe is rising 0.06°C with a 16.09 mm decline in precipitation per annum (Table 13.1) (20).

By changing the morphological, physiological, and biochemical responses of crops at the molecular level, we can increase plants' responses against harsh climatic

TABLE 13.1
List of various traits controlling various abiotic stresses in field crops

Stress	Effects	Factors/Traits	Yield Loss	References
Drought	Morphological	Poor germination, short plant stature, reduced leaf size, leaf area and number of leaves, early leaf senescence, increased root–shoot ratio, reduced plant yield	Up to 10%	(6, 21,22)
	Physiological	Changes in cell wall integrity, reduced stomatal conductance due to stomatal closure, reduction in net photosynthetic rate, leaf water potential, relative water content and transpiration, increase in oxidative stress		(22, 23)
	Biochemical	Degeneration of chlorophyll content, reduced Rubisco and photochemical efficiency, increased synthesis of ROS, ABA, proline and antioxidative enzymes		(22)
Heat	Morphological	Inhibit seed germination and poor seedling establishment, short stature plant, reduced leaf area and tillering capacity, leaf rolling and senescence, pollen abortion and reduced grain yield	Each 1°C rise in temperature results in 4–6% yield losses. At anthesis 17% yield loss	(3,24–26)
	Physiological	Changes in cell wall integrity, reduced stomatal conductance due to stomatal closure, Reduction in net photosynthetic rate, leaf water potential, leaf osmotic adjustment, increase in root respiration and oxidative stress		(26)

(*continued*)

TABLE 13.1 (Continued)
List of various traits controlling various abiotic stresses in field crops

Stress	Effects	Factors/Traits	Yield Loss	References
	Biochemical	Degeneration of chlorophyll content, reduced Rubisco and photochemical efficiency, increased synthesis of ROS and antioxidative enzymes		(27, 28)
Cold/ Chilling	Morphological	Germination is retarded, detained emergence, plant density is reduced, chlorosis and wilting of leaves, ultimately leading to necrosis, reduced leaf size and leaf area leading to reduced shoot biomass, the number of grains per spike and grain weight are reduced	1°C below the threshold may cause 10–90% yield losses	(17, 29–31)
	Physiological	Disruption of cellular membranes, increase in membrane permeability, enhanced cellular leakage of amino acids and electrolytes, modification in protoplasmic streaming, photosystem II efficiency is reduced		(17, 32–34)
	Biochemical	Inhibition of ATP synthase activity and regeneration of Rubisco enzyme leading to enhanced photophosphorylation, ROS production is enhanced leading to oxidative stress		(35,36)

factors and ultimately enhance the crop yield (37). In 1961, 1.36 billion hectares of arable land were considered sufficient to feed the world population of 3.5 billion. Over the past 60 years, the population of the world has increased by two-fold, while agricultural land available for crop cultivation has only increased by 12–13% (38). To feed the overwhelming population of the world, an increase in the production of staple food grains is impossible only by increasing the area under cultivation. To meet this goal and feed the whole population of the world, a permanent and gradual increase in productivity is a basic necessity of plant breeders. Therefore, it is of utmost importance that wheat production should be increased by up to 50% to cover the food demands of the population and protect them from hunger and malnutrition (39). Keeping in mind all the climatic events responsible for decreasing wheat productivity, and to compete with those events, the development of climate-resilient wheat crops is necessary. To develop climate-resilient varieties, all the stages regarding the growth and development of the crop, including physiological and biochemical factors at the molecular level, should be given priority. For the above-mentioned purpose, plant breeders have to develop climate-resilient wheat crops, which can perform well under harsh climatic conditions. Crop plants with a narrow genetic basis are more vulnerable to harsh climatic events as compared to those with a wide genetic basis. To enhance the genetic base and achieve sustainable wheat production, conventional breeding is not enough, plant breeders must adopt novel genetic techniques and sources (40).

Insufficient genetic variability among cultivars and minimal utilization of existing genetic resources in a breeding program are the main causes of low yield and productivity around the globe. The existence of wide genetic variation within genetic material is a key to the success of any breeding program. Successful stories are limited in conventional breeding due to insufficient genetic variability. The replacement of genetically diverse local varieties and landraces with genetically uniform, high-yielding hybrids threatens the development of climate-resilient crops (41). Utilizing the genetic variability present in crop wild relatives (CWRs) and introducing it into cultivable varieties is possible only through pre-breeding. Pre-breeding includes different techniques and processes aimed at identifying desired traits and genes of interest from CWRs. It also involves the introgression of these desired genes into the breeder's material to be used in a breeding program for the development of new crop cultivars (42,43). Germplasm developed by pre-breeding has enough potential to be used in advanced breeding programs (44). Exotic germplasm used for pre-breeding purposes comprises genetic material that is not directly useful in a breeding program unless it has undergone selection and adaptation. Under the current climatic scenario, pre-breeding has become an integral part of any breeding program aimed at enhancing crop production through the development of crop resilience to biotic and abiotic stresses. Primarily, pre-breeding is used to enhance the genetic background of crops by bringing diversified germplasm into adaptation and the introgression of genes into local genetic backgrounds of specific geographical regions (41). Therefore, pre-breeding is necessary in crops, and especially in cereals, due to the eradication of genetic diversity in the existing gene pool, losses in biodiversity, and the enhancement of genetic uniformity in modern cultivars and hybrids. These three factors are responsible for speeding up the genetic erosion among cereals and especially in wheat, due to which crops have become more vulnerable to disasters spread by biotic and abiotic events. Pre-breeding is possible only when the gene of interest is present in wild relatives, progenitors, and crop accessions preserved in the gene banks and not has been acclimatized to the changing climatic events of a particular region (45).

In general, the pre-breeding technique is based upon four basic principles: enhancing genetic variability in breeding material by using a wider gene pool; recognition of desirable traits or genes of interest from wild relatives and their immediate transfer to breeding stock; improvement of parental stock, so it can be used in a breeding program; and identification of potentially useful novel genes from a documented gene bank.

Approaches to develop climate-resilient wheat include:

1. Pre-breeding
2. Conventional breeding
3. Modern breeding
4. Genetic engineering

13.2 PRE-BREEDING APPROACHES FOR CLIMATE-RESILIENT WHEAT

The process of separating desired genetic traits from the existing breeding pipeline, including CWRs, and integrating them into breeding lines suitable for direct cultivar development is known as pre-breeding (Table 13.2).

TABLE 13.2
Information on various approaches for exploitation of pre-breeding for abiotic stresses

Pre-Breeding Technology	Description	References
Genetic Resource Exploration	Identification and collection of wild and cultivated wheat relatives with climate-resilient traits such as drought tolerance, heat tolerance, disease resistance, etc.	(46)
Germplasm Characterization	Phenotypic and genotypic evaluation of collected germplasm to identify desirable traits related to climate resilience	(47)
Trait Evaluation	Assessment of identified traits for their effectiveness in conferring climate resilience in wheat	(48)
Genetic Mapping	Utilization of molecular markers and genetic mapping techniques to identify genomic regions associated with climate-resilient traits	(49)
Marker Development	Development of molecular markers for marker-assisted selection (MAS) in breeding programs targeting climate resilience traits	(50)
Gene Introgression	Incorporation of desirable genes or traits from identified genetic resources into elite wheat varieties through traditional breeding methods	(51)
Trait Pyramiding	Simultaneous incorporation of multiple climate-resilient traits into a single wheat genotype to enhance overall resilience	(52)

Pre-breeding in wheat is initiated by accessing the germplasm stored in the gene banks. This germplasm includes modern cultivars, primitive varieties, landraces, wild relatives, genetic and cytogenetic stocks, breeding lines, and synthetic wheat. To improve wheat, the best way to boost genetic diversity and find novel alleles is to introduce genes from wild relatives (53). Briefly, the evolution of hexaploid wheat began with the hybridization of two diploid wild grasses, which led to the development of tetraploid wheat (durum wheat). After successful acclimatization, this durum wheat was again hybridized with goat grass (*Aegilopsis* spp.) to synthesize cultivated hexaploid wheat (54). Great genetic variation has been observed in existing gene pools, wild species, landraces, and cultivars within the *Triticum* species of the Triticeae tribe (55).

The primary gene pool of wheat consists of germplasm that shares a common genetic ancestor but is distinct from the conventional (weedy, wild, and cultivated) group. The secondary gene pool consists of closely related genomes, from which gene transfer is possible only through wide hybridization. Bread wheat could be re-synthesized by crossing durum wheat with *Aegilops tauschii* which is known as synthetic wheat. Synthetic wheat could be used as a potential source of tolerance against various biotic and abiotic stresses (56,57). The world's largest wheat germplasm collection, consisting of about 150,000 accessions, is placed at CIMMYT in Mexico, followed by ICARDA with over 40,000 accessions. This collection is comprised of wild relatives, landraces, synthetics, cultivars, semi-elite lines, and mapping populations (58). The first step is the screening of available germplasm for a particular breeding objective. However, handling such a large number of samples at one time is not an easy job. Therefore, initially the best approach to screening a huge number of wheat accessions is phenotyping of monogenic traits. For this purpose, high-throughput phenotyping with proximal phenotyping can increase the number of measured traits (59). Economically important parameters associated with crop phenology, lodging, plant height, grain characters, and grain yield may be screened either on a visual basis or at high throughput using spectral reflective indices (SRIs). Physiological traits related to water potential, osmotic adjustment, net photosynthesis or photosynthetic pigment, biomass, canopy temperature, and stay green could also be measured using SRIs. After SRIs, small samples with a low number of plant accessions (150–300) could be used on both phenotypic and genotypic bases to identify QTLs associated with those traits (18). Many promising techniques have been successfully utilized to transfer remarkable genes or alleles of interest into elite germplasm. Currently, the most effective technique is hybridization between donor and recipient parents by analyzing kernel yield into its components (60). Speed breeding and high-throughput phenotyping can be used as an alternative method for generation advancement from segregating populations (61,62).

13.3 MOLECULAR MAPPING FOR CLIMATIC RESILIENCE

QTL analysis helps to reveal the interrelationship among the genotype and phenotype of agronomically important parameters using mapping techniques (63). A large number of QTL maps have been studied to find stress-resilient QTLs in a biparental population of wheat (64). To discover QTLs related to heat tolerance in wheat, substitution lines of durum wheat cultivar Langdon, Chinese Spring, and Hope were used to map heat-tolerant genes (65,66). Mapping results revealed the presence of heat-tolerant genes on chromosomes 3A and 3B. In previous eras, the QTL mapping technique has been extensively used in biparental populations to analyze the genetics of heat resilience, and the results revealed the presence of QTL hotspots on chromosome 3B. Mapping results showed that a prominent number of QTLs for canopy temperature, the quantum efficiency of photosystems, and other heat-tolerant indices along with 1000-grain weight and grain yield have been identified on chromosomes 4A, 5A and 7A (15, 29 and 18 QTLs), 1B, 2B and 7B (17, 16, 15 QTLs), and 2D, 7D (30 and 19 QTLs) (67–69).

Like heat tolerance, QTL mapping of agro-physiological traits in wheat associated with drought tolerance has gained significant consideration (70,71). Among agronomically profound traits most are related to root phenology and performance, such as primary and lateral root length, root elongation rate, root angle, root–shoot ratio, root weight, and biomass, while physiological parameters include canopy temperature, chlorophyll content, photosynthetic rate, cell membrane thermostability, relative water content, and concentration of water-soluble carbohydrates (72–75). A meta-analysis of root-related traits for drought tolerance in wheat revealed the presence of 634 QTLs on a consensus map having 7352 markers. Ninety-four root meta QTL have been identified, out of which 68 gene models were associated with an improvement in root structure for better adaptation against drought stress (76).

The stay green or hampered leaf-dropping phenomenon in wheat is also associated with drought tolerance, as it promotes the process of photosynthesis in plants even under harsh climatic conditions (77,78). Estimations based on reflectance and normalized difference vegetation index (NDVI) have a positive association with stay-green and kernel yield under drought conditions (38,79,80). This interrelationship between kernel yield and NDVI was estimated during early growth and grain-filling stages (79). Results of mapping revealed the presence of major QTL for NDVI on chromosome 5A and pleiotropic QTLs for grain yield and stay green on chromosomes 1B, 3D, 4D, and 7A. The presence of QTLs for NDVI, stay-green, and yield components in such consistency is helpful for the indirect selection of these parameters to enhance and sustain crop yield under a stress environment (80). In the wheat genome, some regions have been identified that possess QTLs for both drought and heat tolerance (38). Meta-QTL information about 81 traits assembled from 30 different studies pinpointed 66 MQTLs allocated throughout the wheat genome. Twenty MQTLs out of 66 belong to drought tolerance and two for heat tolerance, while 43 MQTLs present on chromosomes 1B, 2B, 2D, 4A, 4B, 4D, 5A, and 7A belong to both drought and heat stress tolerance in combination (81).

13.4 GENOME-WIDE ASSOCIATION STUDIES

With the advent of single nucleotide polymorphism (SNP) marker techniques from various sequencing platforms, the genomic marker tool kit has encountered an astonishing change in wheat genomics. The presence of thousands of SNPs has led to the use of genome-wide association (GWAS) techniques for trait selection in wheat. GWAS utilizes a linkage disequilibrium technique to recognize the linkage between trait and genetic loci. This process bypasses the demand to generate mapping panels, hence dispensing a robust substitute for linkage mapping (82). A GC-based linkage mapping in wheat revealed interdependency of SNPs in three CGs, DREB1A, ERA1, and 1-FEH, with several agro-physiological parameters (83). In another CG-based linkage mapping, the TaSnRK2.8 gene (an SNF-1 type serine-threonine protein kinase) was found associated with flag leaf width, water-soluble carbohydrates, and plant height under drought conditions (84).

GWAS comprehensive analysis revealed the existence of marker-trait associations (MTAs) on chromosome 4A for drought tolerance (82). Drought susceptibility index (DSI) and other leaf-associated parameters have been mapped on chromosome 4A, while for grain yield two QTLs have been mapped on chromosome 4A under normal and drought environmental conditions (82,83). High-density genotyping platforms and SNP arrays have enabled the potential exploitation of dense genome-wide SNPs in wheat (9K, 90K, 660K, and 820K SNP arrays, genotyping-by-sequencing, and DArTseq). A haplotype-based GWAS technique has been established for the recognition of stable QTLs in the modern era (85–87). The GWAS approach has been extensively used during the past few years to investigate heat and salinity tolerance in wheat (82,88–91). CG-based QTLs for heat tolerance have been mapped on chromosomes 3B and 4B (92). Fifteen stable haplotypes linked with improved grain yield under heat stress have been marked in the spring wheat germplasm of CIMMYT (62). GWAS is also helpful in the identification of common QTLs for drought and heat resilience in 315 and 105 accessions of spring wheat, respectively (89,90). For combined drought and heat tolerance, QTLs have been marked on chromosome 6A. This 6A locus controls parameters such as the number of grains, 1000-grain weight, and grain yield under drought and heat-stress environments (90). An Affymetrix 660K wheat SNP array was used to investigate a panel of 307 wheat accessions including landraces, obsolete varieties, and cultivars for salt tolerance. The results revealed the presence of three haplotypes on chromosomes 1A (QSt.nwafu-1A), 3B (QSt.nwafu-3B), and 6B (QSt.nwafu-6B) for salinity index (93). Traits and genes of interest identification for salinity indices by GWAS revealed QTLs for salt tolerance and yield-related parameters were found on chromosomes 4A, 5A, 5B, and 7A (94).

13.5 GENOMIC SELECTION FOR ABIOTIC TOLERANCE

Genomic selection (GS) is a widely adopted technique based on genomic-assisted techniques used for selection against abiotic stresses. Genomic estimated breeding values (GEBVs) are estimated by utilizing information obtained from genome-wide markers for individual selection. Estimation of the whole-genome markers is the basic prerequisite of the GS model, so minor and major genetic effects can be easily investigated. Therefore, GS is more robust and useful in comparison with MAS due to bypassing the requirement of QTL discovery for candidate traits (95). It provides the possibility to speed up the genetic gain in harmonization with enhanced selection accuracy and intensity of breeding cycles as compared to conventional breeding approaches.

Over the past few decades, enormous statistical models, markers systems and densities, environmental co-variants, and high-throughput imaging data have been extensively used to predict the nature of complex parameters like stress resilience in wheat (87,96,97). Different kinds of bi-parental and tri-parental populations, multi-subpopulations, multi-families, multi-lines, and gene bank accessions have been used as training populations and investigated as a source population for such investigations (98–100). These investigations are helpful in the construction of the following important results: large-sized training populations are good for better forecasting; a higher interrelationship between test and training populations helps to enhance precision in forecasting; harmonization of G × E interactions into genomic selection models increases precision in prediction; assimilation of pedigree information into the GS model is as productive as a marker system. The efficacy of forecasting models harmonized with G × E interaction is enhanced for trait behavior forecasting under multi-environments or unstable environments as compared to single or univariate environment forecasting models (101,102). The addition of environment-dependent variables, along with marker data into the G × E model, enhanced the forecasting efficacy by 11% for a candidate trait (109). When a multi-aspects random variance–covariance-dependent marker system along with environmental covariates was used, it aided the forecasting efficacy by 17–34% in comparison with forecasting models based on the main effects (103). The same technique was applied to forecast the performance of the CIMMYT wheat lines under erratic climatic conditions in South Asia (104).

Machine learning (ML) combined with deep learning techniques is gaining enormous popularity among scientists nowadays to perform multiple analyses (105). ML is a computer science-based field that observes the qualities of targeted patterns using algorithms and information on existing samples (106). The use of ML techniques is limited for forecasting yield and related parameters under marginal lands. The three most popular ML techniques, counter-propagation artificial neural networks (CP-ANN), XY-fused networks (XY-F), and supervised Kohonen networks (SKN), which utilize satellite-based data, NDVI, and soil traits as input, were used to forecast wheat yield. The average precision of the three models, CP-ANN, XY-F, and SKN, was 78.3%, 80.92%, and 81.65%, respectively (107). GS modeling is creating a paradigm shift in present-day research by integrating stable loci or major genes recognized in GWAS as fixed effects (87,108). More successful stories using this technique are found in forecasting disease resilience. However, little research has been conducted to forecast performance under abiotic stress environments, although the results are astonishing. A haplotype-based GWAS technique was used to observe the fixed effects of stable loci, and these were combined with prediction models to enhance the forecasting precision by 9–10% (87).

Imaging platforms concerned with high-throughput phenotyping (HTP) have undergone a remarkable improvement due to which harmonization of HTP data with the GS model got immense popularity in wheat (109). Secondary yield-related traits like canopy temperature (CT) and red and green NDVI were incorporated in the GS model to forecast grain yield in wheat under contrasting stressful environments. Treatments on the prototype were adjusted for CT and NDVI in training and test populations along with grain yield on the training population under multivariate analysis. The results revealed that after making corrections for days to heading, forecasting accuracy for grain yield was improved by 37% (110).

13.6 GENETIC ENGINEERING APPROACHES TO DEVELOP CLIMATE-RESILIENT WHEAT

In wheat, several genetic engineering approaches have shown promising results for sustainable yield under climate change. It is practically the quickest approach to develop cultivars against rapidly evolving biotic and abiotic stresses posed by climate change. These challenges are not only severe but they are also changing and moving the breeding objectives. Conventional breeding programs can keep pace with the nimble requirements of field crops, while genetic engineering can prepare a customized cultivar with high accuracy and efficiency. However, as compared with other field crops like maize and soybean, genetically modified wheat is still facing acceptability challenges. The initial genetic transformation experiments involving gene transfer from other species coupled with virus origin promoters have invited speculation about safety concerns and its complex genome causing unstable transgenic expression across generations. The low response to tissue culture by several genotypes and the landing transgenic cassette in the heterochromatic region have been bottlenecks for genetic transformation endeavors. Initially restricted to gene and promoter characterization in the laboratory, the advancement in transformation protocols, optimizing native promoters and alien-DNA free genetic modifications, has enabled researchers to take genetically engineered wheat genotypes to the field. Argentina took the lead by approving the world's first genetically engineered wheat, HB4, designed against drought tolerance, for cultivation and consumption in 2020. This cultivar encoding a transcription factor from *Helianthus annuus* showed a 40% yield increase under drought conditions as compared to other wheat varieties (111). In 1992, Vasil *et al.* (112) reported the first transgenic wheat developed through microprojectile bombardment of immature embryos. Since then, several DNA/RNA delivery methods have been tested to modify the wheat genome including *Agrobacterium*-mediated transformation, pollen tube pathway technique, permeabilizing cell membranes through electroporation, ultrasonic waves, and polyethylene glycol to insert foreign DNA for expression and genome integration. Among these, the biolistic method and *Agrobacterium*-mediated genetic transformation are the most commonly used by wheat researchers. The biolistic approach is less affected by the target genotype, and it has a high rate and frequency of construct insertion which works well for gene/locus characterization. However, for stable expression and practical utilization, single copy insertions are more favorable, achieved through *Agrobacterium*-mediated transformation. This approach is successful for genotypes with high totipotency, and the multiple rounds of callus culture require more time and plant death due to contamination during the procedure. Several bacterial and plasmid strains have been identified and optimized to deliver the construct according to the result (113).

Besides the construct delivery mechanism, tools to modify the genome were also improved for stability and accuracy across generations. RNA interference commonly, known as RNAi, was extensively deployed for gene function analysis and engineering required phenotypes by tweaking the gene expression. This was demonstrated first by Yan et al. (114) in 2004 as RNAi application in wheat and trait modification by converting winter wheat into spring type and reducing flower time for a month by down-regulating the VRN2 gene. Further research showed plant trait modifications for tiller number, grain yield, quality, and resistance against disease by a reduction in target gene expression (29,30,115). This tool, coupled with expression analysis, is still valid and helpful for studies where complete gene knock-out is not required. The transgenics provide a range of expression suppressions from complete to partial which provides more insights about gene functionality.

The mutation-based engineering of the genome began with random alteration of the genome. However, in the late 20th century, the situation was improved with the development of site-directed nucleases, namely zinc finger nucleases (ZFNs) and TAL effector nucleases (TALENs). These DNA nucleases equipped with sequence identification domains were able to cut DNA at precise locations and create insertions/deletions to knock out underside genes for loss of function. However, these tools required customized DNA-identifying protein domains, which was the major restriction for the application. The discovery and improvements in the bacterial Clustered Regularly Interspaced Short Palindromic Repeats or CRISPR system have revolutionized site-specific deletions by the Cas9 enzyme which is guided by small RNA. This guide RNA (gRNA) can be synthesized for any sequence and hence provides the possibility to disrupt any region in the genome. Several studies have demonstrated the potential to improve abiotic stress resistance and grain yield improvements in wheat via the CRISPR-Cas9 system (32,33,116) and so far more than 80 wheat genes have been modified and tested through this system (117).

REFERENCES

1. Ahmed HGMD, Zeng Y, Shah AN, Yar MM, Ullah A, Ali M. Conferring of drought tolerance in wheat (Triticum aestivum L.) genotypes using seedling indices. *Frontiers in Plant Science*. 2022 Jul 22;13:961049.
2. Giraldo P, Benavente E, Manzano-Agugliaro F, Gimenez E. Worldwide research trends on wheat and barley: A bibliometric comparative analysis. *Agronomy*. 2019 Jul;9(7):352.
3. Asseng S, Foster I, Turner NC. The Impact of Temperature Variability on Wheat Yields. *Global Change Biology*. 2011;17(2):997–1012.
4. Iqbal MJ, Shams N, Fatima K, Iqbal MJ, Shams N, Fatima K. Nutritional Quality of Wheat. In: *Wheat – Recent Advances [Internet]*. IntechOpen; 2022 [cited 2025 Feb 19]. Available from: www.intechopen.com/chapters/81564

5. World Bank [Internet]. [cited 2025 Feb 13]. *Annual Report 2024*. Available from: www.worldbank.org/en/about/annual-report
6. Dhakal A. Effect of drought stress and management in wheat – a review. *Food & Agribusiness Management (FABM)*. 2021;2(2):62–6.
7. Erenstein O, Jaleta M, Mottaleb KA, Sonder K, Donovan J, Braun HJ. Global trends in wheat production, consumption and trade. In: Reynolds MP, Braun HJ, editors. *Wheat Improvement: Food Security in a Changing Climate [Internet]*. Cham: Springer International Publishing; 2022 [cited 2025 Feb 19]. p. 47–66. Available from: https://doi.org/10.1007/978-3-030-90673-3_4
8. Akram S, Ghaffar M, Wadood A, Shokat S, Hameed A, Waheed MQ, et al. A GBS-based genome-wide association study reveals the genetic basis of salinity tolerance at the seedling stage in bread wheat (*Triticum aestivum* L.). *Frontiers in Genetics*. 2022 Sep 27;13:997901.
9. World Bank [Internet]. [cited 2025 Feb 17]. *Food Security | Food Insecurity Statistics & Solutions*. Available from: www.worldbank.org/en/topic/agriculture/brief/food-security-update
10. www.fao.org [Internet]. [cited 2025 Feb 13]. *The State of Food and Agriculture* 2020. Available from: www.fao.org/interactive/state-of-food-agriculture/2020/en/
11. Bhandari R, Gnawali S, Nyaupane S, Kharel S, Poudel M, Panth P. Effect of drought & irrigated environmental condition on yield & yield attributing characteristic of bread wheat-a review. *Reviews in Food and Agriculture (RFNA)*. 2021;2(2):59–62.
12. Mansour MMF, Salama KHA. Proline and Abiotic Stresses: Responses and Adaptation. *In:* Hasanuzzaman, M. (eds.). *Plant Ecophysiology and Adaptation under Climate Change: Mechanisms and Perspectives II*. Springer: Singapore 2020;357. https://doi.org/10.1007/978-981-15-2172-0_12
13. Ahmad Z, Waraich EA, Akhtar S, Anjum S, Ahmad T, Mahboob W, et al. Physiological responses of wheat to drought stress and its mitigation approaches. *Acta Physiologiae Plantarum*. 2018 Apr;40(4):80.
14. Kajla M, Yadav VK, Khokhar J, Singh S, Chhokar RS, Meena RP, et al. Increase in wheat production through management of abiotic stresses: A review. *Journal of Applied and Natural Science*. 2015 Dec 1;7(2):1070–80.
15. Augspurger CK. Reconstructing patterns of temperature, phenology, and frost damage over 124 years: Spring damage risk is increasing. *Ecology*. 2013;94(1):41–50.
16. Yadav SK. Cold stress tolerance mechanisms in plants. A review. *Agronomy for Sustainable Development*. 2010 Sep 1;30(3):515–27.
17. Hassan MA, Xiang C, Farooq M, Muhammad N, Yan Z, Hui X, et al. Cold stress in wheat: Plant acclimation responses and management strategies. *Frontiers in Plant Science*. 2021 Jul 8;12:676884.
18. Yue Y, Zhou Y, Wang J, Ye X. Assessing wheat frost risk with the support of GIS: An approach coupling a growing season meteorological index and a hybrid fuzzy neural network model. *Sustainability*. 2016 Dec;8(12):1308.
19. Sharma K, Dhingra M, Kaur R, Singh S, Kaur A, Kaur S, et al. Evaluation of Triticum durum–Aegilops tauschii derived primary synthetics as potential sources of drought stress tolerance for wheat improvement. *Cereal Research Communications*. 2022 Dec 1;50(4):1205–16.
20. Paudel B, Zhang Y, Yan J, Rai R, Li L, Wu X, et al. Farmers' understanding of climate change in Nepal Himalayas: Important determinants and implications for developing adaptation strategies. *Climatic Change*. 2020 Feb 1;158(3):485–502.

21. Anjum SA, Ashraf U, Zohaib A, Tanveer M, Naeem M, Ali I, et al. Growth and developmental responses of crop plants under drought stress: A review. *Zemdirbyste-Agriculture [Internet]*. 2017 Aug 7 [cited 2025 Feb 19];
22. Nyaupane S, Poudel MR, Panthi B, Dhakal A, Paudel H, Bhandari R. Drought stress effect, tolerance, and management in wheat – A review. *Cogent Food & Agriculture*. 2024 Dec 31;10(1):2296094.
23. Kou X, Han W, Kang J. Responses of root system architecture to water stress at multiple levels: A meta-analysis of trials under controlled conditions. *Frontiers in Plant Science*. 2022 Dec 9;13:1085409.
24. Khan A, Ahmad M, Ahmed M, Iftikhar Hussain M. Rising atmospheric temperature impact on wheat and thermotolerance strategies. *Plants*. 2021 Jan;10(1):43.
25. Djanaguiraman M, Narayanan S, Erdayani E, Prasad PVV. Effects of high temperature stress during anthesis and grain filling periods on photosynthesis, lipids and grain yield in wheat. *BMC Plant Biology*. 2020 Jun 9;20(1):268.
26. Ni Z, Li H, Zhao Y, Peng H, Hu Z, Xin M, et al. Genetic improvement of heat tolerance in wheat: Recent progress in understanding the underlying molecular mechanisms. *The Crop Journal*. 2018 Feb 1;6(1):32–41.
27. Lai CH, He J. Physiological performances of temperate vegetables with response to chronic and acute heat stress. *American Journal of Plant Sciences*. 2016 Sep 29;7(14):2055–71.
28. Qaseem MF, Qureshi R, Shaheen H. Effects of pre-anthesis drought, heat and their combination on the growth, yield and physiology of diverse wheat (Triticum aestivum L.) Genotypes varying in sensitivity to heat and drought stress. *Scientific Reports*. 2019 May 6;9(1):6955.
29. Rinalducci S, Egidi MG, Karimzadeh G, Jazii FR, Zolla L. Proteomic analysis of a spring wheat cultivar in response to prolonged cold stress. *Electrophoresis*. 2011;32(14):1807–18.
30. Valluru R, Link J, Claupein W. Consequences of early chilling stress in two Triticum species: Plastic responses and adaptive significance. *Plant Biology*. 2012;14(4):641–51.
31. Ji H, Xiao L, Xia Y, Song H, Liu B, Tang L, et al. Effects of jointing and booting low temperature stresses on grain yield and yield components in wheat. *Agricultural and Forest Meteorology*. 2017 Sep 15;243:33–42.
32. Los DA, Mironov KS, Allakhverdiev SI. Regulatory role of membrane fluidity in gene expression and physiological functions. *Photosynthesis Research*. 2013 Oct 1;116(2):489–509.
33. Cvetkovic J, Müller K, Baier M. The effect of cold priming on the fitness of *Arabidopsis thaliana* accessions under natural and controlled conditions. *Scientific Reports* 2017 Mar 9;7(1):44055.
34. Khan TA, Fariduddin Q, Yusuf M. Low-temperature stress: Is phytohormones application a remedy? *Environmental Science and Pollution Research International*. 2017 Sep 1;24(27):21574–90.
35. Liu C, Pinto F, Cossani CM, Sukumaran S, Reynolds MP. Spectral reflectance indices as proxies for yield potential and heat stress tolerance in spring wheat: heritability estimates and marker-trait associations. *Frontiers of Agricultural Science and Engineering*. 2019 Sep 15;6(3):296–308.
36. Hussain S, Shaukat M, Ashraf M, Zhu C, Jin Q, Zhang J, et al. Salinity Stress in Arid and Semi-Arid Climates: Effects and Management in Field Crops. In: Saddam Hussain (ed.). *Climate Change and Agriculture [Internet]*. IntechOpen; 2019 [cited 2025 Feb 19]. Available from: www.intechopen.com/chapters/68075

37. Singh V. *Environmental Plant Physiology: Botanical Strategies for a Climate Smart Planet*. Boca Raton: CRC Press; 2020. 230 p.
38. FAO. Agricultural production statistics 2000–2022. *FAOSTAT Analytical Brief 79*. Rome: FAO; 2023.
39. Poudel MR, Ghimire S, Pandey MP, Dhakal KH, Thapa DB, Poudel HK. Evaluation of wheat genotypes under irrigated, heat stress and drought conditions. *Journal of Biology and Today's World*. 2020;9(1):212.
40. Building Climate Resilience in Agriculture: *Theory, Practice and Future Perspective | SpringerLink [Internet].* [cited 2025 Feb 17]. Available from: https://link.springer.com/book/10.1007/978-3-030-79408-8
41. Jain SK, Omprakash. Pre-breeding: A Bridge between genetic resources and crop improvement. *International Journal of Current Microbiology and Applied Sciences*. 2019 Feb 10;8(2):1998–2007.
42. Kumar V, Shukla YM. *Pre-breeding: Its applicat*. [cited 2025 Feb 17]; Available from: www.researchgate.net/profile/Vinay-Kumar-74/publication/325083861_Pre-breeding_its_applications_in_crop_improvement/links/5b03d07baca2720ba09959d5/Pre-breeding-its-applications-in-crop-improvement.pdf
43. Singh K, Kumar S, Kumar SR, Singh M, Gupta K. Plant genetic resources management and pre-breeding in genomics era. *Indian Journal Of Genetics and Plant Breeding*. 2019 Apr 25;79(Sup-01):117–30.
44. Abebe A, Tafa Z. Pre-breeding concept and role in crop improvement. *International Journal for Research in Applied Sciences and Biotechnology*. 2021 Mar 31;8(2):275–9.
45. Sharma S, Upadhyaya HD, Varshney RK, Gowda CLL. Pre-breeding for diversification of primary gene pool and genetic enhancement of grain legumes. *Frontiers in Plant Science*. 2013 Aug 20;4:309.
46. Sharma M, Punya, Gupta BB. Role of Wild Relatives for Development of Climate-Resilient Varieties. In: Salgotra RK, Zargar SM, editors. *Rediscovery of Genetic and Genomic Resources for Future Food Security [Internet*]. Singapore: Springer; 2020 [cited 2025 Feb 19]. p. 303–14. Available from: https://doi.org/10.1007/978-981-15-0156-2_11
47. Abadya S, Shimelis H, Pasupuleti J, Mashilo J, Chaudhari S, Manohar SS. Assessment of the genetic diversity of groundnut (*Arachis hypogaea* L.) genotypes for kernel yield, oil and fodder quantity and quality under drought conditions. *Crop Science*. 2021;61(3):1926–43.
48. Ober ES, Alahmad S, Cockram J, Forestan C, Hickey LT, Kant J, et al. Wheat root systems as a breeding target for climate resilience. *Theoretical and Applied Genetics*. 2021 Jun 1;134(6):1645–62.
49. Choudhary M, Wani SH, Kumar P, Bagaria PK, Rakshit S, Roorkiwal M, et al. QTLian breeding for climate resilience in cereals: Progress and prospects. *Functional & Integrative Genomics*. 2019 Sep 1;19(5):685–701.
50. Song L, Wang R, Yang X, Zhang A, Liu D. Molecular markers and their applications in marker-assisted selection (MAS) in Bread Wheat (*Triticum aestivum* L.). *Agriculture*. 2023 Mar;13(3):642.
51. Sharma S, Schulthess AW, Bassi FM, Badaeva ED, Neumann K, Graner A, et al. Introducing beneficial alleles from plant genetic resources into the wheat germplasm. *Biology*. 2021 Oct;10(10):982.
52. Wani SH, Khan H, Riaz A, Joshi DC, Hussain W, Rana M, et al. Chapter Six – Genetic Diversity for Developing Climate-Resilient Wheats to Achieve Food Security Goals. In: Sparks DL, editor. *Advances in Agronomy* [Internet]. Academic Press; 2022 [cited

2025 Feb 19]. p. 255–303. Available from: www.sciencedirect.com/science/article/pii/S0065211321000857

53. Gorafi YSA, Kim JS, Elbashir AAE, Tsujimoto H. A population of wheat multiple synthetic derivatives: An effective platform to explore, harness and utilize genetic diversity of *Aegilops tauschii* for wheat improvement. *Theoretical and Applied Genetics*. 2018 Aug 1;131(8):1615–26.
54. Marcussen T, Sandve SR, Heier L, Spannagl M, Pfeifer M, The International Wheat Genome Sequencing Consortium, et al. Ancient hybridizations among the ancestral genomes of bread wheat. *Science*. 2014 Jul 18;345(6194):1250092.
55. Hammer K, Knüpffer H. Genetic Resources of Triticum. In: Ogihara Y, Takumi S, Handa H, editors. *Advances in Wheat Genetics: From Genome to Field*. Tokyo: Springer Japan; 2015. p. 23–31.
56. Gadimaliyeva G, Aminov N, Jahangirov A, Hamidov H, Abugalieva A, Shamanin V, et al. Productivity and disease resistance of primary hexaploid synthetic wheat lines and their crosses with bread wheat. *Cereal Research Communications*. 2018 Jun 1;46(2):355–64.
57. Naz AA, Dadshani S, Ballvora A, Pillen K, Léon J. Genetic analysis and transfer of favorable exotic QTL alleles for grain yield across D genome using two advanced backcross wheat populations. *Frontiers in Plant Science*. 2019 Jun 4;10:711.
58. Sukumaran S, Rebetzke G, Mackay I, Bentley AR, Reynolds MP. Pre-Breeding Strategies. In: Reynolds MP, Braun HJ, editors. *Wheat Improvement: Food Security in a Changing Climate [Internet]*. Cham: Springer International Publishing; 2022 [cited 2025 Feb 17]. p. 451–69. Available from: https://doi.org/10.1007/978-3-030-90673-3_25
59. Reynolds M, Chapman S, Crespo-Herrera L, Molero G, Mondal S, Pequeno DNL, et al. Breeder friendly phenotyping. *Plant Science*. 2020 Jun 1;295:110396.
60. Reynolds MP, Pask AJD, Hoppitt WJE, Sonder K, Sukumaran S, Molero G, et al. Correction to: Strategic crossing of biomass and harvest index—source and sink—achieves genetic gains in wheat. *Euphytica*. 2018;214(1):9.
61. Speed breeding is a powerful tool to accelerate crop research and breeding | *Nature Plants [Internet]*. [cited 2025 Feb 18]. Available from: www.nature.com/articles/s41477-017-0083-8
62. Singh D, Wang X, Kumar U, Gao L, Noor M, Imtiaz M, et al. High-throughput phenotyping enabled genetic dissection of crop lodging in wheat. *Frontiers in Plant Science*. 2019;10:394.
63. Leng P fei, Lübberstedt T, Xu M liang. Genomics-assisted breeding – A revolutionary strategy for crop improvement. *Journal of Integrative Agriculture*. 2017 Dec 1;16(12):2674–85.
64. Mondal S, Sallam A, Sehgal D, Sukumaran S, Farhad M, Navaneetha Krishnan J, et al. Advances in Breeding for Abiotic Stress Tolerance in Wheat. In: Kole C, editor. *Genomic Designing for Abiotic Stress Resistant Cereal Crops [Internet]*. Cham: Springer International Publishing; 2021 [cited 2025 Feb 18]. p. 71–103. Available from: https://doi.org/10.1007/978-3-030-75875-2_2
65. Sun QX, Quick JS. Chromosomal Locations of Genes for Heat Tolerance in Tetraploid Wheat. *Cereal Research Communications*. 1991;19(4):431–7.
66. Ruqiang X, Qixin S, Shuzhen Z. Chromosomal location of genes for heat tolerance as measured by membrane thermostability of common wheat cv. *Hope. Yi chuan = Hereditas*. 1996;18(4):1–3.
67. Bhusal N, Sarial AK, Sharma P, Sareen S. Mapping QTLs for grain yield components in wheat under heat stress. *PLOS ONE*. 2017 Dec 19;12(12):e0189594.

68. Sharma DK, Torp AM, Rosenqvist E, Ottosen CO, Andersen SB. QTLs and potential candidate genes for heat stress tolerance identified from the mapping populations specifically segregating for Fv/Fm in wheat. *Frontiers in Plant Science*. 2017 Sep 27;8:1668.
69. QTL on wheat (*Triticum aestivum* L.) Chromosomes 1B, 3D and 5A are associated with constitutive production of leaf cuticular wax and may contribute to lower leaf temperatures under heat stress | *Euphytica [Internet]*. [cited 2025 Feb 19]. Available from: https://link.springer.com/article/10.1007/s10681-014-1193-2
70. Gupta PK, Balyan HS, Gahlaut V. QTL analysis for drought tolerance in wheat: Present status and future possibilities. *Agronomy*. 2017 Mar;7(1):5.
71. Tshikunde NM, Mashilo J, Shimelis H, Odindo A. Agronomic and physiological traits, and associated quantitative trait loci (QTL) affecting yield response in wheat (*Triticum aestivum* L.): A review. *Frontiers in Plant Science*. 2019 Nov 5;10:1428.
72. Bharti S, Balyan HS, Gupta PK. Quantitative trait loci analysis for some root traits in bread wheat (triticum aestivum L.). *International Journal of Agricultural Sciences*. 2014;4(7):214–21.
73. Czyczyło-Mysza I, Marcińska I, Skrzypek E, Cyganek K, Juzoń K, Karbarz M. QTL mapping for germination of seeds obtained from previous wheat generation under drought. *Open Life Sciences*. 2014 Apr 1;9(4):374–82.
74. Ayalew H, Liu H, Yan G. Quantitative analysis of gene actions controlling root length under water stress in spring wheat (*Triticum aestivum* L.) genotypes. *Crop and Pasture Science*. 2016 May 31;67(5):489–94.
75. Voss-Fels KP, Robinson H, Mudge SR, Richard C, Newman S, Wittkop B, et al. *VERNALIZATION1* Modulates root system architecture in wheat and barley. *Molecular Plant*. 2018 Jan 8;11(1):226–9.
76. Soriano JM, Alvaro F. Discovering consensus genomic regions in wheat for root-related traits by QTL meta-analysis. *Scientific Reports*. 2019 Jul 22;9(1):10537.
77. Borrell AK, van Oosterom EJ, Mullet JE, George-Jaeggli B, Jordan DR, Klein PE, et al. Stay-green alleles individually enhance grain yield in sorghum under drought by modifying canopy development and water uptake patterns. *New Phytologist*. 2014;203(3):817–30.
78. Thomas H, Ougham H. The stay-green trait. *Journal of Experimental Botany*. 2014 Jul 1;65(14):3889–900.
79. Hazratkulova S, Sharma RC, Alikulov S, Islomov S, Yuldashev T, Ziyaev Z, et al. Analysis of genotypic variation for normalized difference vegetation index and its relationship with grain yield in winter wheat under terminal heat stress. *Plant Breeding*. 2012;131(6):716–21.
80. Shi J, Gao H, Wang H, Lafitte HR, Archibald RL, Yang M, et al. ARGOS8 variants generated by CRISPR-Cas9 improve maize grain yield under field drought stress conditions. *Plant Biotechnology Journal*. 2017;15(2):207–16.
81. Acuña-Galindo MA, Mason RE, Subramanian NK, Hays DB. Meta-analysis of wheat QTL regions associated with adaptation to drought and heat stress. *Crop Science*. 2015;55(2):477–92.
82. Sehgal D, Autrique E, Singh R, Ellis M, Singh S, Dreisigacker S. Identification of genomic regions for grain yield and yield stability and their epistatic interactions. *Scientific Reports*. 2017 Feb 1;7(1):41578.
83. Edae EA, Byrne PF, Manmathan H, Haley SD, Moragues M, Lopes MS, et al. Association mapping and nucleotide sequence variation in five drought tolerance candidate genes in spring wheat. *The Plant Genome*. 2013;6(2):plantgenome2013.04.0010.

84. Zhang H, Cui F, Wang L, Li J, Ding A, Zhao C, et al. Conditional and unconditional QTL mapping of drought-tolerance-related traits of wheat seedling using two related RIL populations. *Journal of Genetics*. 2013;92(2):213–31.
85. Bhatta M, Morgounov A, Belamkar V, Baenziger PS. Genome-wide association study reveals novel genomic regions for grain yield and yield-related traits in drought-stressed synthetic hexaploid wheat. *International Journal of Molecular Sciences*. 2018 Oct;19(10):3011.
86. Afzal F, Li H, Gul A, Subhani A, Ali A, Mujeeb-Kazi A, et al. Genome-wide analyses reveal footprints of divergent selection and drought adaptive traits in synthetic-derived wheats. *G3 Genes|Genomes|Genetics*. 2019 Jun 1;9(6):1957–73.
87. Sehgal D, Rosyara U, Mondal S, Singh R, Poland J, Dreisigacker S. Incorporating genome-wide association mapping results into genomic prediction models for grain yield and yield stability in CIMMYT spring bread wheat. *Frontiers in Plant Science*. 2020 Mar 4;11:197.
88. Maulana F, Ayalew H, Anderson JD, Kumssa TT, Huang W, Ma XF. Genome-wide association mapping of seedling heat tolerance in winter wheat. *Frontiers in Plant Science*. 2018 Sep 4;9:1272.
89. Qaseem MF, Qureshi R, Muqaddasi QH, Shaheen H, Kousar R, Röder MS. Genome-wide association mapping in bread wheat subjected to independent and combined high temperature and drought stress. *PLOS ONE*. 2018 Jun 27;13(6):e0199121.
90. Schmidt J, Claussen J, Wörlein N, Eggert A, Fleury D, Garnett T, et al. Drought and heat stress tolerance screening in wheat using computed tomography. *Plant Methods*. 2020 Feb 13;16(1):15.
91. Oyiga BC, Sharma RC, Baum M, Ogbonnaya FC, Léon J, Ballvora A. Allelic variations and differential expressions detected at quantitative trait loci for salt stress tolerance in wheat. *Plant, Cell & Environment*. 2018;41(5):919–35.
92. Maulana F, Ayalew H, Anderson JD, Kumssa TT, Huang W, Ma XF. Genome-wide association mapping of seedling heat tolerance in winter wheat. *Frontiers in Plant Science [Internet]*. 2018 Sep 4 [cited 2025 Feb 18];9. Available from: / www.frontiersin.org/journals/plant-science/articles/10.3389/fpls.2018.01272/full
93. Yu S, Wu J, Wang M, Shi W, Xia G, Jia J, et al. Haplotype variations in QTL for salt tolerance in Chinese wheat accessions identified by marker-based and pedigree-based kinship analyses. *The Crop Journal*. 2020 Dec 1;8(6):1011–24.
94. Hu P, Zheng Q, Luo Q, Teng W, Li H, Li B, et al. Genome-wide association study of yield and related traits in common wheat under salt-stress conditions. *BMC Plant Biology*. 2021 Jan 7;21(1):27.
95. Nakaya A, Isobe SN. Will genomic selection be a practical method for plant breeding? *Annals of Botany*. 2012 Nov 1;110(6):1303–16.
96. Juliana P, Singh RP, Singh PK, Crossa J, Rutkoski JE, Poland JA, et al. Comparison of models and whole-genome profiling approaches for genomic-enabled prediction of septoria tritici blotch, stagonospora nodorum blotch, and tan spot resistance in wheat. *The Plant Genome*. 2017;10(2):plantgenome2016.08.0082.
97. Kristensen PS, Jahoor A, Andersen JR, Cericola F, Orabi J, Janss LL, et al. Genome-wide association studies and comparison of models and cross-validation strategies for genomic prediction of quality traits in advanced winter wheat breeding lines. *Frontiers in Plant Science*. 2018 Feb 2;9:69.
98. Joukhadar R, Daetwyler HD, Bansal UK, Gendall AR, Hayden MJ. Genetic diversity, population structure and ancestral origin of australian wheat. *Frontiers in Plant Science*. 2017 Dec 12;8:2115.

99. Dong H, Wang R, Yuan Y, Anderson J, Pumphrey M, Zhang Z, et al. Evaluation of the potential for genomic selection to improve spring wheat resistance to fusarium head blight in the pacific northwest. *Frontiers in Plant Science.* 2018 Jul 3;9:911.
100. Ladejobi O, Mackay IJ, Poland J, Praud S, Hibberd JM, Bentley AR. Reference genome anchoring of high-density markers for association mapping and genomic prediction in european winter wheat. *Frontiers in Plant Science.* 2019 Nov 8;10:1278.
101. Pérez-Rodríguez P, Crossa J, Rutkoski J, Poland J, Singh R, Legarra A, et al. Single-step genomic and pedigree genotype × environment interaction models for predicting wheat lines in international environments. *The Plant Genome.* 2017;10(2):plantgenome2016.09.0089.
102. Lopez-Cruz M, Crossa J, Bonnett D, Dreisigacker S, Poland J, Jannink JL, et al. Increased prediction accuracy in wheat breeding trials using a marker × environment interaction genomic selection model. *G3 (Bethesda).* 2015 Feb 6;5(4):569–82.
103. Jarquín D, Crossa J, Lacaze X, Du Cheyron P, Daucourt J, Lorgeou J, et al. A reaction norm model for genomic selection using high-dimensional genomic and environmental data. *Theoretical and Applied Genetics.* 2014 Mar 1;127(3):595–607.
104. Pérez-Rodríguez P, Burgueño J, Montesinos-López OA, Singh RP, Juliana P, Mondal S, et al. Prediction with Big Data in the Genomic and High-Throughput Phenotyping Era: A Case Study with Wheat Data. In: Manjit S. Kang (ed.). *Quantitative Genetics, Genomics and Plant Breeding [Internet].* 2020 [cited 2025 Feb 19]. p. 213–26. (CABI Books). Available from: www.cabidigitallibrary.org/doi/10.1079/9781789240214.0213
105. González-Camacho JM, Ornella L, Pérez-Rodríguez P, Gianola D, Dreisigacker S, Crossa J. Applications of machine learning methods to genomic selection in breeding wheat for rust resistance. *The Plant Genome.* 2018 Jul;11(2):170104.
106. Gonzalez-Sanchez A, Frausto-Solis J, Ojeda-Bustamante W. Predictive ability of machine learning methods for massive crop yield prediction. *Spanish Journal of Agricultural Research.* 2014 Apr 29;12(2):313–28.
107. Pantazi XE, Moshou D, Alexandridis T, Whetton RL, Mouazen AM. Wheat yield prediction using machine learning and advanced sensing techniques. *Computers and Electronics in Agriculture.* 2016 Feb 1;121:57–65.
108. Sarinelli JM, Murphy JP, Tyagi P, Holland JB, Johnson JW, Mergoum M, et al. Training population selection and use of fixed effects to optimize genomic predictions in a historical USA winter wheat panel. *Theoretical and Applied Genetics.* 2019 Apr 1;132(4):1247–61.
109. Crain J, Mondal S, Rutkoski J, Singh RP, Poland J. Combining high-throughput phenotyping and genomic information to increase prediction and selection accuracy in wheat breeding. *The Plant Genome.* 2018;11(1):170043.
110. Rutkoski J, Poland J, Mondal S, Autrique E, Pérez LG, Crossa J, et al. Canopy temperature and vegetation indices from high-throughput phenotyping improve accuracy of pedigree and genomic selection for grain yield in wheat. *G3 Genes|Genomes|Genetics.* 2016 Sep 1;6(9):2799–808.
111. Ayala F, Fedrigo GV, Burachik M, Miranda PV. Compositional equivalence of event IND-ØØ412-7 to non-transgenic wheat. *Transgenic Research.* 2019 Apr;28(2):165–76.
112. Vasil V, Castillo AM, Fromm ME, Vasil IK. Herbicide resistant fertile transgenic wheat plants obtained by microprojectile bombardment of regenerable embryogenic callus. *Bio/Technology.* 1992 Jun;10(6):667–74.

113. Ye X, Shrawat A, Moeller L, Rode R, Rivlin A, Kelm D, et al. Agrobacterium-mediated direct transformation of wheat mature embryos through organogenesis. *Frontiers in Plant Science*. 2023 May 31;14:1202235.
114. Yan L, Loukoianov A, Blechl A, Tranquilli G, Ramakrishna W, SanMiguel P, et al. The wheat VRN2 gene is a flowering repressor down-regulated by vernalization. *Science*. 2004 Mar 12;303(5664):1640–4.
115. Halder K, Chaudhuri A, Abdin MZ, Majee M, Datta A. RNA Interference for improving disease resistance in plants and its relevance in this clustered regularly interspaced short palindromic repeats-dominated Era in terms of dsRNA-based biopesticides. *Frontiers in Plant Science*. 2022;13:885128.
116. Hamada H, Liu Y, Nagira Y, Miki R, Taoka N, Imai R. Biolistic-delivery-based transient CRISPR/Cas9 expression enables in planta genome editing in wheat. *Scientific Reports*. 2018 Sep 26;8(1):14422.
117. Zhou X, Zhao Y, Ni P, Ni Z, Sun Q, Zong Y. CRISPR-mediated acceleration of wheat improvement: Advances and perspectivess. *Journal of Genetics and Genomics*. 2023 Nov 1;50(11):815–34.

14 Use of Molecular Breeding to Develop Climate-ready Crops

Zeshan Hassan, Azhar Abbas Khan, Umbreen Shehzad, Asad Azeem, Sadia Zainab, Mahlka Mukhtar, Zeba Shehnaz, and Zahid Manzoor

14.1 INTRODUCTION

Climate change has shown a drastic effect on agricultural crop production (1). Owing to changes in climatic and pest patterns, global food insecurity has increased (2). The world is currently facing significant challenges that threaten global food security. Population projections estimate a rise to more than 8 billion by 2030 and more than 9.9 billion by 2050, increasing demand for food resources (3). However, climate change is a problem that needs urgent attention as droughts, heat waves, and floods are upsetting the balance necessary for proper farming operations with crop failures due to food shortages. Due to the severity of extreme weather events, crop production is severely affected. The profound effects of climate change are being perceived by developing nations like Pakistan, worsening food insecurity. Agricultural productivity is disrupted by unpredictable weather patterns and destructive floods, resulting in crop failures and dwindling food supplies (4). Developing climate-ready crops (CRCs) is the only reliable and authentic strategy to tackle the current scenario. Climate-ready crops ensure the constancy of food availability and agricultural productivity even in fluctuating environmental conditions, viz., altered precipitation and temperature swings (5). Many studies have been conducted to ameliorate the effects of climate change on crop production using advanced molecular breeding techniques. Molecular breeding (MB) has played a vital role in pinpointing the attributes of interest in the crop genomes (6). Molecular breeding can selectively produce plants with desirable characteristics, understanding the genetic basis of their traits, which saves time and resources compared to conventional breeding. Innovations in genomics have transformed crop improvement by making innovative plant breeding methods. Modern genomics technologies can accelerate trait mapping, gene identification, marker development, and molecular breeding procedures, in addition to increasing crop productivity (7). It is possible to identify accessions with low-frequency variations that may be dependable for essential traits such as yield components, abiotic stress tolerance, or disease resistance by integrating data from genome-wide sequencing with precise phenotypic variation. Next-generation sequencing and other

DOI: 10.1201/9781032720142-18

molecular tools such as genotyping, marker creation, high-density genetic linkage maps, QTL (quantitative trait loci) identification, ESTs (expressed sequence tags), SNPs (single nucleotide polymorphisms), and transcription factors combined with current breeding processes have made precise and rapid transmission of information possible (8).

By integrating advantageous characteristics, MB is able to develop climate-ready crops that can perform even against challenging weather conditions. Traditional methods have given significant results over time, and molecular breeding has opened new doors to precision agriculture. Molecular breeding has made this shift from conventional approaches to precision agriculture possible. Identifying specific traits is facilitated by utilizing genetic markers, which saves time (9). This shift in approach enhances the precision of trait selection and accelerates the breeding process (10). MB unlocks new opportunities for agricultural development by clarifying the genetic foundations of climate resistance. Through marker-assisted selection, researchers can target specific traits in the genomic regions associated with climatic stress resistance. Cutting-edge MB techniques further increase the efficiency of crop enhancement programs. Utilizing these techniques enhances breeding effectiveness, enabling the rapid development of crops with high-stress tolerance, stability in yields, and adaptability in changing climate (11). Understanding the biochemical mechanisms governing the plant responses to the stresses is important. This understanding can open up how breeders and researchers develop targeted techniques, boosting crop productivity and resilience in the face of climate change. Molecular breeding has revolutionized agriculture by improving crop architecture, which has advanced features that allow it to survive even in stressful and changing environmental conditions. A thorough SWOT analysis of these approaches sheds light on their strengths, weaknesses, opportunities, and threats (12). By adopting advanced molecular breeding techniques, scientists can combat climate change and develop climate-ready varieties that can withstand freezing and scorching conditions. Although MB has many plus points, scientists face some challenges in molecular breeding techniques, which must be addressed. The potential benefits of molecular breeding in developing CRCs are much more significant than the potential drawbacks (13). In conclusion, utilizing MB presents a paradigm shift in crop improvement programs, expressing a focused and efficient method to enhance crop resilience to climate change through genome utilization. Molecular breeding is also hopeful in addressing the problems posed by climate change and ensuring crop productivity and food security stability.

14.2 CLIMATE CHANGE AND ITS IMPACTS ON CROP PLANTS

Crop plants suffer significantly from environmental change because of higher greenhouse gas levels, altered phenology, increased water stress, and increased frequency of severe weather conditions (14). Elevated temperatures cause havoc with crucial developmental phases, while changes in rainfall patterns exacerbate waterlogging and shortages (15). Severe weather events exacerbate food insecurity by causing bodily harm. The increasing threats of pests and diseases challenge crop defense mechanisms. Pollination services, essential for crop reproduction, are hampered by disruptions in plant–pollinator interactions. Arable land is further reduced by salinization and soil degradation, which impair fertility. Creating crop varieties that are tolerant to climate change, implementing sustainable practices, improving irrigation efficiency, and advancing agroecology are examples of mitigation and adaptation tactics (16). Creating robust agricultural systems requires time to guarantee food security in the face of climate unpredictability (17).

14.3 DEVELOPMENTAL JOURNEY FROM CONVENTIONAL TO MOLECULAR CROP BREEDING

Crops have the most tremendous potential to contribute the renewable energy necessary for the survival of all living things (18). Crop productivity is often greatly affected by rapidly changing biotic and abiotic stresses (19). In these adverse conditions, finding other ways to increase yields to support the growing world population is important. Since farmers prefer healthy saplings for farming, we always know how to find the best variety. Thanks to Gregor Johann Mendel's discovery of Mendelian inheritance, we can now use proven and traditional research methods to increase yields. Therefore, action is needed to solve this problem. Modern biotechnological methods, tools, and techniques have gained importance in recent years. These technologies allow researchers to perform tasks such as whole-genome sequencing, annotation, prediction and validation of gene expression, identification of patterns of nucleic acid sequences (motifs) present in promoter regions and their protein–DNA interactions, and predicting potency (20). Functions and their participation in different metabolic processes also ensure the production of high-quality seeds and achieving optimal yields even under challenging conditions. We also strive to produce excellent seed quality and yield good results despite challenging conditions.

Global food security is at risk due to climate change causing crop losses worldwide. As the world's population is expected to surpass 10 billion soon, lawmakers and stakeholders are concerned about food security. Future food needs can be met by combining modern cultivation techniques, good agricultural practices, advances in microbiome-based solutions, and enhanced crop yields achieved through improved crop management strategies that reduce the risk of crop failure. To address future challenges in global food security, in this review, we describe next-generation breeding tools that can be used to increase crop yields by improving weather-tolerant genotypes.

Recent advances in genome-assisted breeding (GAB) have enabled the production of multiple pan-genome crops that provide the entire landscape of genetic diversity (GD) and restore extinct species to gene pool technologies (21). Pan-genomes offer a new way to use rare genes or genomes to improve breeding programs (22). With the emergence of next-generation CRISPR/Cas technologies—such as prime editing, base editing, and homolog-specific editing—the concept of genome editing has evolved into a powerful and organized approach for improving crop yields. In addition, the efficiency of modified Cas orthologs (e.g., Cas9, Cas12, Cas13, and Cas14) has also been increased. Today, CRISPR/Cas systems are widely used in agricultural research and work well for significant crops to develop tolerance to biotic and abiotic stresses (23). Agriculture is moving toward automation or digitalization using high-throughput phenotyping technology and big data analysis tools such as artificial intelligence (AI) and machine learning (ML) (24). Crop improvements can be accelerated by combining genomic and phenomics technologies with rapid breeding, allowing rapid genetic analysis. Additionally, the integration of joint ventures can pave the way for producing crops that are resilient to climate change and contribute to world food security.

14.4 GLOBAL CLIMATE CHANGE AND FOOD SECURITY SCENARIO

The biggest challenge facing the population remains food security. The goal of a world without hunger can only be achieved by developing productive agriculture (25). More than 815 million people face chronic hunger, and nearly 2 billion people face malnutrition. New data show that malnutrition has increased in developing countries in West Asia and Africa since 2014. The severe acute respiratory syndrome coronavirus 2 (SARS-CoV-2) outbreak has worsened the situation by negatively affecting the food chain (26), and the impact on global food security remains unclear.

Food security fundamentally depends on economic development, which plays a crucial role in addressing chronic hunger and poverty, particularly for vulnerable populations and developing countries (27). However, this will not be enough to reduce hunger and malnutrition, as many factors affect global food security. The two biggest challenges to food security are population growth and climate change. The world population is growing rapidly, demand for food is increasing, and dependence on other resources, such as agricultural land, is increasing (28). It is estimated that the population will exceed 8 billion by the end of 2030 and 9 billion by the end of 2050 (29). Climate conditions are changing rapidly, causing biotic and abiotic stress and causing loss of agricultural lands (30). Crop rotation under various climate stresses is a challenging goal for breeding because the future forecast and impact of climate change cannot be predicted (31). The problem is further exacerbated as temperature and annual precipitation changes affect crop growth and increase crop resilience. The yield of staple crops such as rice, wheat, and maize has been reported to decrease by 10–25% at all temperatures (32). Although the main staple foods that account for two-thirds of the calories humans consume are wheat, rice, corn, and soybeans, their increased production in recent years has been reported to be low. At the same time, the population is growing faster than previously estimated. According to the latest estimate, there will be approximately 9.9 billion people by 2050 (33). Population growth and climate trends are expected to worsen crop yields, widening the gap between food production and demand. These issues will significantly impact the future development of global food security.

14.4.1 Plant Breeding as The Redemptor

About 10,000 years ago, plant breeding became an important concept in agriculture. It involves the continued use of wild relatives for the needs of many generations to maximize yield. After a century of continuous research, scientists discovered a genomic region that later identified genes controlling plant agricultural diseases. To meet the increasing demand for food in the 1960s, the Green Revolution increased the yields of many important crops, such as wheat and rice. For this reason, attacks of different insects and diseases are frequent, and plants cannot withstand heat, drought, salinity, and flood (34).

14.5 INNOVATIVE MOLECULAR BREEDING METHODS TO DEVELOP CLIMATE-READY CROPS

Crop breeders now have tremendous potential to use genomics-enabled technology to improve crop yield and growth in response to environmental changes. This technology uses DNA-based markers to select the best hybrids through marker selection services (Table 14.1). Wild or cultivated varieties with biotic/abiotic stress tolerance or agronomic characteristics should be selected for the success of genome-assisted breeding. Aggressive selection to develop climate-adapted crops should be based on genotyping using specific markers to link phenotypes. High-density DNA markers of commercially essential crops can be used for DNA fingerprinting to correlate plant phenology with crop physiology and discover phenotypes with specific traits (35).

Many studies have shown that combining GWAS with next-generation sequencing technology improves map resolution, precise locus identification, and QTL evidence/genes/alleles. For example, GWAS found 90 social markers related to abiotic stress, grain yield, and other agronomic traits in drought-stressed synthetic hexaploid rice (36). Additionally, integration with MAS facilitates the selection of the best QTL response, enabling the process of genomic selection of the best varieties in breeding programs. GWAS also helped identify the relationships between date fruit traits and various variables such as fruit firmness and fruit size. This study identified 21 candidate genes for genetic selection and breeding to improve jujube plants (37). Broadly, cultivation refers to the changes humans make to plants due to their use; this could be breeding using molecular tools or the introduction of deliberately unmodified agriculture. Plants selected on the basis of natural variation present within a species are referred to as plants selected based on observed phenotypic variation. In controlled mating plant breeding, a plant is chosen for the desired genetic recombination from more than one parent. Finally, selecting specific genes or markers and using molecular techniques to monitor intragenic diversity are important components of plant-based integrated monitoring.

14.5.1 SPEED BREEDING

Food security is seriously threatened by the increasing global population and ongoing environmental changes (70). The long planting cycle is one reason for the slow progress of current planting programs. One solution to the problem is using speed breeding methods that shorten generations, support multiple crop generations yearly, and improve agronomy. Australian scientists first used the term ‘accelerated

TABLE 14.1
Specific genes improved/edited through molecular approaches to develop climate-ready crops

Crop	Study/Technique Employed	Key Findings	References
Wheat	Marker-assisted gene stacking	Molecular marker-assisted gene stacking for disease resistance and quality genes in the dwarf mutant of an elite common wheat cultivar Xiaoyan22	(38)
	Expanded base editing	Demonstrated expanded base editing in rice and wheat using a Cas9-adenosine deaminase fusion	(39)
	CRISPR/Cas9 system	Simultaneous editing of three homoeoalleles in hexaploid bread wheat confers heritable resistance to powdery mildew	(40)
	Fluorescence in situ hybridization (FISH)	BAC-FISH in wheat identifies chromosome landmarks consisting of different types of transposable elements	(41)
Rice	Molecular breeding strategy	Discusses molecular breeding strategies and challenges for improving blast disease resistance in rice	(42)
	qRT-PCR, DGE analysis	ZmPIF1 enhances drought tolerance by inducing stomatal closure and improves grain yield	(43)
	TALENs	Discussed the use of TALENs as customizable molecular DNA scissors for genome engineering in plants	(44)
	CRISPR/Cas9-targeted mutagenesis	Enhanced rice blast resistance by CRISPR/Cas9-targeted mutagenesis of the ERF transcription factor gene OsERF922	(45)
	Molecular breeding, phylogenetic analysis, etc.	Variation in NRT1.1B contributes to nitrate-use divergence between rice subspecies	(46)
	Mutagenesis	Natural variation at the DEP1 locus enhances grain yield in rice	(47)
	CRISPR/Cas9 system	The CRISPR/Cas9 system produces specific and homozygous targeted gene editing in rice in one generation	(48)
	Marker-assisted breeding	Development of submergence-tolerant, bacterial blight-resistant, and high-yielding near isogenic lines of popular variety, 'Swarna' through marker-assisted breeding approach	(49)

	CRISPR/Cas9 system: *Gn1a*, *DEP1*, *GS3*, and *IPA1*	Reassessed the function of yield-related genes *Gn1a*, *DEP1*, *GS3*, and *IPA1* in rice using a CRISPR/Cas9 system	(50)
	TALEN-based gene editing	Demonstrated high-efficiency TALEN-based gene editing producing disease-resistant rice	(51)
	Genome editing	Achieved broad-spectrum resistance to bacterial blight in rice using genome editing	(52)
	TALEN technology	Created fragrant rice by targeted knockout of the *OsBADH2* gene using TALEN technology	(53)
	QTL editing	QTL editing conferred opposing yield performance in different rice varieties	(54)
	CRISPR/Cas9 system/ OsNramp5 gene knockout	Knockout of OsNramp5 using the CRISPR/Cas9 system produces low Cd-accumulating indica rice without compromising yield	(55)
	Marker-assisted breeding/ QTLs for drought, salinity, and submergence tolerance were studied	Pyramiding QTLs controlling tolerance against drought, salinity, and submergence in rice through marker-assisted breeding	(56)
	Phylogenetic and expression analysis/NPR1-like gene family	Identified and analyzed the phylogenetic and expression patterns of the NPR1-like gene family in *Persea americana*	(57)
Chickpea	SNP array development	Developed and evaluated a high-density SNP array for genetic mapping and breeding applications in chickpea	(58)
	QTL-based comparative genome mapping and transcript profiling	Identified a seed weight-regulating candidate gene through a comprehensive QTL-based comparative genome mapping and transcript profiling approach	(59)
	Whole-genome re-sequencing	Revealed genome-wide variations among parental lines of 16 mapping populations in chickpea	(60)

(continued)

TABLE 14.1 (Continued)
Specific genes improved/edited through molecular approaches to develop climate-ready crops

Crop	Study/Technique Employed	Key Findings	References
	Phosphorylation of RAV1 transcription factor by SnRK2 kinases: ABI3, ABI4, ABI5 were studied	Phosphorylation of RAV1 by SnRK2 kinases regulates expression of ABI3, ABI4, and ABI5 during seed germination and early seedling development	(61)
	CRISPR/Cas9 system	High-efficiency genome editing in *Arabidopsis* using YAO promoter-driven CRISPR/Cas9 system	(62)
Arabidopsis	Phosphorylation of RAV1 transcription factor by SnRK2 kinases ABI3, ABI4, ABI5	Phosphorylation of RAV1 by SnRK2 kinases regulates expression of ABI3, ABI4, and ABI5 during seed germination and early seedling development	(61)
	Marker-assisted breeding	Development of diagnostic molecular markers for marker-assisted breeding against bacterial wilt in tomato	(63)
Soybean	Molecular breeding	Provides an update on molecular breeding strategies to overcome biotic stresses in soybean	(64)
	CRISPR/Cas9-mediated targeted mutagenesis	Altered plant height and internode length in soybean through CRISPR/Cas9-mediated targeted mutagenesis of GmLHY genes	(65)
	CRISPR/Cas9-mediated genome editing	Described a method for the production and screening of CRISPR/Cas9-mediated non-transgenic mutant plants	(66)
	CRISPR/Cas9-mediated genome editing	Demonstrated successful genome editing in soybean hairy roots using CRISPR/Cas9 technology	(67)
	Direct stacking of sequence-specific nuclease-induced mutations	Produced high oleic and low linolenic soybean oil by stacking sequence-specific nuclease-induced mutations	(68)
	Targeted mutagenesis	Improved soybean oil quality by targeted mutagenesis of the fatty acid desaturase 2 gene family	(69)

growth' in 2003, inspired by experiments conducted by NASA during space operations. This study provides an overview of the processes, available resources, and constraints associated with preterm birth within this specific context. Rapid integration with other breeding methods can shrink the cycle and improve selection for specific traits. Speed breeding to ensure food security is a new and promising method that can accelerate the development and launch of new crops. Speed breeding (SB) uses control methods to promote fast and rapid growth from the growing stage to the maturity stage in high-yielding crops (71). Selection methods (such as pure lines, repeat, batch, group, and strain selection) are not considered suitable for use in SB as they involve a long and drawn-out selection and breeding process. These traditional methods are incompatible with SB's growing needs and reality on modern farms. On the other hand, it is worth noting that some breeding methods follow SB standards, including single seed (SSD) and single plant selection (SPS). These technologies are suitable because they require less farmland and less work for the first few generations. This simple method allows researchers to raise good offspring under controlled growth and HDP conditions in small nurseries (excluding surrounding greenhouse environments). In this process, the reproductive cycle is accelerated, and the allocation of resources is optimized to increase crop yields. This method can also be used in the field (72).

Speed breeding is an advanced technique that uses optimum temperature and light to accelerate the growth cycle of crop plants. This method is efficient as many case studies have proved. For example, a case study at the University of Queensland and the University of Sydney researched wheat breeding using speed breeding, accelerating wheat varietal development. Typically, wheat varieties require 7–8 years to develop; however, this method reduces the development time to just 2–3 years. As a result of speed breeding, faster selection of stress-tolerant and high-yielding varieties was made possible. Another case study was for tomatoes and peas, where researchers experimented with speed breeding, resulting in faster selection of disease-resistant traits, varieties with good flavor, and larger fruit sizes. Tomatoes gave six generations per year, and peas gave four generations per year through speed breeding.

14.5.2 Marker-assisted Breeding

This method facilitates breeding using molecular markers associated with desired traits (6). This strategy accelerates our understanding of beneficial genes related to climate protection, such as disease resistance and water efficiency. This aims to shorten the breeding cycle and accelerate the growth of plants by using controlled environments such as growth rooms with improved lighting, quality, and temperature (73). Because of the rapid changes in this approach, growers can better plan for climate change. This technique allows breeders to test for climate-adaptive traits more successfully because of the rapid generation turnover (Figure 14.1).

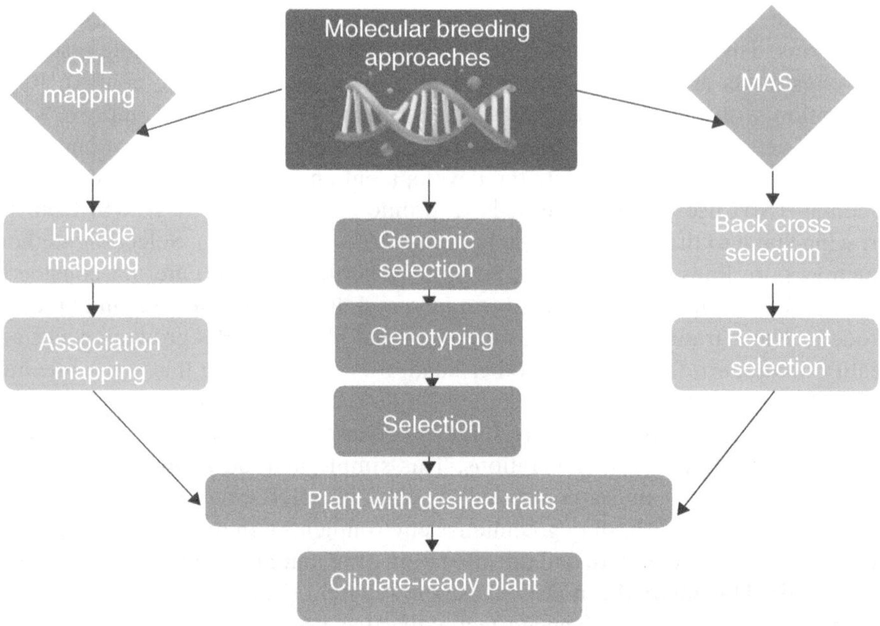

FIGURE 14.1 Illustration of molecular breeding approaches to develop climate-resilient plants.

14.6 MOLECULAR MECHANISMS UNDERLYING GENETIC ADAPTATIONS IN CROPS FOR CLIMATE RESILIENCE

14.6.1 Introgression of Traits

To improve crops' ability to cope with climate change, agriculture faces a critical challenge that demands creative solutions. One standard method involves transferring beneficial traits and genes from wild plants to cultivated crops (74). This approach takes advantage of the genetic variety found in wild plants, which have adapted to different environments over time. Doing so helps crops become more resilient to the difficulties brought about by changing climates (75).

Desirable traits like drought tolerance, pest resistance, and higher yield can be added to crops by transferring genes through introgression. Wild crop relatives often have genetic differences that cultivated varieties lack and provide valuable adaptive qualities, making them better suited to changing environments. By introducing these traits into cultivated crops, breeders can expand the genetic diversity of crop populations, ensuring their long-term sustainability and food security (74). This method helps create crops resilient to climate change and preserves genetic diversity in agriculture. Utilizing introgression from wild relatives holds promise for improving crops and building resilient agricultural systems to combat the effects of climate change (76).

14.6.2 Omics-facilitated Improvement

By tapping into the genetic reservoirs of wild relatives and underutilized species, novel approaches in crop improvement leverage omics technologies to bolster crop resilience (77). Within the genetic makeup and functional attributes of crops, omics techniques encompass a range of methodologies, such as genomics, transcriptomics, proteomics, and metabolomics, which provide comprehensive insights. Researchers can identify and harness specific genes and pathways by employing these advanced tools and molecular markers associated with desirable traits, including stress tolerance, disease resistance, and enhanced productivity (78). Due to their rich diversity and unique adaptive traits, wild relatives and underutilized species of crops serve as invaluable genetic resources. By identifying novel alleles and genomic regions that confer resilience to environmental challenges through omics-guided analyses, scientists can explore the untapped potential of these genetic reservoirs (79). Researchers can introduce beneficial traits into cultivated crops, enhancing their adaptability and sustainability in the face of climate change and other stressors (80). Integrating genetic information from wild relatives and underutilized species into breeding programs broadens the genetic base of cultivated varieties. It contributes to developing climate-smart crop cultivars capable of withstanding fluctuating environmental conditions while ensuring global food security.

14.6.3 Stress Tolerance Enhancement

To enable stress tolerance, scientists use two main methods: changing genes directly and traditional breeding. These methods make crops stronger against stress, such as harsh weather. Scientists can tweak specific genes related to stress tolerance (81) with gene editing, such as CRISPR/Cas9. This helps create crops that can handle harsh conditions better. Traditional breeding is also crucial. Gradually making plants better at handling stress, it mixes different plants to obtain new ones with valuable traits. These methods work well together. Scientists can speed up making stress-resistant crops by combining gene editing with traditional breeding (82). This way, they can create crops that can survive specific environmental challenges enabling farmers to adapt to climate changes. Using both high-tech gene editing and older breeding methods is key to making sure proper achievements are made (83).

14.6.4 Gene Editing for Targeted Adaption

Gene editing technologies such as CRISPR-Cas9 gene editing for targeted adaptation have revolutionized crop improvement by enabling precise modifications in plant genomes. These tools allow researchers to enhance nutrient use efficiency by targeting and modifying specific genes like drought tolerance and disease resistance. Gene editing provides a faster and more accurate means to develop climate-resilient crops. Researchers have used CRISPR to enhance crop tolerance by editing genes involved in stomata regulation. This approach minimizes unintended genetic changes, ensuring crops remain safe and productive while accelerating breeding.

14.6.5 Epigenetics and Crop Resilience

In regulating the plant response to environmental stress, epigenetic modifications such as DNA methylation and histone acetylation play a crucial role. For improving crop resilience, these heritable changes in gene expression do not involve alterations in the DNA sequence itself, making them a promising tool. For example, epigenetic changes have shown that rice plants exposed to drought conditions can pass on enhanced drought tolerance to their offspring. By harnessing these mechanisms, researchers aim to develop crops capable of adapting to fluctuating environmental conditions without requiring extensive genetic modification. The use of epigenetics in breeding programs represents a novel avenue to enhance the adaptability and sustainability of agriculture systems.

14.6.6 Integrating Multiple Strategies

Researchers are increasingly adopting integrated approaches that combine traditional breeding, molecular techniques, and ecological methods to address the complex challenges posed by climate change. Enabling the development of crops that are not only resilient but also high-yielding and resource-efficient, for example, by combining omics technologies with gene editing and microbiome engineering, can provide synergistic benefits. Scientists can create holistic solutions that ensure agriculture sustainability and global food security in the face of climate change.

14.7 PHYSIOLOGICAL RESPONSES FOR CLIMATE-READY CROPS

14.7.1 Increased Heat and Drought Tolerance

Increased temperatures and extended drought conditions have led to the need for climate-ready crops using advanced mechanisms. Enhancing water use efficiency enables plants to optimize their water uptake and utilization, thereby reducing the impact of drought stress (84). By increasing the production of heat shock proteins, these crops enhance their resilience, as these proteins act as molecular chaperones that prevent protein misfolding and help maintain cellular homeostasis under high-temperature conditions. This protein protection is essential for growth and enables plants to withstand heat stress. (85). Climate-ready crops may exhibit altered stomatal behavior as part of their adaptive response. Stomata, the microscopic pores on plant surfaces, are pivotal in regulating gas exchange and transpiration, ensuring efficient carbon dioxide uptake for photosynthesis through genetic mechanisms (86). These crops adjust stomatal openings to minimize water loss while facilitating gas exchange. Climate-ready crops can bloom in harsh environmental conditions by combining these physiological adaptations, offering hope for sustainable agriculture in the face of climate change. Specific genetic mechanisms underlying increased heat and drought tolerance involve regulating plants' key genes and molecular pathways. These mechanisms include the following.

14.7.2 Expression of Heat Shock Proteins (HSPs)

Expression of HSPs facilitated proper protein folding and prevents protein denaturation under high temperatures. Heat stress induces the expression of heat shock proteins, which act as molecular chaperones (87).

14.7.3 Activation of Transcription Factors

Activation of transcription factors enhances heat and drought tolerance. Transcription factors such as DREB (dehydration-responsive element-binding) and HSF (heat shock factor) regulate the expression of stress-responsive genes (88).

14.7.4 Synthesis of Osmo-protectants

Proline and glycine betaine help maintain cellular osmotic balance and protect proteins and cellular structures. Aquaporin regulation from damage caused by heat and drought stress genes is involved in synthesizing osmo-protectants. Genes encoding aquaporins control water movement across cellular membranes, influencing water uptake and transpiration rates (89).

14.7.5 Stomatal Regulation

Genes regulate stomatal conductance, influencing water loss and drought tolerance involved in stomatal development and function, such as those encoding stomatal density and aperture.

14.7.6 Altered Germination and Growth

In response to novel environmental pressures induced by climate change, climate-ready crops exhibit altered germination and growth patterns, such as germination success, plant viability, and growth rates. Under climatic conditions, breeders must select traits that promote optimal growth and development. In unpredictable environments, climate-resilient crops may exhibit enhanced germination rates, ensuring successful establishment under fluctuating temperatures and moisture conditions. Enabling plants to thrive in adverse conditions may help them through improved stress tolerance mechanisms and maintain viability throughout their growth stages (90). Climate-ready crops involve the regulation of key genes, molecular pathways, and genetic mechanisms underpinning altered germination and growth. During seed germination and early seedling development, the expression of genes encoding HSPs plays a vital role in maintaining cellular stability under stress. These proteins ensure proper protein folding, prevent cellular damage, and stabilize essential structures when exposed to high temperatures. In addition, HSP-mediated stress responses significantly influence seed germination success and early plant growth, particularly under drought conditions. Transcription factors such as DREB and ABRE (abscisic acid-responsive element) regulate the expression of genes (91). Understanding these genetic mechanisms enables breeders to harness molecular tools for developing climate-resilient crop varieties with improved germination and growth characteristics.

14.7.7 Optimization of Photosynthesis

To enhance carbon assimilation, transcription factors such as DREB (and ABRE regulate the expression of genes (91). Transcription factors and energy generation ensure sustained productivity in varying environmental conditions (92). In photosynthetic mechanisms, climate-ready crops undergo modifications that increase the activity of enzymes like phosphoenolpyruvate carboxylase (PEPC). Genetic modifications targeting photorespiration pathways can minimize energy losses and improve carbon assimilation (93). In climate-resilient crops, strategies such as introducing novel genetic diversity and manipulating expression levels of genes associated with photosystem assembly and chlorophyll biosynthesis improve photosynthetic efficiency. In ensuring sustainable agricultural productivity in the face of climate change challenges, genetic mechanisms play a vital role in the adaptation of crops to changing environmental conditions (94).

14.7.8 Enhanced Stress Resilience

Climate-ready crops develop enhanced stress resilience through physiological adaptations to combat stressors like heat, drought, and salinity. Increased accumulation of osmolytes such as proline and glycine betaine help to maintain cellular water balance and stabilize proteins under stress. By reducing oxidative damage and enhancing stress tolerance, the upregulation of antioxidant enzymes such as superoxide dismutase (SOD) and catalase (CAT) scavenges reactive oxygen species (ROS), thereby preventing electrolyte leakage and maintaining cellular integrity under stress conditions. Alterations in membrane composition and structure improve membrane stability (95). Specific genetic mechanisms underlying enhanced stress resilience involve the activation of stress-responsive genes, which regulate stress-responsive pathways and transcriptionally activate stress-related genes such as DREB and AREB/ABF (ABA-responsive element-binding factor) (96). The introgression of stress-tolerant genes from wild relatives facilitated by next-generation breeding tools enhances stress resilience. These genetic mechanisms, capable of thriving under adverse environmental conditions and physiological adaptations, contribute to developing climate-ready crops.

14.8 SWOT Analysis of Molecular Breeding Approaches to Develop Climate-Ready Crops

SWOT analysis is a tactical approach expressed in the main four classes: Strengths, weaknesses, opportunities, and threats. Within the context of molecular breeding for CRC, a SWOT analysis is extremely valuable for a better understanding of the benefits and obstacles of molecular breeding strategy.

14.8.1 Strengths

Molecular breeding is a powerful technique for targeting genes linked with traits of interest (97). MB significantly limits the required time for trait identification and the improvement of a variety with the help of advanced DNA markers and genomic data. Trait introgression into crop genomes is precisely achieved by targeting modern breeding, which is effective with this focused method (98). Scientists can improve multiple traits simultaneously by knocking in and knocking out genes using MB and developing climate-resilient crops. MB allows scientists to make the specific required climate-ready crops.

14.8.2 Weaknesses

MB has some challenges in agricultural systems owing to limited resources, which may pose serious challenges due to the requirement for advanced equipment, facilities, and expertise (99). Using MB techniques may be hampered by genetically modified organisms such as GM maize and the regulatory environment around genome-editing technology (100). MB technique implementation requires significant expense due to the high cost of the requirements, including equipment. The achievable trait variability and genetic diversity through MB may be restricted by the accessibility of genomic information and the availability of germplasm supplies (101).

14.8.3 Opportunities

MB has an excellent perspective for evolving crop plants resistant against various climatic stressors, such as abiotic (heat, drought, etc.) and biotic (pests, diseases, etc.) (102). Combined energies through research projects and integrated associations may encourage the innovation and development of current MB approaches (103), increasing the scope of agricultural improvement programs such as international wheat and rice breeding. MB deals with the potential for climate-ready crops to achieve consumer mandates for vigorous, sustainable, fully-fledged nutritive foods. Improving resilient varieties and farming climate-ready crop varieties can reduce the adverse impacts of chemicals on agriculture and the overall natural environment (104).

14.8.4 Threats

MB techniques have intensified socioeconomic disparities that exist in areas with inadequate access to capital and equipment (105). Owing to the rapid advancement of MB technologies, there may be limitations in assessing the long-term effects on overall crop genomes and ecosystems (106). Access to genetic resources and intrusions over intellectual property rights may slow down the proper dispersal of reimbursements owing to MB discoveries. In conclusion, SWOT analysis of MB techniques gives a complete understanding of the strengths, weaknesses, opportunities, and challenges of producing climate-ready crops. Therefore, considering the background and results from the SWOT analysis, plant breeders can augment the exploitation of molecular breeding techniques for sustainable agriculture and climate resilience programs.

14.9 CONCLUSION

The climate is changing all over the globe rapidly and affecting crop production, leading to food insecurity. There is an urgent need to improve crops to produce enough food for the growing population and to adapt to the changing and threatening climatic conditions. Conventional breeding plays a critical role in achieving goals, but researchers have to work at a slow pace. Therefore, molecular breeding is a critical tool to develop climate-ready crops with precise targets at a much faster speed. Molecular breeding has validated its revolutionary potential to cope with the prevailing threats of climate change and overall food insecurity. Molecular breeding can potentially improve climate-resilient farming systems coupled with genetic engineering, molecular biology, and advanced biotechnology. Therefore, molecular breeding is a very effective method to improve crop output and safeguard resilience in varying climatic conditions. Breeders have already developed important climate-ready crops such as wheat, rice, maize, and tomato against biotic and abiotic stresses by utilizing modern breeding techniques. Furthermore, SWOT analysis of molecular breeding provides a comprehensive approach to utilize these modern techniques to develop climate-ready crops to meet global challenges.

REFERENCES

1. Azeem A, Ul-Allah S, Khan S, Dullu MUD, Azeem S, Zaman MS, et al. Enhancing crop resilience in the face of a changing climate: Strategies for sustainable agricultural production. *Water-Soil-Plant-Animal Nexus in the Era of Climate Change*: IGI Global; 2024. p. 114–36.
2. Myers SS, Smith MR, Guth S, Golden CD, Vaitla B, Mueller ND, et al. Climate change and global food systems: Potential impacts on food security and undernutrition. *Annual Review of Public Health*. 2017;38:259–77.
3. Kalenska S. Food security and innovative solutions in crop production. *Plant & Soil Science*. 2022;13(2):14-26.
4. Kumar L, Chhogyel N, Gopalakrishnan T, Hasan MK, Jayasinghe SL, Kariyawasam CS, et al. Climate change and future of agri-food production. *Future Foods*: Elsevier; 2022. p. 49–79.
5. Iqbal Z, Azeem A, Ul-Allah S, Sher A, Ahmad MQ, Haider B, et al. Advances in biotechnological tools and their impact on global climate change and food security. *Climate-Resilient Agriculture, Vol 2: Agro-Biotechnological Advancement for Crop Production*: Springer; 2023. p. 667–76.
6. Ul-Allah S, Azeem A, Sher A, Ijaz M, Sattar A, Naeem M, et al. Plant breeding approaches in developing stress tolerance. *Biology and Biotechnology of Environmental Stress Tolerance in Plants*: Apple Academic Press; 2023. p. 167–86.
7. Singh RK, Prasad A, Muthamilarasan M, Parida SK, Prasad M. Breeding and biotechnological interventions for trait improvement: Status and prospects. *Planta*. 2020;252:1–18.
8. Slatko BE, Gardner AF, Ausubel FM. Overview of next-generation sequencing technologies. *Current Protocols in Molecular Biology*. 2018;122(1):e59.
9. Prohens J, Gramazio P, Plazas M, Dempewolf H, Kilian B, Díez MJ, et al. Introgressiomics: A new approach for using crop wild relatives in breeding for adaptation to climate change. *Euphytica*. 2017;213:1–19.

10. Raza A, Chen H, Zhang C, Zhuang Y, Sharif Y, Cai T, et al. Designing future peanut: The power of genomics-assisted breeding. *Theoretical and Applied Genetics*. 2024;137(3):1–30.
11. Azeem A, Haq MZU, Rasheed HU, Uddin S, Dullu MUD, Azeem S, et al. Bridging the gap between climate change and plant biology: Enhancing resilience through genetic diversity and agroecological practices. *Water-Soil-Plant-Animal Nexus in the Era of Climate Change*: IGI Global; 2024. p. 173–93.
12. Gurl E. *SWOT Analysis: A Theoretical Review*; 2017.
13. Khan MHU, Wang S, Wang J, Ahmar S, Saeed S, Khan SU, et al. Applications of artificial intelligence in climate-resilient smart-crop breeding. *International Journal of Molecular Sciences*. 2022;23(19):11156.
14. Kumar A, Nayak A, Das B, Panigrahi N, Dasgupta P, Mohanty S, et al. Effects of water deficit stress on agronomic and physiological responses of rice and greenhouse gas emission from rice soil under elevated atmospheric CO2. *Science of the Total Environment*. 2019;650:2032–50.
15. Sarkar S, Saha S, Ghosh S, Paul SK, Dey S, Moulick D, et al. Abiotic stress sensitivity and adaptation in field crops. *Climate-Resilient Agriculture, Vol 2: Agro-Biotechnological Advancement for Crop Production*: Springer; 2023. p. 319–62.
16. Altieri MA, Nicholls CI, Henao A, Lana MA. Agroecology and the design of climate change-resilient farming systems. *Agronomy for Sustainable Development*. 2015;35(3):869–90.
17. Mrabet R. Sustainable agriculture for food and nutritional security. *Sustainable Agriculture and the Environment*: Elsevier; 2023. p. 25–90.
18. Majeed Y, Khan MU, Waseem M, Zahid U, Mahmood F, Majeed F, et al. Renewable energy as an alternative source for energy management in agriculture. *Energy Reports*. 2023;10:344–59.
19. Gull A, Lone AA, Wani NUI. Biotic and abiotic stresses in plants. *Abiotic and Biotic Stress in Plants*. 2019:1–19. DOI: 10.5772/intechopen.85832
20. Bashir T, Ul Haq SA, Masoom S, Ibdah M, Husaini AM. Quality trait improvement in horticultural crops: OMICS and modern biotechnological approaches. *Molecular Biology Reports*. 2023;50(10):8729–42.
21. Sun C, Hu H, Cheng Y, Yang X, Qiao Q, Wang C, et al. Genomics-assisted breeding: The next-generation wheat breeding era. *Plant Breeding*. 2023;142(3):259–68.
22. Bayer PE, Golicz AA, Scheben A, Batley J, Edwards D. Plant pan-genomes are the new reference. *Nature Plants*. 2020;6(8):914–20.
23. Ghoshal B. A birds-eye-view on CRISPR-Cas system in agriculture. *The Nucleus*. 2024;67(1):89–96.
24. Tripodi P, Nicastro N, Pane C, Cammarano D. Digital applications and artificial intelligence in agriculture toward next-generation plant phenotyping. *Crop and Pasture Science*. 2022;74(6):597–614.
25. Tanima D, Sharma NR, Bohn R , Ergunova O, Ryhtik D, Makarenko E, et al., editors. The parallels of food self-sufficiency and hunger in light of sustainable agriculture: A case of the BRICS countries. *E3S Web of Conferences*; 2024: EDP Sciences.
26. Barcaccia G, D'Agostino V, Zotti A, Cozzi B. Impact of the SARS-CoV-2 on the Italian agri-food sector: An analysis of the quarter of pandemic lockdown and clues for a socio-economic and territorial restart. *Sustainability*. 2020;12(14):5651.
27. Pollard CM, Booth S. Food insecurity and hunger in rich countries—it is time for action against inequality. *International Journal of Environmental Research and Public Health*. 2019;16(10):1804.

28. Hemathilake D, Gunathilake D. Agricultural productivity and food supply to meet increased demands. *Future Foods*: Elsevier; 2022. p. 539–53.
29. Ray DK, Foley JA. Increasing global crop harvest frequency: Recent trends and future directions. *Environmental Research Letters*. 2013;8(4):044041.
30. Oshunsanya SO, Nwosu NJ, Li Y. Abiotic stress in agricultural crops under climatic conditions. *Sustainable Agriculture, Forest and Environmental Management*. 2019:71–100.
31. Mylonas I, Stavrakoudis D, Katsantonis D, Korpetis E. Better farming practices to combat climate change. *Climate change and food security with emphasis on wheat*: Elsevier; 2020. p. 1–29.
32. Deutsch CA, Tewksbury JJ, Tigchelaar M, Battisti DS, Merrill SC, Huey RB, et al. Increase in crop losses to insect pests in a warming climate. *Science*. 2018;361(6405):916–9.
33. Siegel FR, Siegel FR. Economic realities in 2020 populastions: What do they portend for 2050? 2100? *The earth's human carrying capacity: Limitations assessed, solutions proposed*. 2021:135–45.
34. Parashar A, Sharma JD. Sustainable crop production in stress condition: Strategies and management. *Journal of Rural Advancement*. 2024;12(1):75–83.
35. Afzal M, Hindawi SES, Alghamdi SS, Migdadi HH, Khan MA, Hasnain MU, et al. Potential breeding strategies for improving salt tolerance in crop plants. *Journal of Plant Growth Regulation*. 2023;42(6):3365–87.
36. Reddy SS, Saini DK, Singh GM, Sharma S, Mishra VK, Joshi AK. Genome-wide association mapping of genomic regions associated with drought stress tolerance at seedling and reproductive stages in bread wheat. *Frontiers in Plant Science*. 2023;14:1166439.
37. Pan Y, Luo Y, Bao J, Wu C, Wang J, Liu M, et al. Screening candidate genes for fruit size based on QTL-seq in Chinese jujube. *Frontiers in Plant Science*. 2024;15:1361771.
38. Zheng W, Li S, Liu Z, Zhou Q, Feng Y, Chai S. Molecular marker assisted gene stacking for disease resistance and quality genes in the dwarf mutant of an elite common wheat cultivar Xiaoyan22. *BMC Genetics*. 2020 Dec;21:1–8.
39. Li C, Zong Y, Wang Y, Jin S, Zhang D, Song Q, Zhang R, Gao C. Expanded base editing in rice and wheat using a Cas9-adenosine deaminase fusion. *Genome Biology*. 2018 Dec;19:1–9.
40. Wang Y, Cheng X, Shan Q, Zhang Y, Liu J, Gao C, Qiu JL. Simultaneous editing of three homoeoalleles in hexaploid bread wheat confers heritable resistance to powdery mildew. *Nature Biotechnology*. 2014 Sep;32(9):947–51.
41. Zhang P, Li W, Fellers J, Friebe B, Gill BS. BAC-FISH in wheat identifies chromosome landmarks consisting of different types of transposable elements. *Chromosoma*. 2004 Mar;112:288–99.
42. Ashkani S, Rafii MY, Shabanimofrad M, Miah G, Sahebi M, Azizi P, Tanweer FA, Akhtar MS, Nasehi A. Molecular breeding strategy and challenges towards improvement of blast disease resistance in rice crop. *Frontiers in Plant Science*. 2015 Nov 16;6:886.
43. Gao Y, Wu M, Zhang M, Jiang W, Ren X, Liang E, Zhang D, Zhang C, Xiao N, Li Y, Dai Y. A maize phytochrome-interacting factors protein ZmPIF1 enhances drought tolerance by inducing stomatal closure and improves grain yield in Oryza sativa. *Plant Biotechnology Journal*. 2018 Jul;16(7):1375–87.
44. Chen K, Gao C. TALENs: Customizable molecular DNA scissors for genome engineering of plants. *Journal of Genetics and Genomics*. 2013 Jun 20;40(6):271–9.

45. Wang F, Wang C, Liu P, Lei C, Hao W, Gao Y, Liu YG, Zhao K. Enhanced rice blast resistance by CRISPR/Cas9-targeted mutagenesis of the ERF transcription factor gene OsERF922. *PloS One*. 2016 Apr 26;11(4):e0154027.
46. Hu B, Wang W, Ou S, Tang J, Li H, Che R, Zhang Z, Chai X, Wang H, Wang Y, Liang C. Variation in NRT1. 1B contributes to nitrate-use divergence between rice subspecies. *Nature Genetics*. 2015 Jul;47(7):834–8.
47. Huang X, Qian Q, Liu Z, Sun H, He S, Luo D, Xia G, Chu C, Li J, Fu X. Natural variation at the DEP1 locus enhances grain yield in rice. *Nature Genetics*. 2009 Apr;41(4):494–7.
48. Zhang H, Zhang J, Wei P, Zhang B, Gou F, Feng Z, Mao Y, Yang L, Zhang H, Xu N, Zhu JK. The CRISPR/C as9 system produces specific and homozygous targeted gene editing in rice in one generation. *Plant Biotechnology Journal*. 2014 Aug;12(6):797–807.
49. Mohapatra S, Panda AK, Bastia AK, Mukherjee AK, Sanghamitra P, Meher J, Mohanty SP, Pradhan SK. Development of submergence-tolerant, bacterial blight-resistant, and high-yielding near isogenic lines of popular variety,'Swarna'through marker-assisted breeding approach. *Frontiers in Plant Science*. 2021 Jul 27;12:672618.
50. Li M, Li X, Zhou Z, Wu P, Fang M, Pan X, Lin Q, Luo W, Wu G, Li H. Reassessment of the four yield-related genes Gn1a, DEP1, GS3, and IPA1 in rice using a CRISPR/Cas9 system. *Frontiers in Plant Science*. 2016 Mar 30;7:377.
51. Li T, Liu B, Spalding MH, Weeks DP, Yang B. High-efficiency TALEN-based gene editing produces disease-resistant rice. *Nature Biotechnology*. 2012 May;30(5):390–2.
52. Oliva R, Ji C, Atienza-Grande G, Huguet-Tapia JC, Perez-Quintero A, Li T, Eom JS, Li C, Nguyen H, Liu B, Auguy F. Broad-spectrum resistance to bacterial blight in rice using genome editing. *Nature Biotechnology*. 2019 Nov;37(11):1344–50.
53. Shan Q, Zhang Y, Chen K, Zhang K, Gao C. Creation of fragrant rice by targeted knockout of the Os BADH 2 gene using TALEN technology. *Plant Biotechnology Journal*. 2015 Aug;13(6):791–800.
54. Shen L, Wang C, Fu Y, Wang J, Liu Q, Zhang X, Yan C, Qian Q, Wang K. QTL editing confers opposing yield performance in different rice varieties. *Journal of Integrative Plant Biology*. 2018 Feb;60(2):89–93.
55. Tang L, Mao B, Li Y, Lv Q, Zhang L, Chen C, He H, Wang W, Zeng X, Shao Y, Pan Y. Knockout of OsNramp5 using the CRISPR/Cas9 system produces low Cd-accumulating indica rice without compromising yield. *Scientific Reports*. 2017 Oct 31;7(1):14438.
56. Muthu V, Abbai R, Nallathambi J, Rahman H, Ramasamy S, Kambale R, Thulasinathan T, Ayyenar B, Muthurajan R. Pyramiding QTLs controlling tolerance against drought, salinity, and submergence in rice through marker assisted breeding. *PloS One*. 2020 Jan 7;15(1):e0227421.
57. Backer R, Mahomed W, Reeksting BJ, Engelbrecht J, Ibarra-Laclette E, van den Berg N. Phylogenetic and expression analysis of the NPR1-like gene family from Persea americana (Mill.). *Frontiers in Plant Science*. 2015 Apr 29;6:300.
58. Roorkiwal M, Jain A, Kale SM, Doddamani D, Chitikineni A, Thudi M, Varshney RK. Development and evaluation of high-density Axiom® Cicer SNP array for high-resolution genetic mapping and breeding applications in chickpea. *Plant Biotechnology Journal*. 2018 Apr;16(4):890–901.
59. Bajaj D, Upadhyaya HD, Khan Y, Das S, Badoni S, Shree T, Kumar V, Tripathi S, Gowda CL, Singh S, Sharma S. A combinatorial approach of comprehensive QTL-based comparative genome mapping and transcript profiling identified a seed weight-regulating candidate gene in chickpea. *Scientific Reports*. 2015 Mar 19;5(1):9264.

60. Thudi M, Khan AW, Kumar V, Gaur PM, Katta K, Garg V, Roorkiwal M, Samineni S, Varshney RK. Whole genome re-sequencing reveals genome-wide variations among parental lines of 16 mapping populations in chickpea (Cicer arietinum L.). *BMC Plant Biology*. 2016 Jan;16:53–64.
61. Feng CZ, Chen Y, Wang C, Kong YH, Wu WH, Chen YF. Arabidopsis RAV 1 transcription factor, phosphorylated by S n RK 2 kinases, regulates the expression of ABI 3, ABI 4, and ABI 5 during seed germination and early seedling development. *The Plant Journal*. 2014 Nov;80(4):654–68.
62. Yan L, Wei S, Wu Y, Hu R, Li H, Yang W, Xie Q. High-efficiency genome editing in Arabidopsis using YAO promoter-driven CRISPR/Cas9 system. *Molecular Plant*. 2015 Dec 7;8(12):1820–3.
63. Abebe AM, Choi J, Kim Y, Oh CS, Yeam I, Nou IS, Lee JM. Development of diagnostic molecular markers for marker-assisted breeding against bacterial wilt in tomato. *Breeding Science*. 2020;70(4):462–73.
64. Tripathi N, Tripathi MK, Tiwari S, Payasi DK. Molecular breeding to overcome biotic stresses in soybean: Update. *Plants*. 2022 Jul 28;11(15):1967.
65. Cheng Q, Dong L, Su T, Li T, Gan Z, Nan H, Lu S, Fang C, Kong L, Li H, Hou Z. CRISPR/Cas9-mediated targeted mutagenesis of GmLHY genes alters plant height and internode length in soybean. *BMC Plant Biology*. 2019 Dec;19:1–1.
66. Chen L, Li W, Katin-Grazzini L, Ding J, Gu X, Li Y, Gu T, Wang R, Lin X, Deng Z, McAvoy RJ. A method for the production and expedient screening of CRISPR/Cas9-mediated non-transgenic mutant plants. *Horticulture Research*. 2018 Dec 1;5.
67. Cai Y, Chen L, Liu X, Sun S, Wu C, Jiang B, Han T, Hou W. CRISPR/Cas9-mediated genome editing in soybean hairy roots. *PLoS One*. 2015 Aug 18;10(8):e0136064.
68. Demorest ZL, Coffman A, Baltes NJ, Stoddard TJ, Clasen BM, Luo S, Retterath A, Yabandith A, Gamo ME, Bissen J, Mathis L. Direct stacking of sequence-specific nuclease-induced mutations to produce high oleic and low linolenic soybean oil. *BMC Plant Biology*. 2016 Dec;16:1–8.
69. Haun W, Coffman A, Clasen BM, Demorest ZL, Lowy A, Ray E, Retterath A, Stoddard T, Juillerat A, Cedrone F, Mathis L. Improved soybean oil quality by targeted mutagenesis of the fatty acid desaturase 2 gene family. *Plant Biotechnology Journal*. 2014 Sep;12(7):934–40.
70. Penuelas J, Coello F, Sardans J. A better use of fertilizers is needed for global food security and environmental sustainability. *Agriculture & Food Security*. 2023;12(1):1–9.
71. Rai NK, Ravika, Yadav R, Jattan M, Karuna, Rai PS, et al. Speed breeding: A budding technique to improve crop plants for drought and salinity tolerance. *Salinity and Drought Tolerance in Plants: Physiological Perspectives*: Springer; 2023. p. 295–313.
72. Sinha D, Maurya AK, Abdi G, Majeed M, Agarwal R, Mukherjee R, et al. Integrated genomic selection for accelerating breeding programs of climate-smart cereals. *Genes*. 2023;14(7):1484.
73. Sajja SB, Mathew A, Pasupuleti J. Speed breeding to accelerate crop improvement. *Digital Agriculture: A Solution for Sustainable Food and Nutritional Security*: Springer; 2024. p. 425–43.
74. Salgotra RK, Chauhan BS. Genetic diversity, conservation, and utilization of plant genetic resources. *Genes*. 2023;14(1):174.
75. Mondal R, Kumar A, Gnanesh BN. Crop germplasm: Current challenges, physiological-molecular perspective, and advance strategies towards development of climate-resilient crops. *Heliyon*. 2023;9(1).

76. Hafeez U, Ali M, Hassan SM, Akram MA, Zafar A. Advances in breeding and engineering climate-resilient crops: A comprehensive review. *International Journal of Research and Advances in Agricultural Sciences*. 2023;2(2):85–99.
77. Visioni A, Basile B, Amri A, Sanchez-Garcia M, Corrado G. Advancing the conservation and utilization of Barley genetic resources: Insights into germplasm management and breeding for sustainable agriculture. *Plants*. 2023;12(18):3186.
78. Reyes VP. Fantastic genes: Where and how to find them? Exploiting rice genetic resources for the improvement of yield, tolerance, and resistance to a wide array of stresses in rice. *Functional & Integrative Genomics*. 2023;23(3):238.
79. Yasmin F. *Investigating the Genetic Basis and Selection of Diverse Plant Specialized Metabolites in Wild Soybean, Glycine soja*: The University of North Carolina at Charlotte; 2023.
80. Shahzad A, Ullah S, Dar AA, Sardar MF, Mehmood T, Tufail MA, et al. Nexus on climate change: Agriculture and possible solution to cope future climate change stresses. *Environmental Science and Pollution Research*. 2021;28:14211–32.
81. Mahrukh, Hussain K, Lone JK, Bhardwaj R, Mushtaq M. Tweaking CRISPR/Cas for developing salt and drought tolerant crop plants. *Advanced Crop Improvement, Volume 2: Case Studies of Economically Important Crops*: Springer; 2023. p. 513–27.
82. Ahmad M. Plant breeding advancements with "CRISPR-Cas" genome editing technologies will assist future food security. *Frontiers in Plant Science*. 2023;14:1133036.
83. Singh MK, Alam MK, Pandey MV, Singh MS, Kumar M. *Advanced technology of horticulture*. Daya Publishing House; 2023.
84. Leotta L, Toscano S, Ferrante A, Romano D, Francini A. New strategies to increase the abiotic stress tolerance in woody ornamental plants in Mediterranean climate. *Plants*. 2023;12(10):2022.
85. Puppala N, Nayak SN, Sanz-Saez A, Chen C, Devi MJ, Nivedita N, et al. Sustaining yield and nutritional quality of peanuts in harsh environments: Physiological and molecular basis of drought and heat stress tolerance. *Frontiers in Genetics*. 2023;14:1121462.
86. Wang Y, Li P, Sun W, Zhang T. Plant cell walls: Emerging targets of stomata engineering to improve photosynthesis and water use efficiency. *New Crops*. 2024;1:100021.
87. Jeyachandran S, Chellapandian H, Park K, Kwak I-S. A review on the involvement of heat shock proteins (extrinsic chaperones) in response to stress conditions in aquatic organisms. *Antioxidants*. 2023;12(7):1444.
88. Prusty A, Panchal A, Singh RK, Prasad M. Major transcription factor families at the nexus of regulating abiotic stress response in millets: A comprehensive review. *Planta*. 2024;259(5):118.
89. Byrt CS, Zhang RY, Magrath I, Chan KX, De Rosa A, McGaughey S. Exploring aquaporin functions during changes in leaf water potential. *Frontiers in Plant Science*. 2023;14:1213454.
90. Zafar S, Afzal H, Ijaz A, Mahmood A, Ayub A, Nayab A, et al. Cotton and drought stress: An updated overview for improving stress tolerance. *South African Journal of Botany*. 2023;161:258–68.
91. Sadau SB, Liu Z, Ninkuu V, Guan L, Sun X. DREB transcription factors are crucial regulators of abiotic stress responses in Gossypium spp. *Plant Stress*. 2024;11:100350.
92. Chaudhari RS, Jangale BL, Krishna B, Sane PV. Improved abiotic stress tolerance in Arabidopsis by constitutive active form of a banana DREB2 type transcription factor, MaDREB20. CA, than its native form, MaDREB20. *Protoplasma*. 2023;260(3):671–90.

93. Balparda M, Bouzid M, Martinez MdP, Zheng K, Schwarzländer M, Maurino VG. Regulation of plant carbon assimilation metabolism by post-translational modifications. *The Plant Journal*. 2023;114(5):1059–79.
94. Das S, Ray MK, Panday D, Mishra PK. Role of biotechnology in creating sustainable agriculture. *PLOS Sustainability and Transformation*. 2023;2(7):e0000069.
95. Altaf MA, Hao Y, Shu H, Mumtaz MA, Cheng S, Alyemeni MN, et al. Melatonin enhanced the heavy metal-stress tolerance of pepper by mitigating the oxidative damage and reducing the heavy metal accumulation. *Journal of Hazardous Materials*. 2023;454:131468.
96. Kim J-S, Kidokoro S, Yamaguchi-Shinozaki K, Shinozaki K. Regulatory networks in plant responses to drought and cold stress. *Plant Physiology*. 2024;195:kiae105.
97. Ikram M, Rauf A, Rao MJ, Maqsood MFK, Bakhsh MZM, Ullah M, et al. CRISPR-Cas9 based molecular breeding in crop plants: A review. *Molecular Biology Reports*. 2024;51(1):227.
98. Aziz MA, Masmoudi K. Molecular breakthroughs in modern plant breeding techniques. *Horticultural Plant Journal*. 2024.
99. Tian Z, Wang JW, Li J, Han B. Designing future crops: Challenges and strategies for sustainable agriculture. *The Plant Journal*. 2021;105(5):1165–78.
100. Fiaz S, Ahmar S, Saeed S, Riaz A, Mora-Poblete F, Jung K-H. Evolution and application of genome editing techniques for achieving food and nutritional security. *International Journal of Molecular Sciences*. 2021;22(11):5585.
101. Delannay X, McLaren G, Ribaut J-M. Fostering molecular breeding in developing countries. *Molecular Breeding*. 2012;29(4):857–73.
102. Haggag WM, Abouziena H, Abd-El-Kreem F, El Habbasha S. Agriculture biotechnology for management of multiple biotic and abiotic environmental stress in crops. *Journal of Chemistry and Pharmeceutical Research*. 2015;7(10):882–9.
103. Xu Y, Li P, Zou C, Lu Y, Xie C, Zhang X, et al. Enhancing genetic gain in the era of molecular breeding. *Journal of Experimental Botany*. 2017;68(11):2641–66.
104. Rimmer M. Climate-ready crops: Intellectual property, agriculture and climate change. *Intellectual Property and Emerging Technologies*: Edward Elgar; 2012.
105. Njogu JW, Karuku G, Busienei J, Gathiaka JK. Assessing determinants of scaling up pathways for adopted CSA Climate Smart Agricultural practices: Evidence from climate smart villages in Nyando Basin, Kenya. *Cogent Food & Agriculture*. 2024;10(1):2316362.
106. Shende SS, Rajput VD, Ranjan A, Mandzhieva SS, Sushkova SN, Minkina TM. Nanotoxicity to soil ecosystems. *Nanotoxicology for Agricultural and Environmental Applications*: Elsevier; 2024. p. 109–34.

Index

H

N

O

X

Y

Z

For Product Safety Concerns and Information please contact our EU representative GPSR@taylorandfrancis.com Taylor & Francis Verlag GmbH, Kaufingerstraße 24, 80331 München, Germany